RANDOM NUMBER GENERATORS AND SIMULATION

MATHEMATICAL METHODS OF OPERATIONS RESEARCH

Series Editor: A. PRÉKOPA

Vol. 1. Studies on Mathematical Programming
Edited by A. PRÉKOPA

Vol. 2. Combinatorial Methods of Discrete Programming
L. B. KOVÁCS

Vol. 3. Véletlenszám generátorok és alkalmazásuk
I. DEÁK

AKADÉMIAI KIADÓ, BUDAPEST 1990

RANDOM NUMBER GENERATORS AND SIMULATION

by

ISTVÁN DEÁK

AKADÉMIAI KIADÓ, BUDAPEST 1990

This is the revised and updated version of the Hungarian
Véletlenszám generátorok és alkalmazásuk,
published by Akadémiai Kiadó as Vol. 3,
in the Series *Az operációkutatás matematikai módszerei*

Translated by István Deák

ISBN 963 05 5316 3

Printed in Hungary by Akadémiai Kiadó és Nyomda Vállalat

CONTENTS

INTRODUCTION

One of the most important methodologies in operations research, numerical analysis and statistics is the deployment of random experiments for solving problems. Examples of this type can be cited from long ago, one of the best known one makes possible the approximate determination of the value π. The method of the random experiments has undergone since the end of the 1940s a fast development for in those years the rapid growth in the computational capacities of the computers made the executions of the random experiments in really large numbers possible. First the new method was used for solving problems related to the atomic power plants but it soon became apparent that it is a very useful tool in solving a great number of problems in economics, physics and mathematics.

The methods using random experiments are also called Monte Carlo methods, a name given by John Neumann. In this book these methods will be called Monte Carlo methods only in the case if the mathematical foundations and aspects of these procedures are discussed, otherwise generally the solution procedures using randomness are called simulational methods or simulation according to the more usual terminology.

When applying simulational methods two areas emerge as possible sources of problems (1) how the random quantities to be used should be produced and (2) how a model is to be prepared that would describe the reality close enough and be computationally still efficient. The procedures which produce the random quantities are called random number generators. In the first part of the book we describe general and specific methods for manufacturing random quantities that follow a particular distribution, generally several generators are discussed for a distribution. In the second part of the book some possible applications of these random number generators are described, having highlighted the mathematical theory of the simulational methods by exposing the basic techniques some mathematical, engineering and economic problems are discussed where simulation can successfully be applied.

Our style is computational and algorithmic. Besides the mathematically precise description of methods and algorithms special care was taken to emphasize the computational aspects of developing an algorithm and to give the possible fullest details on the procedures. The presented methods are compared and evaluated with respect to their speed, ease of programming, user friendliness, length of code, number of data, etc. The formal description of the algorithms is given in such details that preparing the computer code of an algorithm is possible without fully understanding the nuts and bolts of the algorithm, that was aimed to facilitate the work of computer scientists.

To understand the book only basic mathematical knowledge is required. For the sake of completeness and to help the Reader, the necessary concepts and theorems from probability theory and statistics are briefly outlined in the first chapter.

A great part of the material contained in this book was part of the postgraduate courses the author gave at the Eötvös University, Budapest during the years 1978–1983 and also at Dalhousie University, Halifax, 1984–1986. Thus the emphasis is more that of a textbook, but advanced researchers could also use it as a reference because of its comprehensive range, the references and the bibliographical notes given at the end of each chapter.

This book is a thoroughly revised, updated and extended version of the author's book, "Random number generators and their applications" (in Hungarian), published by the Publishing House of the Hungarian Academy of Sciences. This extended translation has been prepared while the author was enjoying a visit to Dalhousie University, Halifax, arranged by Professor M. A. H. Dempster, to whom the author would like to express his gratitude. Without his support this work would not have come about.

Budapest, 1986

István Deák

CHAPTER 1

BASICS OF PROBABILITY THEORY AND STATISTICS

1.1. Probability and random variables

Frequently in a physical process certain phenomena occur randomly, by some chance. The possible phenomena are called elementary events, the measure of the chance that a given phenomenon occurs is called the probability of the event. More precisely the probability $P(A)$ of the event A is defined in the following way: denoting the number of occurrences of the event A by $\nu(n)$ during n independent realizations of the physical process the probability of A is given by

$$P(A) = \lim_{n \to \infty} \frac{\nu(n)}{n}.$$

The probability of the sure event is 1, that of the impossible event is 0.

In case of throwing a dice the number obtained may be 1,2,. . .,6, these are the elementary events. The individual elementary events occur with equal chance, that is the probability of each elementary event equals 1/6. Of course if the dice has on two facets 5, then the probability of throwing 5 is doubled, it is 1/3. An elementary event may not necessarily be a numerical value; if the facets of the dice are coloured to six different colours, then the set of the elementary events consists of the possibilities, that after throwing the dice we will obtain red, yellow, etc.

Phenomena dependent on the elementary events are called random variables, in other words they are real valued functions defined on the set of elementary events. Staying with our example of throwing a dice with numbered facets we can define the random variable ξ that takes on the value 1 if the result of the throw is even and 0 if the number obtained is odd; this ξ is a function defined on the set of the elementary events 1,2,. . .,6 indeed. Denote that elementary event, where the outcome was i by A_i, $i = 1, \ldots, 6$, then for the probability of the event A_i we have $P(A_i) = 1/6$. Since in case of occurring A_1, A_3 and A_5 the value of ξ is 1, thus for the probability of the occurrence of the event $\xi = 1$ we have $P\{\xi = 1\} = 1/2$, that is the random variable ξ equals 1 with probability 1/2.

We can define other random variables, too, on the elementary events of throwing a dice. Consider the following game, played by two players: if the outcome of the throw is 4 or less, then the first player obtains 2 dollars from the second one, otherwise the first player owes 3 dollars to the second. This defines the random variable η of the first player's profit on the set of the elementary events A_i, $i = 1, \dots, 6$ as

$$\eta(A_i) = \begin{cases} 2, & \text{if } \ i \leqq 4 \\ -3, & \text{if } \ 5 \leqq i \leqq 6. \end{cases}$$

The probability of the first player's winning equals the probability of the event $\eta > 0$, that is the joint probability of the elementary events A_1, A_2, A_3 and A_4:

$$P\{\eta > 0\} = P(A_1) + P(A_2) + P(A_3) + P(A_4) = 2/3.$$

In the examples mentioned so far we had a finite number of elementary events. For a different example consider the temperature in New York at the Times Square. A possible elementary event is any number t between a possible minimum m and maximum M as temperature, giving thus rise to infinitely many events. On these elementary events we can define the random variable p of freeze for instance, which is 1 if t is less than 32°F, otherwise p assumes the value 0.

The probability for the general case is defined by a set of axioms. Let the space of the elementary events Ω be given and consider a σ-algebra $\mathcal{A}$ of the subsets of Ω, $-\mathcal{A}$ is an algebra if the union of two sets from $\mathcal{A}$, the complementer of a set and Ω itself is in $\mathcal{A}$, the notion of σ-algebra implies that the sum of any number (finite or enumerably infinite) of sets from $\mathcal{A}$ is inside $\mathcal{A}$. We define the function $P(A)$ for any subset A of the σ-algebra $\mathcal{A}$ by the following three axioms

(i) $0 \leqq P(A) \leqq 1, \quad A \in \mathcal{A}$.
(ii) $P(\Omega) = 1$.
(iii) If $A_1, A_2, \dots, A_k, \dots$ are subsets of $\mathcal{A}$ such that the intersection of any two of them is empty (among the events $A_1, A_2, \dots, A_k, \dots$ at most only one can occur at a time), then $P(\bigcup_{k=1}^{\infty} A_k) = \sum_{k=1}^{\infty} P(A_k)$.

The function $P(A)$ is the probability of the event A; the space of events Ω, the σ-algebra $\mathcal{A}$ and the function P together is called a probability field.

The probability of the event, that in case of occurring the event B (of positive probability) the event A also occurs is called the conditional probability of the event A provided event B and is denoted by $P(A \mid B)$.

The numerical value assigned to this conditional probability is a quotient: the probability of the joint occurrence of the events A and B divided by the probability of the condition B, that is

$$P(A \mid B) = \frac{P(AB)}{P(B)}. \tag{1.1}$$

Consider a finite number of events $B_1, \ldots, B_n$ forming a complete system of events $(\bigcup_{i=1}^{n} B_i = \Omega)$ and the events B_i are mutually exclusive $(B_i \cap B_j = \emptyset$ if $i \neq j$ and $\emptyset$ denotes the empty set). The following theorem holds for any event A

$$P(A) = \sum_{i=1}^{n} P(A \mid B_i) P(B_i) \tag{1.2}$$

which is called the theorem of complete (or total) probability.

The events A and B are mutually independent if the probability of their occurrence can be given as the product of the probabilities of the individual events

$$P(AB) = P(A)P(B)$$

or in a slightly different, but equivalent form $P(A \mid B) = P(A)$.

1.2. Distribution function, density function

Consider the random variables with real values, that is examine those functions ξ defined on the elementary events ω_i whose range is a set of the real axis. Then we can speak about the probability of the event that a particular random variable $\xi = \xi(\omega_i)$ falls into the interval $[a, b]$, which is the following

$$P\{\xi \in [a, b]\} = P\{\cup \omega_i \mid \xi(\omega_i) \in [a, b]\}.$$

To characterize the random variable ξ it is sufficient to give these probabilities for intervals of type $(-\infty, x]$, since — according to the axiom (iii) of the probability — making use of these intervals the probability of ξ's falling into any interval $[a, b]$ can be expressed as

$$P\{\xi \in [a, b]\} = P\{\xi \in (-\infty, b]\} - P\{\xi \in (-\infty, a]\}.$$

This probability $P\{\xi \in (-\infty, x]\} = P\{\xi < x\}$ is clearly dependent on x, that is it can be expressed in the form

$$F(x) = P\{\xi < x\}. \tag{1.3}$$

The function $F(x)$ thus defined is called the distribution function of the random variable ξ. It is easily seen from the definition that $F(x)$ is monotone non-decreasing and left continuous function. Moreover, this function $F(x)$ completely determines the random variable ξ; as consequence random variables are generally given through their distribution functions. A random variable is called absolutely continuous if there exists a function $f(x)$ for which

(1.4) $$F(x) = \int_{-\infty}^{x} f(z)\mathrm{d}z$$

holds for any x. This function f (if exists) is called the density function of the random variable ξ and of course we have $F'(x) = f(x)$. Since $P\{\xi < \infty\} = 1$ holds thus $F(\infty) = \lim_{x\to\infty} F(x) = 1$, that is the integral of the density function over the whole real axis is always 1 (Fig. 1)

$$\int_{-\infty}^{+\infty} f(z)\mathrm{d}z = 1.$$

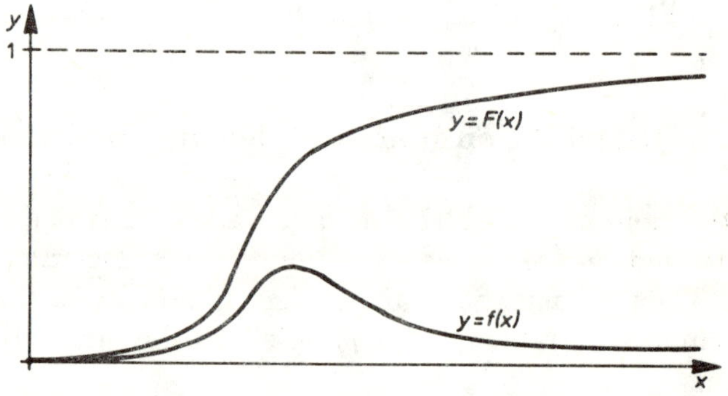

Fig. 1. The density function $f(x)$ and the distribution function $F(x)$ of a continuous random variable

If the random variable η is discrete, that is it takes upon a finite or enumerable many values $x_1, \dots, x_n, \dots$ then more conveniently the probabilities

(1.5) $$P\{\eta = x_i\} = p_i, \quad i = 1, 2, \dots$$

are used instead of the density function. The distribution function of the discrete random variable η is a stepwise constant staircase function (Fig. 2)

$$F(x) = P\{\eta < x\} = \sum_{i \in I(x)} P_i$$

where the summation is carried out for all indices $i \in I(x)$ such that $x_i < x$, that is $I(x) = \{i \mid x_i < x, \ i = 1, 2, \ldots\}$.

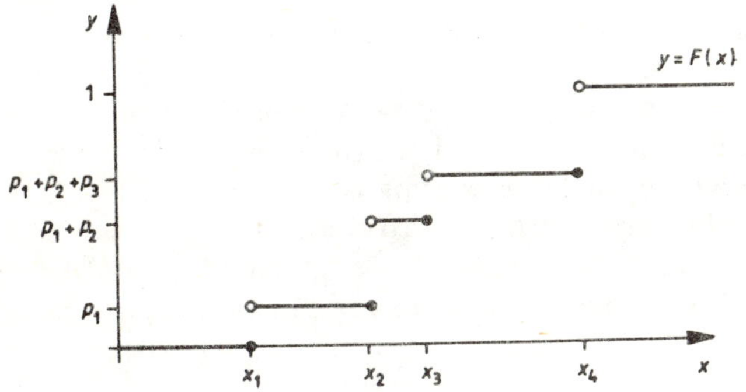

Fig. 2. The distribution function $F(x)$ of discrete random variable ξ given by
$P\{\xi = x_i\} = p_i, i = 1, \ldots, 4$

As an example we discuss the uniform distribution. In the discrete case η is uniformly distributed if it assumes any of the n values $x_1, \ldots, x_n$ with equal probability $p_i = \dfrac{1}{n}, i = 1, \ldots, n$. In the continuous case ξ is uniformly distributed in the interval $[a, b]$ if its density function $f(x) = \dfrac{1}{b-a}$, $x \in [a, b]$ or equivalently $F(x) = \dfrac{x-a}{b-a}$ if $x \in [a, b]$.

The joint distribution function of the random variables ξ and η is the function F defined by the following equation

(1.6) $$F(x, y) = P\{\xi < x, \eta < y\}.$$

If there exists a function $f(x, y)$ which satisfies the equation

(1.7) $$F(x, y) = \int_{-\infty}^{x} \int_{-\infty}^{y} f(u, v) du dv$$

for any values x, y, then this function $f(x, y)$ is called the joint density function of the random variables ξ and η. One obviously has for this function the equality $f(x, y) = \dfrac{\partial^2 F(x, y)}{\partial x \partial y}$. In case of discrete random variables ξ and η the joint probabilities

$$p_{ij} = P\{\xi = x_i, \eta = y_j\}$$

will conveniently characterize the joint distribution of ξ and η.

If the random variables ξ and η are independent, then their joint distribution function $F(x,y)$ can be decomposed into the product of two one-dimensional distribution function

$$F(x,y) = G_1(x)G_2(y)$$

as it follows from the decomposition of the right-hand side of equation (1.6).

The reverse of this assertion is true, too: if the joint distribution function can be decomposed into such a product then the corresponding two random variables are independent of each other.

Projecting the joint distribution to one of the coordinate axis we obtain the marginal distribution of the random variable ξ and η, respectively. These marginal distribution functions can be obtained from the following equations

$$\begin{aligned} G_1(x) &= \int_{-\infty}^{x} \int_{-\infty}^{+\infty} f(u,v)\mathrm{d}v\mathrm{d}u, \\ G_2(y) &= \int_{\infty}^{y} \int_{\infty}^{-\infty} f(u,v)\mathrm{d}u\mathrm{d}v, \end{aligned} \tag{1.8}$$

respectively. In the continuous case the density functions corresponding to these marginal distribution functions are $g_1(x)$ and $g_2(x)$. In the discrete case the marginal distributions are defined by the relationships

$$\begin{aligned} q_i^{(1)} &= \sum_j p_{ij}, \qquad i = 1,2,\ldots \\ q_j^{(2)} &= \sum_i p_{ij}, \qquad i = 1,2,\ldots \end{aligned}$$

where the summation is over all possible indices j and i, respectively.

We can speak about the conditional distribution function $H(x \mid y)$ of the random variable ξ under the condition $\eta = y$, which is defined as follows

$$H(x \mid y) = \lim_{d\to 0} P\{\xi < x \mid y \leqq \eta \leqq y + d\} \tag{1.9}$$

on the condition that the limit in (1.9) exists. This conditional distribution function can easily be derived from the joint distribution function $F(x,y)$ of the random variables ξ and η, since because of definition (1.1) of the conditional probability we have

$$H(x \mid y) = \lim_{d\to 0} \frac{F(x,y+d) - F(x,y)}{G_2(y+d) - G_2(y)},$$

where G_2 is the unconditional (marginal) distribution function of the random variable η. Assuming the differentiability of the function F and G_2 one can obtain

$$H(x \mid y) = \frac{\dfrac{\partial G_1(x)G_2(y)}{\partial y}}{g_2(y)} = G_1(x),$$

that is the conditional distribution function is the usual one-dimensional distribution function of ξ. If $H(x \mid y)$ is differentiable then the function $h(x \mid y) = \dfrac{\partial H(x \mid y)}{\partial y}$ is called the conditional density function of the random variable ξ provided the condition $\eta = y$ holds. The theorem of complete (or total) probability can be given in two special forms for the discrete and the continuous random variables.

Let ξ and η be two discrete random variables (not necessarily independent), taking up values $x_1, \ldots, x_n$ and $y_1, \ldots, y_m$ respectively. Then (1.2) can be rewritten as

$$(1.10) \qquad P\{\xi = x_i\} = \sum_{k=1}^{m} P\{\xi = x_i \mid \eta = y_k\}\, P\{\eta = y_k\}.$$

For the continuous case let ξ and η be continuous random variables with a joint density function $f(x, y)$, their marginal density functions being $g_1(x)$ and $g_2(y)$ respectively and the conditional density function of ξ under the condition $\eta = y$ being $h(x \mid y)$. Then we have the second form of the complete probability theorem as

$$(1.11) \qquad g_1(x) = \int_{-\infty}^{+\infty} h(x \mid y) g_2(y) dy.$$

The expected value $E(\xi)$ of a continuous random variable ξ with density function $f(x)$ is defined by the equation

$$E(\xi) = \int_{-\infty}^{+\infty} x f(x) dx.$$

In other words we can say that the possible values of the random variable are centered about the value $E(\xi)$ as a center of gravity (see Fig. 3).

The expected value of the discrete random variable η given by the probabilities $P\{\eta = x_i\} = p_i$, $i = 1, 2, \ldots, n$ is the sum

$$(1.12) \qquad E(\eta) = \sum_{i=1}^{n} x_i p_i.$$

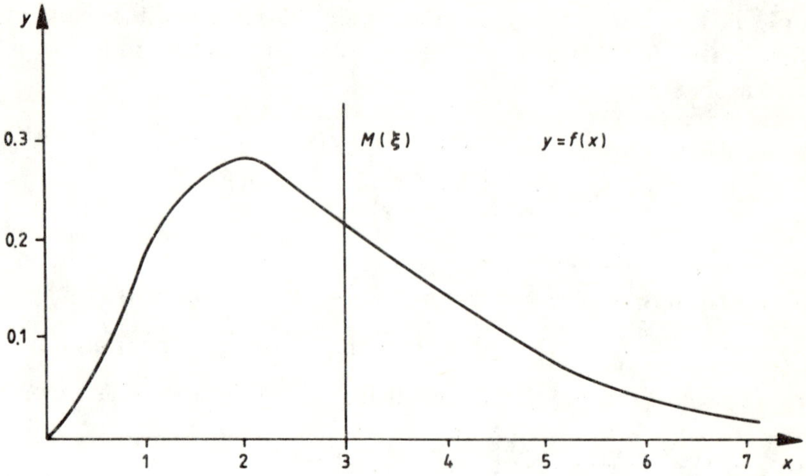

Fig. 3. The expected value $M(\xi) = 3$ of the random variable ξ with density function $f(x) = e^{-x}x^2/2$

By applying the definitions we can derive the following equivalences. The expected value of a sum of two random variables is the sum of the expected values of the individual random variables, that is

$$E(\xi + \eta) = E(\xi) + E(\eta). \tag{1.13}$$

Here it should be stressed that the expected value of a product of random variables will be the product of the expected values if the random variables are independent, but sometimes it may be the case even if the random variables are not independent (see formula (1.16)). Otherwise, if ξ and η are not independent and their joint density function is $f(x, y)$, then we have to use the general expression

$$E(\xi\eta) = \int_{-\infty}^{\infty} \int_{-\infty}^{\infty} xyf(x, y)dxdy. \tag{1.14}$$

Note furthermore, that if b and c are constant, then

$$E(b\xi + c) = bE(\xi) + c.$$

The standard deviation $D(\xi)$ (or the variance $\text{var}(\xi) = D^2(\xi)$) gives us information about the difference between the values of the random variable and the expected value, it is defined by the equation

$$D^2(\xi) = E\left([\xi - E(\xi)]^2\right) = E(\xi^2) - E^2(\xi). \tag{1.15}$$

For a random variable with continuous density function $f(x)$ the variance can be obtained as

$$D^2(\xi) = \int_{-\infty}^{+\infty} x^2 f(x)dx - \left[\int_{-\infty}^{+\infty} x f(x)dx\right]^2.$$

For a discrete random variable η, where $P\{\eta = x_i\} = p_i$, $i = 1, 2, \ldots, n$ the variance is evaluated as

$$D^2(\eta) = \sum_{i=1}^{n} x_i^2 p_i - \left[\sum_{i=1}^{n} x_i p_i\right]^2.$$

For any constants b and c we have

$$D^2(b\xi + c) = b^2 D^2(\xi).$$

Relationship between two random variables is characterized by the covariance and the correlation. The covariance $c(\xi, \eta)$ of the random variables ξ and η is given by

$$c(\xi, \eta) = E[(\xi - E(\xi))(\eta - E(\eta))] = E(\xi\eta) - E(\xi)E(\eta).$$

In case of independent ξ and η the covariance is zero because of the equality $E(\xi\nu) = E(\xi) \cdot E(\eta)$. Note that the independence of the random variables ξ and η is not implied by the equation $c(\xi, \eta) = 0$. Normalizing the covariance with the standard deviations the correlation coefficient $r(\xi, \eta)$ is obtained as

$$r(\xi, \eta) = \frac{c(\xi, \eta)}{D(\xi)D(\eta)} = \frac{E[(\xi - E(\xi))(\eta - E(\eta))]}{D(\xi)D(\eta)}. \tag{1.16}$$

While the covariance between two random variables can be very great (or small) for the correlation coefficient r we always have the inequalities $-1 \leqq r \leqq +1$; this can be deduced from the inequality $E^2(\xi\eta) \leqq \leqq E(\xi^2) \cdot E(\eta^2)$. If $r = +1$ or $r = -1$ then the two random variables are in a close relationship; in fact one of them is a linear function of the other.

The relation $D^2(\xi) = c(\xi, \xi)$ is true, thus we may write for the random variables $\xi_1, \xi_2, \ldots, \xi_n$

$$D^2\left(\sum_{i=1}^{n} \xi_i\right) = \sum_{i=1}^{n}\sum_{j=1}^{n} c(\xi_i, \xi_j),$$

an equation which is sometimes useful. For the variance of the sum of two random variables we have

$$\begin{aligned} D^2(\xi + \eta) &= D^2(\xi) + D^2(\eta) + 2E(\xi\eta) - 2E(\xi)E(\eta) \\ &= D^2(\xi) + D^2(\eta) + 2c(\xi, \eta), \end{aligned} \tag{1.17}$$

that is the equality $D^2(\xi + \eta) = D^2(\xi) + D^2(\eta)$ holds for variables ξ and η with zero covariance only.

1.3. Transformation theorems

When discussing the methods of random number generation the following theorems will be needed to show how upon the transformation of a random variable is its density function changing.

Let ξ be a continuous random variable with density function $f(x)$, distribution function $F(x)$, possible values of ξ are in the interval $[a, b]$. Let the function $y = T(x)$ be a differentiable and monotone increasing (or monotone decreasing) in the interval $[a, b]$, its inverse is denoted by $x = T^{-1}(y)$.

Theorem 1. The density function of the random variable $\eta = T(\xi)$ is

$$g(y) = f(T^{-1}(y))\left|\frac{\partial T^{-1}(y)}{\partial y}\right|.$$

Proof. Assume that $T(x)$ is monotone increasing, the proof of the other case is quite similar. The distribution function of the random variable $\eta = T(\xi)$ can be expressed as

$$G(y) = P(\eta < y) = P(T(\xi) < y) = P(\xi < T^{-1}(y)) = F(T^{-1}(y)).$$

Differentiating the last expression on the right-hand side with respect to the variable y the desired result is obtained.

A similar theorem holds for the multivariable case, too. Let $\xi = (\xi_1, \xi_2, \ldots, \xi_n)$ be a continuous random vector, whose possible values are in the domain D, its density function is $f(\mathbf{x})$ (see Fig. 4).

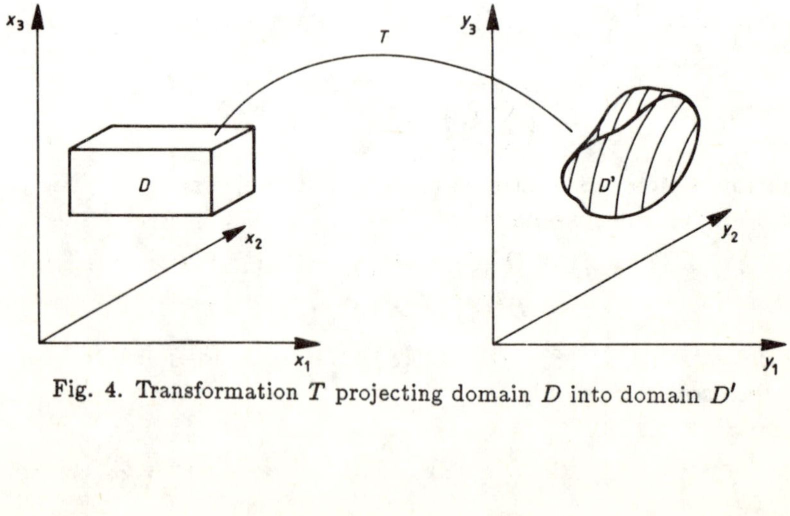

Fig. 4. Transformation T projecting domain D into domain D'

Consider now the one-to-one transformation $\eta_1 = t_1(\xi_1, \ldots, \xi_n)$, $\eta_2 = t_2(\xi_1, \ldots, \xi_n), \ldots, \eta_n = t_n(\xi_1, \ldots, \xi_n)$ given by the functions $t_1, t_2, \ldots, t_n$ which projects the domain D into the domain D', that is using the notation $T = (t_1, \ldots, t_n)$ we have

$$D' = \{\mathbf{y} = T(\mathbf{x}), \mathbf{x} \in D\}.$$

The inverse transformation projecting D' into D is denoted by $\mathbf{x} = = T^{-1}(\mathbf{y})$.

Theorem 2. The density function $g(\mathbf{y})$ of the random vector $\boldsymbol{\eta} = T(\boldsymbol{\xi})$ is

$$g(\mathbf{y}) = f(T^{-1}(\mathbf{y}))\left|\frac{\partial T^{-1}(\mathbf{y})}{\partial \mathbf{y}}\right|,$$

where the expression $\left|\frac{\partial T^{-1}\left(\frac{y}{z}\right)}{\partial \mathbf{y}}\right|$ stands for the Jacobian of the inverse transformation T^{-1}.

Proof. The density function $g(\mathbf{y})$ of the random vector $\boldsymbol{\eta} = T(\boldsymbol{\xi})$ is — by definition — the function, for which the equation

$$P(\boldsymbol{\eta} \in E') = \int_{E'} g(\mathbf{y})d\mathbf{y}$$

holds in case of any domain E' inside D'. On the other hand we can write for the domain $E = T^{-1}(E') \subset D$ the equation

$$P(\boldsymbol{\eta} \in E') = P(T(\boldsymbol{\xi}) \in E') = P(\boldsymbol{\xi} \in T^{-1}(E')) = P(\boldsymbol{\xi} \in E) = \int_E f(\mathbf{x})d\mathbf{x}.$$

Substituting in the integral the variable $\mathbf{y} = T(\mathbf{x})$ we obtain

$$P(\boldsymbol{\eta} \in E') = \int_E f(\mathbf{x})d\mathbf{x} = \int_{E'} f(T^{-1}(\mathbf{y}))\left|\frac{\partial T^{-1}(\mathbf{y})}{\partial \mathbf{y}}\right|d\mathbf{y}$$

which compared to the expression defining the density function $g(\mathbf{y})$ yields the required result.

1.4. Statistical estimators and tests

A set of certain objects together with a numerical value related to them is called a statistical population. Such a statistical population may be constituted by the inhabitants of a city and their heights or a river with the possible water levels during a year. From a mathematical point of view the statistical population can be viewed as a probabilistic field, the numerical value attached to its element as a random variable. In this sense we can speak about the distribution of the statistical population. Assume that the type of the distribution is known but not the numerical values of the parameters. Then the usual procedure in statistics will be taking a sample from the statistical population for determining the unknown parameter(s); these samples are the values $x_1, \ldots, x_n$ and a function $t(x_1, \ldots, x_n)$ of these values is used for estimating the correct value of the parameter in concern. This function t is called the estimator of the parameter sought for.

An estimator t of the parameter p is called unbiased, if its expected value equals the parameter, that is

$$p = E(t) = E(t(x_1, \ldots, x_n)).$$

An estimator is considered better than another one, if it has less variance, which means that it will probably deviate from the expected value p to a lesser extent than the other estimator. If there exists an estimator t_0 amongst the unbiased estimators having the smallest variance it is called an efficient estimator. If for example the parameter p is the expected value of a distribution, then the estimator $t(x_1 \ldots, x_n) = \frac{1}{n}(x_1 + \ldots + x_n)$ is an unbiased estimator of p. Generally the determination of an efficient estimator is a rather difficult task, so frequently one is satisfied with a linear estimator which has the smallest variance, that is only the linear functions are considered amongst the possible functions t. The above estimator of the expected value is exactly of this type, that is out of the unbiased linear estimators $t'(x_1, \ldots, x_n) = b_1x_1 + \ldots + b_nx_n$, $b_1 + \ldots + b_n = 1$, the estimator $t(x_1, \ldots, x_n) = \frac{1}{n}(x_n + \ldots + x_n)$ has the smallest variance.

The deviation between the unbiased estimator t and the looked for parameter value p is of random nature, it is characterized by a confidence-interval. This means that an interval containing p is given and the estimator t falls into this interval with great probability: for any $\varepsilon > 0$, we can give a value h that $P\{p - h \leqq t \leqq p + h\} = 1 - \varepsilon$ holds, in practice the use of $1 - \varepsilon = 0.95$ is frequent and gives satisfactory results in most cases.

If more samples can be taken from the statistical population at the same time, which are dependent on each other, then this property can be exploited in constructing the estimator. Assume that we want to estimate the value of the parameter p, where it is known, that for the samples $x_1, \ldots, x_n$ we have $x_i = b_i p + \delta_i$ where b_i is a given constant (in simple cases it equals 1) δ_i is a random error, whose expected value is 0, furthermore the covariance matrix $\mathbf{C} = (c_{ij})$ is known, $c_{ij} = c(x_i, x_j)$. Then

$$t(x_1, \ldots, x_n) = (\mathbf{b}'C^{-1}\mathbf{b})^{-1}\mathbf{b}C^{-1}\mathbf{x} \tag{1.18}$$

is an unbiased and minimum variance linear estimator of the parameter p, and the variance of this estimator is $(\mathbf{b}'C^{-1}\mathbf{b})^{-1}$. This construction can be extended to cover the multidimensional case. If an estimator is to be constructed for the unknown vector $\mathbf{p}$ and we have the samples x_i as $x_i = \sum_j b_{ij}p_j + \delta_i$ where b_{ij} is a constant obtained from the sampling or determined by the statistical population, δ_i is a random "noise" with expected value 0, $\mathbf{C}$ is the covariance matrix of the errors δ_i, then

$$\mathbf{t}(x_1, \ldots, x_n) = (B'C^{-1}B)^{-1}B^{-1}C^{-1}\mathbf{x}$$

is an unbiased estimator of $\mathbf{p}$ and has the minimum variance $((B'C^{-1}B)^{-1})$ among the linear estimators.

Frequently one encounters situations when the problem is not to determine the numerical value of a parameter but answer the question whether a hypothesis on a parameter of the given distribution is true or the statistical population's distribution is of the assumed type or not. Such questions can be answered by the help of the statistical tests.

All statistical tests are constructed along the following general guidelines. Assume that the population, that is, the samples taken from it are distributed according to a given distribution function. This hypothesis H_0 will serve us as a starting point. If this hypothesis is true, then another property, derived from the hypothesis H_0 will also be true for a function $d(x_1, \ldots, x_n)$ of the samples $x_1, \ldots, x_n$.

We are able to check this derived — secondary by its nature — property only, not the primary hypothesis H_0. Thus a statistical test consists of constructing and testing the existence of such a secondary property. Since we work with random quantities, this testing can be examined and accepted in probabilistic terms only, that is one can check only whether a property of the samples is true with a great enough probability. This latter means, that if we test this particular property several times, that is the function $d(x_1, \ldots, x_n)$ is evaluated several times for different sample-batches, then in most of the cases the hypothesis holds. If we check only once the existence of the property, then according to the previous reasoning we may expect that the more frequent case, i.e. the case of greater probability will happen.

A statistical test is then carried out in the following steps. First, the test is constructed, second by the aid of a sample the soundness of the hypothesis is decided. The construction of the test involves the formulation of the hypothesis H_0 and from it a property of the hypothesis is deduced that is the function d corresponding to this property is defined. Completion of a test means taking a sample $x_1, \ldots, x_n$ from the statistical variety, evaluation of the function value $d(x_1, \ldots, x_n)$ and finally scrutinizing the obtained value whether it possesses the property looked for or not, that is the theory is confronted with practice. In case the practice conforms with the theory, then the test is said to be accepted and the initial hypothesis H_0 on the distribution is judged to be true, while if the required property does not hold then the initial hypothesis is rejected as false.

The essence of constructing and applying statistical tests may be illustrated by the following example. We are in a river on a boat and would like to know whether presently the boat is in deep water or near to the riverbank. The hypothesis H_0 is that we are in the middle of the river. From this it follows that if a pole is pushed downward, until contact, i.e. the riverbed, then the pole should be wet at least on a 4-meter part of it. After setting up this theoretical model we test it: the pole is pushed down until it gives way and pulling back the length of its wet part is measured. If it is longer then 4 meter the hypothesis is accepted, otherwise rejected. (Of course to make this crude example statistically valid, some conditions are to be observed: the boat should be drifting down the river, the pole is to be pushed down several times, etc.)

We have two types of error when applying statistical tests. An error of type I occurs if the hypothesis is rejected but it is true and the error is of type II if it is accepted though it is false. Remaining at the previous example an error may result if in the shallow water we pushed the pole into an underwater pit, and it became wet on a long portion or if in deep water the pole has been stuck in another boat, or rock and it remained dry. In this latter case the error is of type I, while in the first case it was of type II.

In statistical tests the presence of a property is required with a probability $1 - \varepsilon$. The probability of the first-type error is ε, so it can be decreased to an arbitrarily small value. We want stress here the subtle formulation of the description of tests: if we accept a hypothesis it does not mean, that the hypothesis is true, all what we say is that the sample does not contradict the hypothesis and so we have no reason to reject the hypothesis.

1.5. Notes

At the end of each chapter we give some historical remarks and bibliographical notes pertaining to the presented material. On one side we try to show and illustrate the development of certain problem exposures, topics and solutions with some remarks, on the other advanced researchers wishing to get involved deeper in this branch of science are helped by citing references used — generally here can one find the omitted proofs — and giving further relevant papers.

A great many of the ideas presented in the book are of folklore type, that is more persons seem to know them, use them, so it is rather hard to pinpoint the original author. In such cases we either give the author of the earliest publication known to us, or we do not give any references.

For general introduction and overview of the field of random number generators and simulation the valuable book of Hammersley and Handscomb (1964) on Monte Carlo methods, the books of Ahrens and Dieter (1974) on random number generators, the work of Knuth (1969) on computational aspects, Sobol's (1978) book on simulational applications can be recommended. Researchers may use Rubinstein's (1981) monograph, Bratley, Fox, Schrage's (1983) book on simulation, the extensive work of Devroye (1986) on random number generation, two review papers of Schmeiser (1980), (1981); and Sowey's (1986) bibliography which is the most recent one.

The notations introduced and used in Chapter 1 to describe the basics of probability theory are mostly along the lines of Prékopa's book.

CHAPTER 2

GENERATION OF UNIFORMLY DISTRIBUTED RANDOM NUMBERS

2.1. Introduction

In different types of Monte Carlo calculations realizations of random variables are required. These numerical realizations — to be used in actual computations are to be produced some way or other. If these numbers are independent and random realizations of random variables, then the underlying probability theory ensures the applicability of the Monte Carlo methods and the soundness of the results.

In practice three methods are employed to produce uniformly distributed random numbers. (First, we deal with manufacturing uniform random numbers only because on one hand such numbers can be produced in a relatively easy and fast way on the other using some transformations uniformly distributed numbers can be converted into numbers with any desired distributions.) The produced uniform numbers depending on their way of manufacturing are called natural random numbers, quasirandom numbers and pseudorandom numbers. The reader may ask why are we speaking about different types of random numbers. The reason behind this is that there is not a generally accepted exact definition of the concept of random numbers. In probability theory an event is called random, if we do not know all the reasons of the event, so we cannot predict it. For example throwing a dice is a random event, since we do not know the forces effecting it, the distribution of its mass, the density of the air, etc., so we are unable to determine the outcome of a rolling. The three different types of random numbers are three possible ways of bridging the gap between the abstract concept of randomness and the actual real life phenomena.

In case of random numbers the question is the following. Let ξ be a random variable, uniformly distributed in the interval $[0,1)$ and consider a finite or infinite sequence of numbers $x_1, x_2, \ldots, x_n, \ldots$ from the interval $[0,1)$. Can we deem the numbers $x_1, x_2, \ldots, x_n$ be the realizations of the random variable ξ and if so how can we produce similar sequences? Several answers are possible to this question.

An answer is that several mathematically sound definitions can be formulated for the concept of a sequence of random numbers. Choosing one of the definitions we can proceed by checking the assumptions made to determine the randomness of a sequence and produce random sequences. Theoretically this would be the most satisfactory solution to our problem, in actual computations it probably never was applied though.

Another answer may be based on using "real" random numbers produced by a physical gadget or observing the outcomes of a natural physical phenomenon, and these values, called natural random numbers are used in actual computations.

According to a third answer this question is not the proper one; instead of the use of random numbers that give relatively bad results with slow convergence we have to look for carefully selected numbers by the aid of which certain problems can be solved easily, with satisfactory convergence speed. These numbers are called quasirandom numbers and the theoretical background justifying the use of such numbers are called the quasi-Monte Carlo methods.

Finally the most frequent opinion; these numbers do not have to be random numbers, it is quite sufficient if they seem to be random. Numbers of this type are produced (generated) on computers by the aid of algorithms (generators) and they are called pseudorandom numbers. These numbers, similarly to the previously discussed quasirandom numbers, are completely deterministic ones, the only difference is that pseudorandom numbers may and do depend on the initial value and a couple of parameters. The use of pseudorandom numbers in Monte Carlo computation is carried out with a certain common sense and some optimism, that are substitutes for the rigid mathematical justification. Common sense dictates not to test these pseudorandom numbers with full thoroughness whether they are random or not, just a few basic tests are used to check out the initial hypotheses on randomness of these numbers. Afterwards we proceed in an optimistic way, believing that this sequence of numbers would pass any other — theoretically an infinite number of — tests, and what is most important, would solve the problem we have at hand with satisfactory precision.

As a matter of fact we are not really interested in the question, whether the pseudorandom sequence would pass the tests derived to check upon the initial hypothesis on its distribution, and other similar hypotheses, but the solitary most important question is whether our problem can be solved using this sequence or not. In computational practice it is a usual procedure to test the pseudorandom numbers (i.e. the generators) by solving a problem, similar in structure to the one we have to solve whose solution is known or can be evaluated with another technique. If the result obtained through the use of pseudorandom numbers is near enough to the exact result, then these pseudorandom numbers can be used with expectedly similar accuracy to solve problems of this type. This reasoning though theoretically is not exactly sound, in actual computations gives very good results, the practice verifies the inexact theory.

2.2. The concept of random sequence

Several possibilities exist for formulating the concept of a random sequence in exact mathematical definitions. Some of them are elaborated here.

For the sake of simplicity our treatment is restricted to the sequence of numbers uniformly distributed in the $[0, 1)$ interval and integer numbers originating from the discrete uniform distribution defined on the values $1, 2, \dots, m$.

Let the random variable ξ be uniformly distributed in the interval $[0,1)$, $u_1, u_2, \dots, u_n, \dots$ being numbers from the interval $[0,1)$. We aim at defining the random sequence so that on one side it contains as many of the properties of ξ as possible, on the other side there exists at least one sequence, complying with the definition and it is easily computable. Arbitrarily many characteristics of the random variable ξ can be derived, but only a few of them, the most typical ones according to our intuitive feeling of randomness will be required in order to call a sequence $u_1, \dots, u_n, \dots$ random.

According to the definition of uniformity we have

$$P(\xi \in [a, b)) = b - a$$

for any interval $[a, b) \subset [0, 1)$. This property is easily reformulated for a sequence of numbers. An infinite sequence of numbers $u_1, \dots, u_n, \dots$ is called 1-uniform, if

$$\lim_{n \to \infty} \frac{\nu(n)}{n} = b - a$$

holds for any interval $[a, b) \subset [0, 1)$, where $\nu(n)$ denotes the number of the values $u_1, \dots, u_n$ inside the interval $[a, b)$, $\nu(n) = \#\{u_i \mid a \leqq u_i < < b, \quad i = 1, \dots, n\}$, that is the probability of being in an interval should be proportional to the length of the interval. This 1-uniformity does not pose a strong restriction on the sequence of numbers, since if $u_1, \dots, u_n \dots$ is 1-uniform then the sequence $u_1/2,\ 1 - u_2/2,\ u_3/2,\ 1 - u_u/2, \dots$ is

1-uniform, too, even for the fact that in this latter sequence a number in $[0,1/2)$ is always followed by a number in $[1/2,1)$ which contradicts our sense of randomness. The dependency of the successive numbers is at least limited if the sequence is required to satisfy the k-dimensional generalization of the 1-uniformity. This is the following

Definition 1. The infinite sequence $u_1,\ldots,u_n,\ldots$ is called k-uniform, if for the k-dimensional points $\mathbf{x}_1=(u_1,\ldots,u_k)$, $\mathbf{x}_2=(u_{k+1},\ldots,u_{2k})$, $\ldots$, $\mathbf{x}_q=(u_{(q-1)k+1},\ldots,u_{qk})$ we have

$$\lim_{n\to\infty}\frac{\nu_D(n)}{n}=V_D, \tag{2.1}$$

for any domain D inside the k-dimensional unit cube $K=\{\mathbf{y}\mid \mathbf{0}\leqq\mathbf{y}\leqq\mathbf{1}\}$, where V_D is the volume of the domain D and $\nu_D(n)$ is the number of the points $\mathbf{x}_1,\ldots,\mathbf{x}_n$ falling into D, $\nu_D(n)=\#\{\mathbf{x}_i\mid\mathbf{x}_i\in D,\ i=1,\ldots,n\}$.

This definition is the natural extension of the 1-uniformity. A similar definition can be given for a discrete uniform distribution.

Definition 2. The infinite sequence of the digits $j_1,\ldots,j_n,\ldots$ (base b) is called k-uniform, if for any given k digit number $b_1,b_2,\ldots,b_k$ $(0\leqq j_i,b_i<b)$ the relation

$$\lim_{n\to\infty}\frac{\mu(n)}{n}=\frac{1}{b^k} \tag{2.2}$$

holds, where $\mu(n)$ denotes the number of those k digit numbers $j_{ik+1}\cdots j_{(i+1)k}$, $i=0,\ldots,n-1$, that are equal to $b_1\ldots b_k$, that is $\mu(n)=$ $=\#\{j_{ik+1}\cdots j_{(i+1)k})\mid j_{ik+1}\cdots j_{(i+1)k}=b_1\ldots b_k,\quad i=0,\ldots,n-1\}$.

This definition has the advantageous property that once a sequence is k-uniform, it is $(k-1)$-uniform at the same time. Logically from a random sequence one expects high order uniformity. This necessitates in formulating

Definition 3. An infinite sequence $u_1,\ldots,u_n$ is ∞-uniform for any natural number k.

The ∞-uniformity, of a sequence in itself ensures the existence of several desirable properties what are expected to be attributed to a random sequence. For example it can be shown that an ∞-uniform sequence behaves in many statistical tests as random, i.e. if in the test the initial hypothesis is the uniform distribution then it is, because of the

∞-uniformity, accepted. Examples of such statistical tests are the permutation test, the pairwise correlation, the sequence, the t-maximum, the runs up and down, the gap and the poker test (see Section 2.8). On the other side it can be proved, that a random uniform sequence (a sequence of the realizations of a random variable) is always ∞-uniform (with probability 1), which means that the randomness and the ∞-uniformity is almost equivalent. Thus the following definition of randomness is reasonable.

Definition 4. An ∞-uniform infinite sequence is called random.

Based on our intuitive perception of a random sequence one would expect that if the infinite sequence $u_1, \dots, u_n, \dots$ is random than its subsequences should also be random. But the ∞-uniformity of a sequence does not imply the ∞-uniformity of its subsequences. Therefore Definition 4 is to be modified in such a way that the sequence $u_1, \dots, u_n, \dots$ is called random if almost all of its subsequences are also ∞-uniform (subsequences obtained via "simple" rules from the original sequence — we refer the precise formulation for sake of simplicity).

It is difficult to see how can we interpret the randomness of a finite sequence. Since in simulation almost always we work with finite sequences, such a definition is very desirable. Since the limits in expressions (2.1) and (2.2) do not have sense in this case, we have to seek other ways. Two possibilities are mentioned here how the concept of randomness can be defined for finite sequences.

Consider a finite sequence of digits $j_1, \dots, j_N$ in base b, $0 \leqq j_i < b$, $i = 1, \dots, N$ and let $\nu(N)$ denote the number of occurrences of the event A which is defined for $j_1, \dots, j_N$.

Definition 5. The event A has an approximate probability p if

$$\left|\frac{\nu(N)}{N} - p\right| \leqq \frac{1}{\sqrt{N}},$$

and for this we use the notation $P(A) \approx p$.

This inequality is in accordance with the definition of probability; the longer is the sequence the smaller is the deviation of the relative frequency from the probability. Using this definition we can formulate the k-uniformity for finite sequences as

Definition 6. The sequence $j_1, \dots, j_N$ is k-uniform if

$$P(j_{qk+1} \dots j_{(q+1)k} = b_1 \dots b_k) \approx \frac{1}{b^k}$$

for any k digit number $b_1 \dots b_k$ in base b, $0 \leqq b_i < b$.

Contrary to the situation encountered in connection with infinite sequences, if a finite sequence is k-uniform, it does not follow that the sequence is $(k-1)$-uniform. Nevertheless the randomness is formulated on the base of k-uniformity.

Definition 7. A finite sequence of digits $j_1, \dots, j_N$ is base b is called random, if it is k-uniform for any $k \leqq \log_b N$.

According to this definition the longer is the sequence the higher uniformity is required from the sequence in order to be able to call it random.

Example 1. Let $b = 2$, $N = 3$, then $k \leqq \log_2 3$. Thus a sequence of 3 binary bits j_1, j_2, j_3 has to satisfy the 1-uniformity only and this is the condition requiring

$$P(j_i = b_j) \approx \frac{1}{2}.$$

From this one can obtain the inequality

$$\frac{1}{2} \cdot N - \sqrt{N} \leqq \nu(N) \leqq \frac{1}{2} \cdot N + \sqrt{N}, \text{ that is } -0.2 \leqq \nu(N) \leqq 3.2$$

for the frequency $\nu(N)$ which means no restrictions on j_1, j_2, j_3. Consequently any sequence of 3 bits is called random according to Definition 7.

Example 2. Let $b = 2$, $N = 6$, then the sequence $j_1, \dots, j_6$ is to be 1-uniform and 2-uniform. From the 1-uniformity similarly to the above we obtain

$$1 \leqq \nu_1(N) \leqq 5$$

while the 2-uniformity gives the meaningless $0 \leqq \nu_2(N) \leqq 3$ restriction ($\nu_1(N)$ means here the number of the bit b_i among the bits $j_1, \dots, j_6$, $\nu_2(N)$ means the number of occurrences of the pair $b_i b_j$). Consequently from the sequences of six bits those are random in the sense of Definition 7 which contain at least one 0 and at least one 1, this altogether gives us $2^6 - 2$ random sequences of length 6.

The randomness of a finite sequence can be defined in the light of the remark after Definition 4. Consider a set B of algorithms, that determine a subset $i_1, \dots, i_m$ of the indices $1, 2, \dots, N$.

Definition 8. A sequence of digits $j_1, \ldots, j_N$ in base b is called (n, ε) random with respect to the algorithm set B if for any index sequence $i_1, \ldots, i_m$ determined by an algorithm from B we have either

(i) $m < n$, or

(ii) $\left|\frac{1}{m}\nu_c(j_{i_1}, \ldots, j_{i_m}) - \frac{1}{b}\right| \leqq \varepsilon,$

where $\nu_c(j_{i_1}, \ldots, j_{i_m})$ denotes the number of occurrences of the digit c among the numbers $j_{i_1}, \ldots, j_{i_m}$ $(0 \leqq c < b)$.

Actually, if the sequence is long enough the definition requires that it be 1-uniform. According to this definition of randomness there exist random sequences, if the algorithms in B are not too numerous and sophisticated, compared to the value of N.

Example 3. Consider the following rules for selecting a subsequence of indices

(i) Take the whole sequence

(ii) Select the odd indices

(iii) Take the indices from the first element to the half of the sequences and one more (the next one).

Consider the sequence $j_1, \ldots, j_8$ of binary bits $(b = 2)$. It will be $(4, \frac{1}{8})$ random with respect to the algorithm set ((i), (ii) and (iii)) if the following conditions hold

1. $\left|\frac{1}{8}(j_1 + \ldots + j_8) - \frac{1}{2}\right| \leqq \frac{1}{8}$, which means that there are three, four or five zeros in the sequence.
2. $\left|\frac{1}{4}\left(j_1 + j_3 + j_5 + j_7) - \frac{1}{2}\right)\right| \leqq \frac{1}{8}$, that is exactly two elements with odd indices are zero.
3. $\left|\frac{1}{5}(j_1 + \ldots + j_5) - \frac{1}{2}\right| \leqq \frac{1}{8}$, which means that among the first five elements, there should be two or three zeros.

There are sequences satisfying the restrictions 1, 2 and 3, for example the sequence 10011001 is $(4, \frac{1}{8})$ random.

If the number of selecting algorithms in the set B is increased (or their sophistication), then for any given N we can reach a point where no (n, ε) random sequence exists. On the other hand according to a theorem we know that if the number of algorithms in B is less than $\frac{1}{2}\exp\{2n\varepsilon^2(1 - \varepsilon)\}$, then there exists an (n, ε) random sequence always.

2.3. Natural random numbers

Assignment of a numerical value to a state of a physical system or a natural phenomenon may provide us with a source of natural numbers. As a basis of this procedure of course we choose a phenomenon exhibiting random behaviour, for example it may be a "noisy" telecommunication device and its noise level, cosmic radiation or radioactive disintegration.

Special gadgets were manufactured in order to directly produce decimal digits (electronic rulett), or throwing an icozaeder (a polyeder with 20 facets, where each of the digits $0, 1, \ldots, 9$ was written on two facets) and recording the outcomes also can serve as a source of random numbers. We will investigate only the case of producing a random sequence of binary digits, that is we want to produce an infinite sequence $b_1, b_2, \ldots b_n \ldots$ of zeros and ones, such that the two digits are occurring with equal probability $1/2$, and b_{j+1} is independent from b_i, $i = 1, \ldots, j$. Writing k digits from such a sequence one after the other as $u_i = 0,\ b_{i+1}, \ldots, b_{i+k}$, this number in base 2 will be uniformly distributed in the interval $[0, 1)$. Two devices' structure will be outlined here that may produce a random sequence $b_1, \ldots, b_n, \ldots$.

Consider a noisy telecommunication device, such as a phone line, or a transistor, that can be augmented by a measuring and signalling device. This device is capable of registering the number of the events, whenever the noise climbed over a critical noise level c during a fixed time period Δt (Fig. 5). If the number is even, we will register a 0, otherwise a 1. By a correct choice of Δt and c a good random sequence can be produced. If simultaneously k devices are operating in a parallel way, then during the time Δt we can obtain a random number with k bits, uniformly distributed in $[0, 1)$.

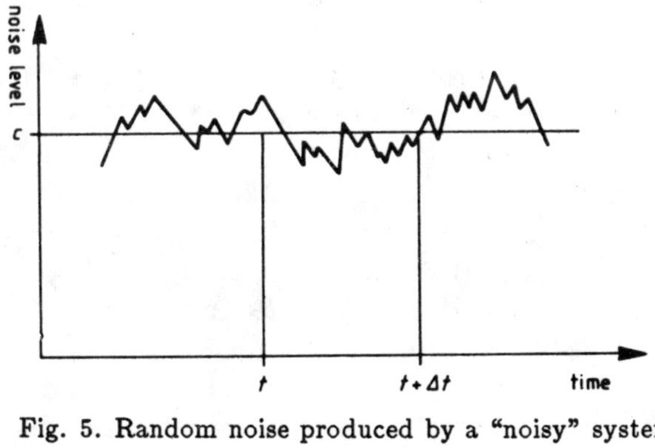

Fig. 5. Random noise produced by a "noisy" system

Another, more sophisticated equipment for generating natural random numbers was a gamma particle emanating source, a gamma particle meter, 4000 detectors and a clock generator giving signals in Δt time apart (Fig. 6).

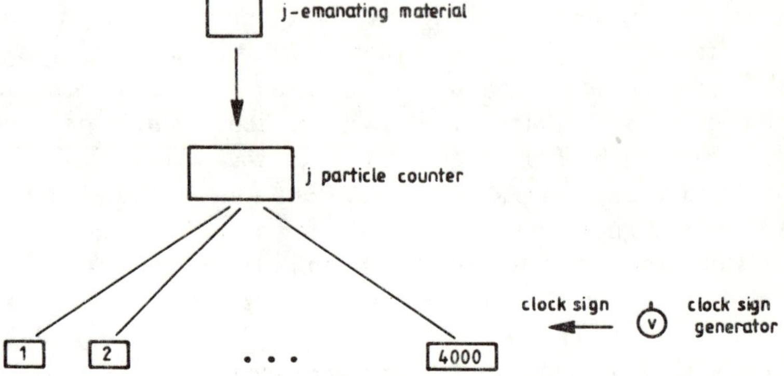

Fig. 6. Natural random number generator based on radioactive material

The number of emanated gamma particles from the first clock signal to the second clock signal is registered by the first detector, from the second clock signal to the third one the number of particles are registered by the second detector, the third detector counts the particles up to the fourth clock signal and so forth, S_i will denote the number of the registered gamma particles by the ith detector during the relevant time-period Δt $(i = 1, \dots, 4000)$. The whole procedure is repeated from the beginning, yielding the numbers S_i', and a third time going through the same routine, the numbers S_i'' are gained. Consider now the last (least significant) bit b_i in the binary form of the sum $S_i + S_i' + S_i''$ as a random bit, then this procedure gives us 4000 random bits in one step (in $3\Delta t$ time).

Up to now we tacitly assumed, that the subsequent bits produced by this or some other machinery are independent of each other and the probability of occurrence of the zeros and ones is equal. If this second assumption does not hold, that is the probability of zero's occurrence is p, the probability of a 1 is $1 - p$ we can still remedy the situation in the following way. Both pairs (0,1) and (1,0) occur with equal probability $p(1 - p)$, provided the subsequent bits b_i and b_{i+1} are independent of each other. Split up the original bit sequence $b_1, b_2, \dots$ into the pairs $(b_1, b_2), (b_3, b_4), \dots$ and simply delete the pairs (0,0) and (1,1) whenever they occur. Define now $b_i' = 0$ if the ith pair is (0,1) and $b_i' = 1$ otherwise.

In this new sequence $b'_1, b'_2, \ldots, b'_n, \ldots$ the probability of 0-s and 1-s is equal.

Several objections can be raised against the use of natural random numbers. The sequence cannot be repeated though it is frequently desirable when comparing results of tests or debugging programs. The physical devices should be kept under constant surveillance to ensure that the produced sequence is still random and has the desired distribution or its distribution has been distorted because of the device's wearing out. It is quite possible furthermore that the distribution of the produced sequence is slightly different from the desired one because of a defect in the device but this distortion can be detected only after a longer time period, when the distortion has caused some fatal defect in the computation that is hardly curable if at all. A partial remedy to these problems can be achieved by carefully checking out a sequence and storing the same on a magnetic tape.

Natural random numbers comprised in tables were published in several books (a small table is reproduced in Chapter 5). These are generally natural random numbers thoroughly tested. In case of computer use these numbers are stored on magnetic tapes or disks. The tables can be used for small examples in hand calculations, the numbers stored on magnetic tapes are of practical importance. A frequent objection to their use is that the sequence of numbers is not infinite but very limited. If more runs are to be performed, or calculations with heavy demand on random numbers are done then the resources of a tableau or a magnetic tape can easily be exhausted. 10^6–10^7 numbers that are contained in most tableaus are frequently required in practical computations. Furthermore we do not want to use all the tabulated numbers but only 20–30% of them in one calculation in order to retain randomness. The randomness and the number of random quantities can be increased by using different "reading" rules: every third number is used only, or some bits are deleted, etc. As an advantage of natural random numbers stored on a magnetic tape, we have to note they can be tested very thoroughly by several statistical tests in advance, discarding the nonrandomly behaving portions of the sequence.

2.4. Quasirandom numbers

Realizations of random variables can be substituted by quasirandom numbers. As later we will see, all Monte Carlo problems can be described as multidimensional integrals, sometimes with a very high dimensionality, though. If our problem is the evaluation of an integral whose integration domain is an n-dimensional rectangle then we can use the well-developed theory of the quasirandom numbers' application, that is described in this point.

Consider the problem of evaluating the integral

$$I = \int_0^1 \dots \int_0^1 f(\mathbf{x}) d\mathbf{x}. \tag{2.3}$$

The value I is approximated by the sum

$$\theta = \frac{1}{n}(f(\xi^{(1)}) + \dots + f(\xi^{(N)})), \tag{2.4}$$

where the points $\xi^{(i)}$, $i = 1, \dots, N$ are located in the n-dimensional unit cube $K = \{\mathbf{x} \mid \mathbf{0} \leqq \mathbf{x} \leqq 1\}$. If the points $\xi^{(1)}, \dots, \xi^{(N)}$ are random vectors, uniformly distributed in K, then it can be proved easily, that the error of approximating the integral is of order $1/\sqrt{N}$, that is

$$|I - \theta| < c \cdot \frac{1}{\sqrt{N}} \tag{2.5}$$

holds with some constant c (see Section 6.4). The point of applying quasirandom points is that for certain fixed point sequences $\xi^{(1)}, \dots, \xi^{(N)}$ (quasirandom points) the magnitude of error is of $(\log N)/N$ instead of the $\frac{1}{\sqrt{N}}$ characteristic for Monte Carlo method and the random points. To present the theory some new definitions are to be introduced.

Consider the n-dimensional brick $B = \{\mathbf{x} \mid \mathbf{0} \leqq \mathbf{x} \leqq \mathbf{b}\}$ inside the cube K, given by the vector $0 \leqq \mathbf{b} \leqq 1$, its volume is $b_1 b_2, \dots, b_n$. Let $S : (\xi^{(1)}, \dots, \xi^{(N)})$ be a fixed point sequence in cube K, the number of points inside B will be denoted by $\mu(\mathbf{b})$, i.e. $\mu(\mathbf{b}) = \#\{\xi^{(i)} \mid \xi^{(i)} \in B, i = 1, \dots, N\}$. Then the uniformity of the sequence S in the cube K can be measured by the difference

$$\frac{\mu(b)}{N} - b_1 b_2 \dots b_n, \tag{2.6}$$

which gives the difference between the relative frequence and the probability of the event that a point falls into B. This difference (2.6) is also called local discrepancy.

Definition 1. The (global) discrepancy of the sequence $S = (\xi^{(1)}, \dots, \xi^{(N)})$ is the quantity

$$(2.7) \qquad D_N(S) = \sup_{\mathbf{b} \in K} \frac{1}{N} |\mu(\mathbf{b}) - N b_1 b_2 \dots b_n|.$$

The smaller is the discrepancy of a sequence, the more evenly is the sequence spread in the cube K.

The variation of the function $g(x)$ of one variable is the smallest constant $V(g)$ which is an upper bound for the sum

$$(2.8) \qquad \sum_{i=1}^{m} |g(x_i) - g(x_{i-1})|$$

for any m and splitting points $0 < x_1 < \dots < x_m < 1$. If $V(g)$ is finite, then the function g is called of finite variation. This definition can be extended to the multidimensional case; we give the definitions for the two-dimensional function $f(x, y)$ only for the sake of simplicity.

Consider the two-dimensional unit square $K = \{(x, y) \mid 0 \leqq x, y < 1\}$ and divisions of sides given by the point sequences $x_1, x_2, \dots, x_m$ and $y_1, y_2, \dots, y_k$, where $x_i < x_{i+1}$, $y_j < y_{j+1}$.

Definition 2. The two-dimensional Vitali variation of the function $f(x, y)$ is

$$V^2(f(x, y)) = \limsup \sum_{i=1}^{m} \sum_{j=1}^{k} |f(x_{i+1}, y_{j+1}) - f(x_{i+1}, y_j) - f(x_i, y_{j+1}) + f(x_i, y_j)|.$$

If $V^2(f)$ is finite, then $f(x, y)$ is of finite (Vitali) variation.

A function with finite variation is expected to be "smooth". Functions twice differentiable are of finite variation. On the other hand, one can easily construct an example that has finite variation but may be of very unpleasant shape; consider $f(x, y) = h(x)$, that is f does not depend on y, it implies $V^2(f) = 0$ no matter what shape $h(x)$ has. To exclude such erratic behaviour it seems to be expedient to introduce the following slightly more stringent definition.

Definition 3. The function $f(x, y)$ is called of finite variation in the sense of Hardy–Krause if the two- and one-dimensional variations $V^2(f(x, y)), V(f(x, 1)), V(f(1, y))$ in the sense of Vitali are finite.

A sufficient condition to ensure the finite variation of a function in the sense of Hardy–Krause is if the existence of the continuous second-order derivatives is ensured. Very simple example can be given to be of no finite variation. Consider the function

$$f(x,y) = \begin{cases} 1 & \text{if} \quad x \geqq y \\ 0 & \text{otherwise.} \end{cases}$$

This function is not of bounded variation; a function of bounded variation may only have enumerable discontinuous points on the hyperplanes parallel to the coordinate axis.

Now we present the theorem without proof that provides us with the theoretical basis for employing quasirandom points in multidimensional integration.

Theorem 1. If the function $f(x,y)$ is of bounded variation in the sense of Hardy–Krause, a sequence S of points in the cube K is the product of the one-dimensional sequences S_x and S_y, where

$$\begin{aligned} S &= \{(x_i, y_j)\,, & i &= 1,\ldots,m, \quad j = 1,\ldots,k\}\,, \\ S_x &= \{x_i, & i &= 1,\ldots,m\}\,, \\ {}_y &= \{y_j, & j &= 1,\ldots,k\}\,, \end{aligned}$$

then using the notation $N = mk$

$$\left| \int\int f(x,y)\mathrm{d}x\mathrm{d}y - \frac{1}{N}\sum_{i=1}^{m}\sum_{j=1}^{k} f(x_i,y_j) \right| \leqq V^2(f(x,y))D_{mk}(S) + \\ + V(f(x,1))D_m(S_x) + V(f(1,y))D_k(S_y).$$

Based on this theorem one is inclined to construct point sequences S with low discrepancy for numerical integration of functions (there is nothing we can do about the variations since it depends on the function). Some of the known constructions will be described next.

In the one-dimensional case the sequence $S = (x_m = \{m\alpha\},\ m = 1,2,\ldots,N\}$ is of interest, where α is an irrational constant, $\{\cdot\}$ denotes the fraction part function, e.g. $\{3.14\} = 0.14$. According to a well-known number theoretical proof we have

$$\lim_{N\to\infty} D_{-N}(S) = 0.$$

The result can easily be modified for the n-dimensional case. Consider the sequence S of the points $\mathbf{x}_m$, for which

$$\mathbf{x}_m = (\{m\alpha_1\},\ \{m\alpha_2\},\ldots,\{m\alpha_n\}) \tag{2.9}$$

where $\alpha_1,\ldots,\alpha_n$ are independent irrational numbers. (The numbers $\alpha_1,\ldots,\alpha_n$ are called independent irrational numbers if they are the roots of an at most $(n+1)$-st degree polynomial with integer coefficients, but the equality $c_1\alpha_1 + \ldots + c_n\alpha_n = 0$ with integer coefficients c_i can hold for $c_i = 0$, $i = 1, \ldots, n$ only.) Such a sequence S may be the one having the points $\mathbf{x}_m = (\{m\sqrt{2}\}\ \{m\sqrt{3}\}, \ldots, \{m\sqrt{p_n}\})$, where p_n is the nth prime number. The following theorem is presented without proof, too.

Theorem 2. If the Fourier series of the function f is absolutely convergent and $\mathbf{x}_m$ is the sequence given in (2.9), then with a constant c_2 we have

$$\left| \int f(\mathbf{x})\, dx - \frac{1}{N} \sum_{i=1}^{N} f(\mathbf{x}_i) \right| \leqq c_2 \frac{1 + \log N}{N}.$$

Another quasirandom sequence of low discrepancy, giving good approximations to integrals can be constructed in the following way. Let the positive integer r be given in base b as $r = (a_1 a_2, \ldots, a_m)_b$ that is

$$r = a_m + b a_{m-1} + \ldots + b^{m-1} a_1.$$

The function $F_b(r)$ is defined on these integers as

$$F_b(r) = a_m b^{-1} + a_{m-1} b^{-2} + \ldots + a_1 b^{-m},$$

that can be perceived as reflecting the number r to its "decimal" point. Let now $p_1, \ldots, p_{n-1}$ be the first $n-1$ number and denote the sequence of the points

$$\mathbf{x}_m = \left(\frac{m}{N}, F_{p1}(m), F_{p2}(m), \ldots, F_{pn-1}(m) \right) \tag{2.10}$$

by S'. Then the following theorem holds.

Theorem 3. We have with a constant c_3 for S' the inequality

$$D_N(S') \leqq c_3 \frac{(\log N)^{n-1}}{N}.$$

A lower bound can be established for a quantity closely related to the discrepancy. Consider the square discrepancy given by

$$D_N^2(S) = \int_K (\mu(\mathbf{b}) - N b_1 \ldots b_n)^2\, d\mathbf{b}$$

for which it can be shown that the inequality

$$D_N^2(S) \geqq c_{4,n} (\log N)^{n-1} \tag{2.11}$$

is true with any sequence S ($c_{4,n}$ is a constant, dependent on n). This inequality gives a lower bound on the integration error that arises when using quasirandom sequences. In some dimensions sequences can be constructed that have faster convergence rates than the ones in Theorem 2 and Theorem 3, but of course satisfying (2.11), too.

Example 1. The first couple of points in sequence (2.10) will be determined here in $n = 3$ dimensions working with binary numbers ($b = 2$).

Table 1

The first quasirandom number by (2.10)

m	m	m	$F_2(m)$	$F_2(m)$	$F_3(m)$	$F_3(m)$
	binary	trinary	binary	decimal	trinary	decimal
1	1	1	0.1	0.5	0.1	0.333
2	10	2	0.01	0.25	0.2	0.666
3	11	10	0.11	0.75	0.01	0.111
4	100	11	0.001	0.125	0.11	0.444
5	101	12	0.101	0.625	0.21	0.777
6	110	20	0.011	0.375	0.02	0.222

Thus the first six points from the sequence S are:

$$\begin{aligned}
x_1 &= (\,0.166,\ 0.5,\ 0.333),\\
x_2 &= (\,0.333,\ 0.25,\ 0.666),\\
x_3 &= (\,0.5,\ 0.75,\ 0.111),\\
x_4 &= (\,0.666,\ 0.125,\ 0.444),\\
x_5 &= (\,0.833,\ 0.625,\ 0.777),\\
x_6 &= (\,1.000,\ 0.375,\ 0.222),
\end{aligned}$$

Summarizing previous remarks the following assessment of quasirandom numbers can be made. Using random numbers for evaluating integrals the error is of $O(N^{-1/2})$, where N denotes the number of points employed, while the use of quasirandom numbers gives a substantially faster diminishing error $O(\log(N)^{n-1}/N)$. As a disadvantage of quasirandom numbers we can mention that the constants figuring in the error estimates cannot be evaluated generally, and sometimes they are very great rendering the application of quasirandom numbers impractical. If great batches of quasirandom numbers are needed then precision of a number representable on a computer may be a problem (or alternately the additional computational work required to carry out the calculations with increased accuracy e.g. in double precision) — this may be the case with the sequences of type $\{m\alpha\}$, where for numbers with $m > 10^6$, six digits of α will be lost. A further drawback is that quasirandom points can only be applied for certain class of problems, integration over an n-dimensional rectangle. There are no known results concerning construction of point sequences over a hypersphere or other domains, neither a result corresponding to Theorem 1. But on the other hand for specific dimensions ($n = 3$–8) very good results were obtained in evaluating integrals by quasirandom sequences.

2.5. Congruential generators

Pseudorandom numbers are produced by algorithms on digital computers. Numbers so obtained are completely deterministic, since the fixed algorithm and parameter values together with a possible choice of an initial starting value determine the whole sequence; in practice we are satisfied with the semblance that these numbers appear as if they came from a random source.

Several algorithms can be used for producing uniformly distributed pseudorandom numbers. The most frequently used type is called the congruential generator, this will be considered below in detail. We will limit ourselves to deal with the production of numbers u_i, uniformly distributed in the interval $[0,1)$, for numbers uniformly distributed in any interval $[c,d)$ can be obtained by the linear transformation $(d-c)u_i + c$ from the numbers u_i.

Consider the recursive relationship

$$z_{i+1} = bz_i + r \pmod{m} \quad i = 0, 1, 2, \ldots \tag{2.12}$$

where b and m are positive integers, $r \geqq 0$ is integer. Having fixed a nonnegative integer z_0 as an initial value the numbers $z_1, z_2, \ldots, z_n, \ldots$ can recursively be evaluated by (2.12) that in turn are used to obtain the values

$$u_i = z_i/m, \quad i = 1, 2, \ldots \tag{2.13}$$

in the interval $[0,1)$. The random number generator based on the recursions (2.12) and (2.13) is called congruential generator. (In the sequel, if it is not explicitly stated otherwise the term "random number" will refer to pseudorandom number, similarly the term random number generator means an algorithm capable of producing pseudorandom numbers.) The corresponding computer algorithm may be the following. Assume that the variable z holds the previously generated (last) value z_i or before the first activization of the algorithm z contains the initial value z_0.

UC Uniform, congruential

1. Set $z \leftarrow bz + r$.
2. Compute the residue $z \leftarrow z \pmod{m}$.
3. Deliver the value $u \leftarrow z/m$ as a number uniformly distributed in the interval $[0, 1)$.

In the description of algorithms the expression $x \leftarrow y$ means transfer of the value, in FORTRAN the corresponding formula is $x = y$, in ALGOL the expression $x := y$ has the same meaning. The word "deliver" signifies the end of the algorithm (in FORTRAN it would be replaced by RETURN), the calculations necessary to produce the number with the desired distribution are completed there.

The constant b is the multiplier of the generator, r is the increment and m is the modulus. If the increment is zero ($r = 0$), then the generator is called simple or pure congruential generator, if $r \neq 0$ then it is a mixed congruential generator. The advantage of a pure generator over a mixed one is, that in evaluating the random number we need one addition less.

The properties of the sequence u_i are dependent on the constants b, r, m and z_0. In the sequel we will scrutinize the constraints under which the sequence u_i will have desirable properties. For the sake of simplicity the sequence of numbers z_i will be investigated instead of the sequence u_i.

The most important property of the sequence is the number of different numbers appearing in the sequence. Any number z_{i+1} is dependent on the previous number z_i thus if one value already evaluated recurs, it will mean repeating the same sequence, beginning with the recurring value.

Definition 1. The smallest integer number n, for which the values z_i recur, that is $z_i = z_{i+n}$ is called the period of the generator.

Example 1. Consider the congruential generator

$$z_{i+1} = 3z_i + 7 \pmod{15}.$$

Using the initial value $z_0 = 1$ the generator produces the sequence $1, 10, 7$, 13, 1,... that is the period is 4. If the initial value $z_0 = 4$ is used, then the sequence is $4, 4, \ldots$, the period of the generator is 1 only.

A congruential generator can produce the numbers $0, 1, \ldots, m - 1$ at most, which implies that the period of a generator is m or less than m. What choice of the constants b, r, m and z_0 will ensure that the maximum period is actually attained and what is the maximum period length for a given m (or b, etc.)? The random number generator $z_{i+1} \equiv z_i + r$ corresponding to the choice of $b = 1$ is of no interest since the resulting sequence can hardly be accused of random behaviour, thus the case $b = 1$ will be excluded from the further considerations. Furthermore we assume for the sake of simplicity, that z_0 is repeated first, i.e. $z_0 = z_{0+n} \pmod{m}$.

First it will be shown that only the generators with prime modulus should be of concern to us. Namely if m can be given as the product of two relative primes then the period belonging to the modulus m can be obtained through the following

Theorem 1. Let $m = m_1 m_2$, where m_1 and m_2 are relative primes. Denote the period length of the generator with modulus m by $n(m)$, then we have

$$n(m) = \mathrm{lcm}(n(m_1), n(m_2)). \tag{2.14}$$

Proof. Let the sequences $x_i, y_i, z_i, \quad i = 1, 2, \ldots$ be given by the recurrence relationships

$$\begin{aligned} x_{i+1} &\equiv bx_i + r && \pmod{m_1} \\ y_{i+1} &\equiv by_i + r && \pmod{m_2} \\ z_{i+1} &\equiv bz_i + r && \pmod{m} \end{aligned}$$

using the same initial value $x_0 = y_0 = z_0$. According to the last congruence we have

$$z_{i+1} = bz_i + r + km$$

with some integer constant k, and similarly $x_1 \equiv z_1 = bz_0 + r + km \pmod{m_1}$ which by induction implies

$$x_{i+1} \equiv z_{i+1} \pmod{m_1} \tag{2.15}$$

m_1 being a divisor of m. Similarly we can derive

$$y_{i+1} \equiv z_{i+1} \pmod{m_2}. \tag{2.16}$$

Using the notations $n_1 = n(m_1)$, $n_2 = n(m_2)$, $n = n(m)$ for the period length of the generators we can write

$$\begin{aligned} z_0 = x_0 &\equiv x_{0+n_1} \equiv z_{0+n_1} && \pmod{m_1} \\ z_0 = y_0 &\equiv x_{0+n_2} \equiv z_{0+n_2} && \pmod{m_2} \end{aligned}$$

where the previous remarks and the definition of the period length were employed. Denote the least common multiple of n_1 and n_2 by q, $q = \mathrm{lcm}(n_1, n_2)$ then the equations

$$z_{0+q} = z_0 + k_1 m_1$$
$$z_{0+q} = z_0 + k_2 m_2$$

hold with some integer constants k_1 and k_2 which in turn implies $k_1 m_1 = k_2 m_2$. According to our assumption m_1 and m_2 are relative primes thus m_2 is a divisor of k_1 which means

$$z_{0+q} \equiv z_0 \pmod{m}, \tag{2.17}$$

that is the period length $n = n(m_1 m_2)$ is a divisor of the least common multiple $q = \mathrm{lcm}(n(m_1), n(m_2))$ since q is an index for which the sequence recurs in light of (2.17).

On the other side according to the definition we have

$$z_{0+n} \equiv z_0 \pmod{m},$$

that is $z_{0+n} - z_0 = k_0 m_1 m_2$ with some integer constant k_0. It follows evidently, that

$$z_{0+n} - z_0 \equiv 0 \pmod{m_1},$$
$$z_{0+n} - z_0 \equiv 0 \pmod{m_2},$$

that is n is a repetition number for both the x_i and the y_i sequences. Both n_1 and n_2 are divisors of n, since the period length is the smallest repetition number, which implies that $q = \mathrm{lcm}(n_1, n_2)$ is a divisor of n, too. This proves the assertion of the theorem.

An immediate consequence of this theorem is

Theorem 2. Let the modulus m of the congruential generator be of the form $m = p_1^{e_1} p_2^{e_2} \dots p_s^{e_s}$, where $p_1, \dots, p_s$ are relative primes, then the period length is

$$n(m) = \mathrm{lcm}\left(n\left(p_1^{e_1}\right), n\left(p_2^{e_2}\right), \dots, n\left(p_s^{e_s}\right)\right).$$

Consider now the possible choices of the modulus m. A very comfortable choice is $m = 2^e$, where e is the number of the bits of the greatest integer in the computer, since in this case if the product bz_i is greater than m, then the greater part is immediately lost as a result of automatic overflow and the remaining part is exactly the product (mod m). One has nevertheless to exercise a certain amount of caution while using modulus of this type, $m = 2^e$. Consider the generator given by $z_{i+1} \equiv bz_i + r \pmod{m}$, let d be a divisor of m and let $x_{i+1} \equiv bx_i + r \pmod{d}$ a related generator. Then similarly to the argumentation of Theorem 1 one can show that $x_i \equiv z_i \pmod{d}$. This implies that the period of the numbers x_i is at most d. In other words if the modulus $m = 2^e$ was chosen then a divisor of m is $d = 2^3$, which means that the last three bits of the generated numbers z_i will form a sequence with period length 8 at most (the last five bits will form a sequence with period length 32 in the most favourable case). The bottom line of this is that the lower bits of a sequence produced by a congruential generator with $m = 2^e$ tend to be less random, and it has to be given attention in applications, especially if the lower bits of the numbers are separately used (e.g. for producing random indices).

Another possible choice is if $m = 2^e - 1$ is used: in this case the nonrandom behaviour of the less significant digits can be avoided too. In Table 2 we present the factorization of the modulus $m = 2^e - 1$ for the most frequently occurring word lengths e. Finally the m modulus can be the greatest prime representable in the computer. In this case evaluation of the remainder is not so easy as in the previous case but the lower bits of the number will behave in a random manner and the period length is generally greater than the period length for modulus $m = 2^e$ as we will see. The greatest primes in a computer with word length e are given in Table 3 for the most frequent values of e.

Table 2

Prime factors of the modulus $m = 2^e - 1$

e	$2^e - 1$
16	$3 \cdot 5 \cdot 17 \cdot 257$
23	$47 \cdot 178481$
24	$3^2 \cdot 5 \cdot 7 \cdot 13 \cdot 17 \cdot 241$
31	2147483647
32	$3 \cdot 5 \cdot 17 \cdot 257 \cdot 65537$
35	$31 \cdot 71 \cdot 127 \cdot 122921$
36	$3^2 \cdot 5 \cdot 7 \cdot 13264529$
47	$2351 \cdot 4513 \cdot 13264529$
48	$3^2 \cdot 5 \cdot 7 \cdot 13 \cdot 17 \cdot 97 \cdot 241 \cdot 257 \cdot 673$
59	$179951 \cdot 3203431780337$
60	$3^2 \cdot 5^2 \cdot 7 \cdot 11 \cdot 13 \cdot 31 \cdot 41 \cdot 61 \cdot 151 \cdot 331 \cdot 1321$
63	$7^2 \cdot 73 \cdot 127 \cdot 337 \cdot 92737 \cdot 649657$
64	$3 \cdot 5 \cdot 17 \cdot 257 \cdot 641 \cdot 65537 \cdot 6700417$

The period length attains the maximum m if only all the numbers between 0 and $n - 1$ will appear once, including 0 too. Assume $z_0 = 0$. Repeatedly applying the recursive relation one can obtain

$$\mathbf{z}_i \equiv \frac{b^j - 1}{b - 1} \cdot r \qquad (\text{mod } m)$$

as the ith term in the sequence. Hence to achieve maximum period length the numbers m and r should be relative primes. This and other necessary conditions to ensure maximum period length are summarized in the following theorem given without proof.

Table 3

The greatest prime numbers p, for which $p < 2^e$

e	p
16	65 521
23	8 388 593
24	16 777 213
31	2 147 483 649
32	4 294 967 291
35	34 359 738 337
36	68 719 476 731
47	140 737 488 355 213
48	281 474 976 710 597
59	576 460 752 303 423 433
60	1 152 921 504 606 846 883
63	9 223 372 036 854 775 783
64	18 446 744 073 709 551 557

Theorem 3. The random number generator given by $z_{i+1} \equiv bz_i + r \pmod{m}$ ($r \neq 0$, $b \neq 1$, $m = p^e$, where p is prime, $e > 0$ integer, $p^e > 2$) has maximum period length $n(m) = m$ if and only if

(i) r and m are relative primes
(ii) $b \equiv 1 \pmod{p}$
(iii) $b \equiv 1 \pmod{4}$ if $p = 2$.

Consider now the pure congruential generator where $r = 0$. The maximum period length may be $m - 1$ at most, since if 0 turns out once, the generator produces only zeros afterwards. Assuming the initial value z_0 and applying the recursive relation we have

$$z_i \equiv b^i z_0 \qquad \pmod{m}$$

which means that z_0 and b should be relative prime to m. We need some more definitions to describe the necessary and sufficient conditions for maximum period length.

Definition 2. The smallest number λ for which

$$b^{\lambda} \equiv 1 \pmod{p^{e-f}}$$

holds, where λ is dependent on $m = p^e$ and b as well, and $p^f = \gcd(b, m)$ defines f, — is called the order of the number b with respect to the modulus m.

Definition 3. Those numbers b whose order with respect to the modulus m will be the maximum $\lambda = \lambda(m)$ are called primitive elements.

Since the order is the same as the period length of the generator thus all we need is the primitive elements and their order. In the two subsequent theorems, the function $\lambda(m)$ is given without proof and the existence of the primitive elements is ensured.

Theorem 3. The maximum order with respect to modulus $m = p^e$ is denoted by $\lambda(m)$ and is given as

$$\lambda(p^e) = p^{e-1}(p-1) \quad \text{if} \quad p \text{ is an odd prime,}$$

$$\lambda(2^e) = \begin{cases} 1 & \text{if} \quad e = 1 \\ 2 & \text{if} \quad e = 2 \\ 2^{e-2} & \text{if} \quad e > 2. \end{cases}$$

Theorem 4. For any modulus m there exists a primitive element b, whose order is $\lambda(m)$ and there are not numbers whose order is greater than $\lambda(m)$.

The next theorem gives properties of the primitive elements, that can be used towards finding primitive elements.

Theorem 5. The number b is a primitive element with respect to the modulus $m = p^e$ if

(i)
$$\begin{aligned} b &\equiv 1 \mod 2 \quad \text{if} \quad p^e = 2 \\ b &\equiv 3 \mod 4 \quad \text{if} \quad p^e = 4 \\ b &\equiv 3 \text{ or } 5 \mod 8 \quad p^e \geqq 8,\ p = 2, \end{aligned}$$

(ii) $b^{(p-1)/q} \not\equiv 1 \pmod{p}$ for any prime divisor q of the number $(p-1)$ if p is odd, $e = 1$ and $b \not\equiv 0 \pmod{p}$, or

(iii) b satisfies (ii) furthermore $b^{p-1} \not\equiv 1 \pmod{p^2}$ if $e > 1$, p odd.

Summarizing all what has been said about the maximum period length we can state

Theorem 6. The maximum period length of a random number generator $z_{i+1} \equiv bz_i \pmod{m}$ ($b \neq 1, m = p^e, p$ prime, e integer) is $\lambda(m)$ and it is attained if and only if

(i) z_0 and m are relative primes
(ii) b is a primitive element with respect to modulus m.

In the frequently encountered case of modulus $m = 2^e$ (where e denotes the word length in bits on the computer) the attainable maximum period length is 2^{e-2} and the necessary condition to have this maximum is that $b = 8k \pm 3$ with any integer k. Scrutinizing the relationship between consecutive numbers, that is their correlation it can be shown that the correlation will be the smallest if b is near to $2^{e-2}\xi$, where $\xi = \frac{1}{2}(\sqrt{5}-1) \sim 0.618033$ is the number from the golden section rule. This generator (with modulus 2^e) will supply us with 2^{e-2} numbers of the possible 2^e numbers; to have them evenly distributed on the set of 2^e notches between 1 and 2^e one needs the choice $b \equiv 5 \pmod 8$ for it gives a better result than the one resulting from the other possible choice $b \equiv 3 \pmod 8$. The reason for this difference is the following. If $b \equiv 3$ (mod 8) then starting with $z_0 = 1$ all the produced numbers will be of form $8k+1$ and $8k+3$, thus the distance between neighbouring points u_i in interval $[0,1)$ will be $2/(2^e)$ and $6/(2^e)$ alternately. With the choice $b \equiv 5 \pmod 8$, $z_0 = 1$, the generated numbers z_i will be of form $8k+1$ and $8k+5$, consequently the distance between the points u_i thus produced will always be $4/(2^e)$ — that is the set of the possible 2^e numbers is evenly covered.

Table 4

Multipliers: b of form $8k+5$ that are near $2^{e-2}\xi$

	$b \sim 2^{e-2}\frac{1}{2}(\sqrt{5}-1)$
24	2 592 221
31	331 804 469
32	663 608 941
35	5 308 871 541
36	10 617 743 077
47	21 745 137 823 725
48	43 490 275 647 445

```
SUBROUTINE RANDU (U,R)
DATA FACTOR /.../
R=DMOD (R*FACTOR, 1.0D0)
U=SNGL(R)
RETURN
END
```

For some frequent word lengths the multiplicator b, satisfying the above conditions, is given in Table 3. Note further that the computer program or a pure (or multiplicative) congruential generator is very simple, actually it contains only two executable lines in FORTRAN where prior to the first call the variable FACTOR is given the multiplier value b and R is given $1/(2^e)$. Note that this version requires R and FACTOR to be declared as of double precision. Also one can replace these two lines with the single line U=AMOD(U*FACTOR, 1.0) but in this case one has to be content with numbers of poorer quality since deformations are likely to occur due to roundoff errors.

2.6. Linear recursive generators

Uniformly distributed random bit sequences are produced by the linear recursive generators or by another name shift register generators. Let the numbers $c_1, \ldots, c_p$ be constants, $c_i = 0$ or $c_i = 1$, but $c_p = 1$, and consider the p-fold recursive congruency

$$x_k \equiv c_1 x_{k-1} + c_2 x_{k-2} + \ldots + c_p x_{k-p} \pmod 2 \tag{2.18}$$

and the bit sequence $x_1, x_2, \ldots$ produced by the same. Provided we are given an initial p-tuple of bits $x_1, \ldots, x_p$ the generation of the sequence is straightforward. The generator based on the recurrence (2.18) is called linear recursive generator (the procedure termed congruential generator is both linear and recursive, though) since the sequence produced $x_1, \ldots, x_n, \ldots$ is called linear recursive sequence in coding theory.

The period length of the sequence of the generated bits will be investigated next. For a fixed set of constants $c_1, \ldots, c_p$ the bit x_k is dependent only on the previous p-tuple $x_{k-1}, \ldots, x_{k-p}$, hence the period length of the linear recursive generator is the smallest integer, for which the p-tuple $x_{k-1}, \ldots, x_{k-p}$ is repeated. (For the sake of simplicity assume, that the first p-tuple repeated is the initial p-tuple $x_p, x_{p-1}, \ldots, x_1$.) If we have a p-tuple of zeros then any subsequent bit will be 0, that is the period length may be $2^p - 1$ at most.

Theorem 1. The period attains its maximum ($n = 2^p - 1$) if and only if the polynomial

$$f(x) = 1 + c_1 x + c_2 x^2 + \ldots + c_p x^p$$

is primitive over the finite field of the polynomials with coefficients 0 and 1 (over GF (2)).

If the polynomial corresponding to a generator is primitive then all $n = 2^p - 1$ possible p-tuples consisting of 0-s and 1-s will be generated, that is $\frac{1}{n}\sum_{k=1}^{n} x_k = \frac{1}{2}$, which is a desirable property of the generator. The algebraic background of linear recursive sequences is outlined in the Appendix.

Example. Consider the primitive trinomial $f(x) = x^5 + x^3 + 1$ and the corresponding generator. (Note here that if $f(x)$ is a primitive polynomial of order p, then $x^p f(x^{-1})$ is also primitive.) The generator is based on the recurrence

$$x_k \equiv x_{k-3} + x_{k-5} \pmod 2. \tag{2.19}$$

If our initial sequence is $x_1 = 1, \quad x_2 = 1, \quad x_3 = 0, \quad x_n = 1, \quad x_5 = 0$ then the generated sequence is as follows:

11010 10000 10010 11001 11110 00110 1

and from this bit the sequence will be repeated.

The advantage of the linear recursive generator over the congruential generator is that instead of performing a multiplication, in case of a trinomial an "exclusive or" is needed only, which is generally faster than the multiplication (Fig. 7).

$\oplus$	0	1
0	0	1
1	1	0

Fig. 7. The "exclusive or" $\oplus$ operation

This "exclusive or" operation is denoted by $\oplus$ and one can generally find it among the assembly level instructions of a computer. Besides this "exclusive or" realisation of the algorithm of the generator requires only shifts that can be executed in very short time too. Generators of this type can be hardware implemented easily, that is a unit can be built with very simple circuits, that will produce the random sequence. This is why in the engineering practice the linear recursive generators (though in essence they are also congruential generators, see the Appendix) are especially welcomed.

Example. Consider the generator based on the previously discussed congruency (2.19). The wiring realizing the generator can be seen in Fig. 8.

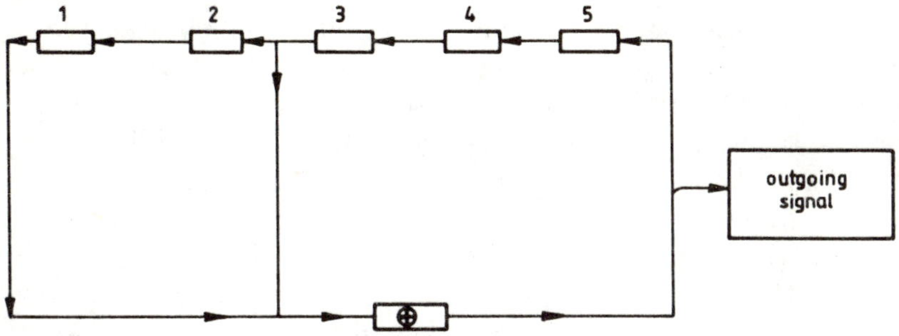

Fig. 8. Wiring model corresponding to the generator based on the primitive trinomial $x^5 + x^3 + 1$

The units $1, 2, \ldots, 5$ are memory cells (registers) consisting the previous 5-tuple (or the initial five bits, respectively). During the time interval Δt the content of each memory unit is shifted to the unit on the left, furthermore the contents of the registers 1 and 3 are summed up in the unit with sign $\oplus$, which executes an addition mod 2 and the sum is transferred into register 5. After four more repetition the obtained set of five bits is the generated bit sequence, the new 5-tuple.

Consider the primitive trinomial $x^p + x^q + 1$. We describe the corresponding computer algorithm for generating a random bit sequence. Assume, that the word length in the computer is p bits and also denote $m = p - q$. The shift operation in the computer is executed by erasing the outshifted bits and filling in zeros in empty spaces, that is from the number 11111 after left-shifting by two bits we obtain 11100, such a shift operation can generally be found among the machine coded operations of the computers. As a preparation fill in the initial p-tuple of bits into register A.

US Uniform, shifted

1. [The previous p-tuple is in register A.] Set $B \leftarrow A$ and shift the content of B by m bits to the left.
2. Compute $A \leftarrow A \oplus B$.

3. Replace $B \leftarrow A$ and right shift register B by q bits.
4. Sum up the two registers once more $A \leftarrow A \oplus B$ and deliver the number $u \leftarrow A2^{-p}$ as uniformly distributed in $[0,1)$.

Generally this algorithm is faster than the congruential generator UC, — in both cases it is assumed, that the computer program is written in low level (assembly, machine code, etc.) language. A further disadvantage of the algorithm US — (besides the inconvenience of using assembly language — this is the only expedient way of implementation) is that the period length of the generator in this form is $2^p - 1$ at most, where p denotes the word length of the computer p as well, that is p fixed. Dividing the bit sequence into p bits the resulting numbers will have a period of $(2^p - 1)/p$; the subsequent numbers even if they do not repeat exactly the first numbers of the sequence, nevertheless they will show strong correlation with the beginning of the sequence. The $(2^p - 1)/p$ different numbers in the case of a computer with $p = 32$ word length means about 2^{27} numbers which may not be a sufficient supply always so we need an enhancement of the basic method.

Example. The algorithm US corresponding to expression (2.19) performs the following operations using the initial sequence 11010. After the first step the content of register A which is denoted by (A) is 11010, that of B is $(B) = 01000$, after the second step $(A) = 10010$. In the third step (A) is not changed, but we get $(B) = 00010$ while at the end of the algorithm in register A we have the required result, 10000, which is the next five bits in the sequence.

Let $w_1, w_2 \ldots, w_p$ be p empty words, which are filled in by the first bits of the sequence $x_1, \ldots, x_n, \ldots$ produced by the generator of the polynomial $x^p + x^q + 1$. The way of filling in the bits is the following. In the first bit of each word the first p bits of the sequence are placed, the first bit of w_1 will hold x_1, the first bit of w_2 will hold x_2, etc., in the first bit of w_p we have x_p. Fix an integer k which is the delay index and in the second bit of the words $w_1, w_2 \ldots, w_p$ put the bits $x_{1+k}, x_{2+k}, \ldots, x_{p+k}$. In the third bit of the words $w_1, \ldots, w_p$ the bits $x_{1+2k}, x_{2+2k}, \ldots, x_{p+2k}$ are written, etc. This procedure is repeated until the words $w_1, \ldots, w_p$ with word length e are not filled in. Then the generalized shift algorithm for producing random *words* is the following.

As a preparation set the indices i and j to the initial values $i \leftarrow 0$, $j \leftarrow p - q$.

UGS Uniform, generalized shift

1. Increase the index i as $i \leftarrow i+1$. If $i > p$ then reset $i \leftarrow 1$.
2. Evaluate index $j \leftarrow i+q$. If $j > p$ then decrease the index by p as $j \leftarrow j-p$.
3. [Evaluate the "exclusive or" sum of the two words selected.] Set $w_i \leftarrow w_i \oplus w_j$ and deliver the number $u \leftarrow w_i 2^{-e}$ as uniformly distributed in [0, 1).

Assume that the words w_i are p bit long. Then it can be shown that if the initial words $w_1, \ldots, w_p$ are independent (the matrix containing these words has linearly independent rows) then the length of the period of the words w_i is $2^p - 1$. Such a system of independent words can be easily constructed. (To choose the first word w_1 we have $2^p - 1$ possibilities, there are $2^p - 2$ possibilities for choosing w_2, $2^p - 4$ possibilities for w_3 — since no linear combination of the $\mathbf{0}$, w_1 and w_2 words is permitted — $2^p - 8$ possibilities for w_4, etc.) Thus in case of using a trinomial of order $p = 100$ the period length may be very great (around 10^{30}). The implementation of the computer algorithm UGS is essentially independent of p, we need to store p words only, so we can implement a version with arbitrarily great power p. A generator having the value $p = 500$ will produce a sequence of numbers with period $2^{500} \sim 10^{150}$, that even by using a fast computer (generating 10^6 numbers in 1 sec) cannot be exhausted in 10^{130} years.

If the overall shift pk in constructing the initial p-tuple is less than the full period $2^p - 1$, then $w_1, \ldots, w_p$ are linearly independent. (Independence is ensured if k is relative prime to $2^p - 1$ but in an intuitive way one feels that in case of $pk > 2^p - 1$ very strong correlations occur among w_i's.) If the words w_i are only of e bit long, $e > p$, what is generally the case than we will encounter the same words several times among the $2^p - 1$ generated words; nevertheless the period length remains $2^p - 1$, thus the period length and the choice of p effectively becomes independent of the length of the words in the computer.

A convenient way to initialize the words $w_1, \ldots, w_p$ is the following. Set the initial p-tuple to all ones, that is $x_1 = \ldots = x_p = 1$. Run the original recurrence relation $x_{p+i} \equiv x_i + x_{i+q} \mod 2$ l times ($l = 5000\,p$ should be a sufficient choice) which is aimed to decrease the effect of the all one p-tuple on the randomness. Start to fill in the bits $x_l, x_{l+1}, \ldots$ into the words $w_1, \ldots, w_p$ in the way described, using a delay $k = 100\,p$.

A linear recursive (or shift register or Tausworthe) generator with a p such that $2^p - 1$ is also a prime (these are called Mersenne primes) is of special interest. In this case irreducibility of the polynomial induces primitivity of the polynomial and the delay k will evidently be relative prime to $2^p - 1$, consequently for $k < 2^p - 1$ the initial matrix composed of words $w_1, \ldots, w_p$ will have full rank, that is $w_1, \ldots, w_p$ are independent.

Some computer experimentations with statistical tests of random numbers produced by UGS-type generators revealed if primitive trinomials with small q or with q near to $(p-1)/2$ are used, the sequences produced will behave somewhat nonrandomly, hence the use of such trinomials is to be avoided. The parameters of some primitive trinomials are displayed in Table 5, the row 111 10, 49 stands for $x^{111} + x^{10} + 1$ and $x^{111} + x^{49} + 1$ which are both primitive trinomials.

Table 5

Some exponents of the primitive trinomials $x^p + x^q + 1$. Those denoted by asterisk are Mersenne primes, that is $2^p - 1$ is a prime, too

p	q
5*	2
9	4
11	2
23	5
31*	3, 6, 7, 13
47	5, 14, 20, 21
49	15
89*	38
95	11, 17
98	11, 27
111	10, 49
124	37
127*	1, 7, 15, 30, 63
170	23
250	103
380	47
476	15, 141
521*	32, 48, 158, 168
532	37
1279*	216, 418
2281*	715, 915, 1029
9689*	84, 471, 1836, 2444, 4187

Example. Consider the primitive trinomial $x^5 + x^3 + 1$ once more. Let the first five bits be 11010 again, and choose a delay of $k = 6$. For $e = 3$ bits words we will have the following results:

$\longleftarrow$	$w_1 = 101$	$w_9 = 001$
	$w_2 = 101$	$w_{10} = 011$
$\oplus\longleftarrow$	$w_3 = 011$	$w_{11} = 100$
	$w_4 = 111$	$w_{12} = 011$
	$w_5 = 001$	$w_{13} = 001$
$\longrightarrow$	$w_6 = 110$	$w_{14} = 101$
	$w_7 = 010$	$w_{15} = 000$
	$w_8 = 010$	$w_{16} = 101$

2.7. The distribution of the generated numbers

The period length is but one of the basic properties of the sequence of numbers produced by random number generators. Some other characteristics of the numbers will be discussed and possible enhancement will be provided here.

Assume that the periodicity of a sequence of numbers is n, then the numbers are expected to be uniformly distributed in the interval $[0,1)$ with an accuracy $1/n$, that is if all the numbers are generated and displayed then the distance between two neighbouring points should be $1/n$. We have seen it to be true for a pure congruential generator provided for a modulus $m = 2^e$ we have $b \equiv 5 \pmod 8$ as the multiplier of the generator.

Consider now the multidimensional uniformity of the generated sequence. In particular if one side of the unit square is split by n dividing points then we obtain n^2 gridpoints. A sequence of number u_i with a period length n — when used to form pairs from subsequent numbers as (u_i, u_{i+1}) that are viewed as two-dimensional points in the unit square — may produce n different gridpoints at most, so it is certain that $n^2 - n$ points will be missing. Obviously this reasoning can be repeated in any k-dimensions: of the n^k gridpoints, theoretically possible in a k-dimensional unit-cube only n can be produced by a congruential generator with period length n, $n^k - n$ gridpoint will never be generated. An additional stumbling-block is created by the fact that even these n generated k-tuples are placed according to some regularities to be described next.

Consider the two-dimensional unit square and a generator $z_{i+1} \equiv bz_i + r \pmod m$ with period length n. For some integer l we have

$$z_{i+1} = bz_i + r + l \cdot m,$$

that is for the generated numbers u_i we obtain

$$u_{i+1} = bu_i + \frac{r}{m} - l, \tag{2.20}$$

that is u_{i+1} is a linear function of u_i, the straight line $y = bx + \frac{r}{m}$ is "broken" into the unit square (Fig. 9). Consequently the generated pairs (u_i, u_{i+1}) are located on $b+1$ parallel lines — parametrized by $l = 0, 1, \ldots, b$, in the unit square. In case of $b = 1$ this yields only two lines accounting for the poor behaviour of the generator $z_{i+1} \equiv z_i + r \pmod{m}$.

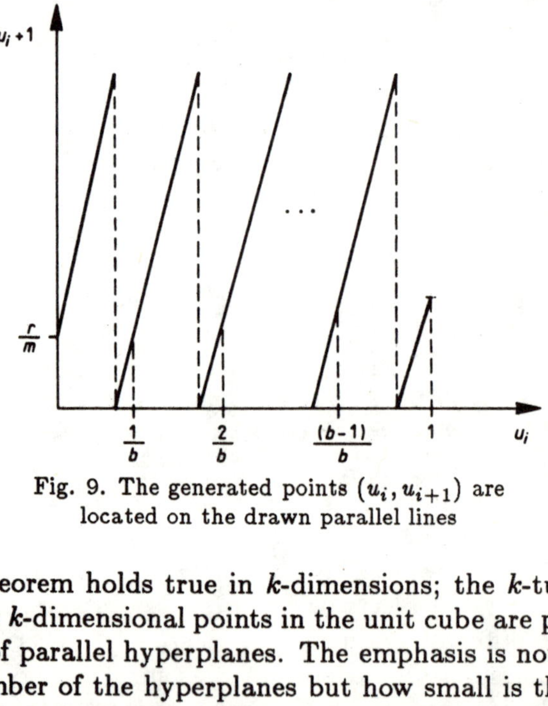

Fig. 9. The generated points (u_i, u_{i+1}) are located on the drawn parallel lines

A similar theorem holds true in k-dimensions; the k-tuples $(u_i, u_{i+1}, \ldots, u_{i+k-1})$, as k-dimensional points in the unit cube are positioned on a finite number of parallel hyperplanes. The emphasis is not on the finiteness of the number of the hyperplanes but how small is this number: in a computer with 32 bit long words the 3-dimensional points produced by a congruential generator with modulus $m = 2^{32}$ lie on at most 2953 parallel hyperplanes, the 10-dimensional points lie on 41 parallel hyperlanes at most. For a computer with 48 bit words and a generator with modulus $m = 2^{48}$ the upper bound for the number of the hyperplanes is 126 in $n = 10$ dimensions.

At the linear recursive generators — beside the substantially longer period — another good property can be observed. The generalized shift's algorithm UGS based on the primitive trinomial $x^p + x^q + 1$ produces

$2^p - 1$ different word. If the computer's word length is e bits, then the $(2^e)^k$ k-dimensional points representable in the computer are all produced provided $(2^e)^k \leqq 2^{p-1}$, that is $e^k \leqq p$. Taking a usual primitive trinomial of degree $p = 100$ on an $e = 32$ bits computer we get complete uniformity up to $k = 3$ dimensions, that is all the gridpoints are produced (in case of $p = 500$ complete uniformity is achieved up to $k \leqq 16$ dimensions). This argument also justifies the preference of linear recursive generators.

To enhance the multidimensional behaviour of the generated numbers without increasing actually the generator's period length the produced numbers are to be used not in the order of generation, but in a randomly shuffled order. Let $\mathbf{T}$ be a vector with k components (possible choices for k are $2^5, 2^6$ or 2^7) and consider the following two congruential generators

$$u_{i+1} \equiv b_1 u_i + r_1 \pmod 1$$
$$v_{i+1} \equiv b_2 v_i + r_2 \pmod 1$$

being called u-generator and v-generator, respectively.

As a preparation the numbers u_i are generated by the u-generator and filled into vector $\mathbf{T}$ as $T(i) = u_i, \quad i = 1, \ldots, k$.

UT Uniform, table look-up

1. [Generate a random index.] Generate v by the v-generator and compute the index $j \leftarrow [kv] + 1$ (where [ ] stands for the integer part).
2. [Interchange the jth element.] Generate u by the u-generator and let $x \leftarrow T(j), \quad T(j) \leftarrow u$.
3. Deliver the value x as uniformly distributed in [0,1).

The disadvantage of this algorithm that two uniform random numbers are used to produce one random number, halving the efficiency of generation. This can, partially be improved by the following, usual technical trick according to which one splits up the number into an index and the remainder.

UM Uniform, modified table look-up

1. [In vector $\mathbf{T}$ the previously generated numbers are stored.] Generate a number u by the u-generator, set $j \leftarrow [ku] + 1$.
2. Let $x \leftarrow T(j), \quad T(j) \leftarrow ku - j - 1$.
3. Deliver x as a uniformly distributed sample.

Such shuffling procedures may yield a very satisfactorily behaving sequence from two sequences exhibiting poor performance, hence if quality of the generated numbers is of greater concern than speed the use of algorithm UT (or UM) is recommended.

2.8. Statistical tests

As we mentioned some properties of pseudorandom numbers are tested to check if they behave in a truly random way or not. This test is necessary in most cases as the distribution of the generated sequence cannot be described in a mathematical way. This verification is carried out by statistical tests, mostly some types of χ^2 text are being used, that is a function of the produced numbers is considered that has supposedly a χ^2 distribution. The theoretical background of the χ^2 tests is given in the next theorem.

Let be given a random variable ξ, its values are in the set X, which is split up into the mutually disjunct subsets $X_1, \ldots, X_r$, $X_i \cap X_j = \emptyset$ if $i \neq j$. The probability of ξ having its value in the subset x_j is denoted by p_j, that is

$$P\{\xi \in X_j\} = p_j, \quad j = 1, \ldots, r,$$

where $p_1 + \ldots + p_r = 1$. Consider now N realizations $\xi_1, \ldots, \xi_N$ of the random variable ξ and denote the number of realizations falling into the subset X_j by ν_j which is a random variable itself, obviously $E(\nu_j) = p_j N$. The relative frequencies realized by the actual realizations $\xi_1, \ldots, \xi_N$ differ from the theoretically precise probabilities p_j. To measure this deviation the values $(\nu_j - Np_j)^2$ or more precisely their weighted sum are used. The statistical behaviour of this weighted sum is described by the next theorem.

Theorem 1. The distribution of the random variable

$$\chi_N^2 = \sum_{j=1}^{r} \frac{(\nu_j - Np_j)^2}{Np_j}$$

for $N \to \infty$ converges to the χ^2 distribution with $r-1$ degree of freedom, that is

$$\lim_{N \to \infty} P\{\chi_N^2 < y\} = \int_0^y g_{r-1}(x)\mathrm{d}x,$$

where $g_r(x) = [2^{r/2}\Gamma(r/2)]^{-1} x^{r/2-1} e^{-x/2}$.

Based on this theorem tests can be built to investigate the distribution of numbers produced by random number generators. We emphasize that the theorem is valid for scalar and vector random variables ξ equally well and any decomposition consisting of subsets χ_j with probabilities $p_j > 0$.

According to the general description of statistical tests we have to go through the following steps in order to evaluate a random number generator. Assume that the numbers $x_1, \ldots, x_N$ were produced by the generator, they are situated in interval $[a, b]$ and supposedly they have a common density function $f(x)$. Decompose the interval $[a, b]$ into r parts by the splitting points $c_0 = a, c_1, \ldots, c_{r-1}, c_r = b$ and denote the jth subinterval by $X_j, X_j = (c_{j-1}, c_j), j = 1, \ldots, r$. Let $p_j = \int\limits_{c_{j-1}}^{c_j} f(x)dx$ and the number of the points $x_1, \ldots, x_N$ in the subinterval X_j be denoted by y_j. Choose a reliability level β ($\beta = 0.95$ is usual) to the test and compute the quantity

$$k_N^2 = \sum_{i=1}^{n} \frac{(y_j - Np_j)^2}{Np_j}. \tag{2.21}$$

(The quantity $\alpha = 1 - \beta$ is the probability of type I error.)

According to the assumption on the density function of the sample and the previous theorem the distribution of k_N^2 is approximately χ^2 distribution with density function $g_{r-1}(x)$, so we expect the probability

$$1 - p = \int\limits_{k_N^2}^{\infty} g_{r-1}(l)dx$$

to be not very small. (This value $1 - p$ is called the level of significance of the test.) A decision how small is the quantity $(1 - p)$ can be made in the following way. To the reliability level β the threshold value $\chi^2(r - 1, \beta)$ is evaluated, for which one has

$$\int\limits_{0}^{\chi^2(r-1,\beta)} g_{r-1}(x)dx = \beta.$$

If for the number k_N^2 evaluated from the sample the inequality $k_N^2 < < \chi^2(r - 1, \beta)$ holds then $1 - p$ is greater than $1 - \beta$ and in this case we can state that the experiment does not contradict the initial hypothesis. On the other hand if $k_N^2 > \chi^2(r - 1, \beta)$ then the hypothesis is rejected at the reliability level β. We speak about the probability of the statistical test, this is the value $p = \int\limits_{0}^{k_N^2} g_{r-1}(x)dx$. The condition of accepting a hypothesis can be formulated using the notion of the test's probability: if $p \geqq \beta$, then the hypothesis is rejected at a reliability level β.

It can easily be seen, that if the χ^2 test is repeatedly executed, that is if the value k_N^2 in (2.21) (or the probability p of the test) is evaluated for different sequences $x_1, \ldots, x_N$, then sometimes we may encounter large values of k_N^2 (or great probabilities p), but the hypothesis can be accepted only in the case if, for instance, for nine sequences out of ten we have $k_N^2 < \chi^2(r-1,\beta)$ (or what is the same $p < \beta$) for a reliability level β fixed in advance.

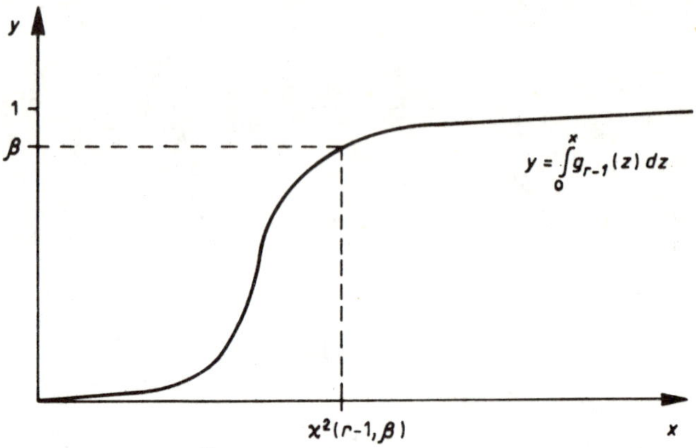

Fig. 10. A ξ^2 distribution function, β reliability level and its $\xi^2(r,\beta)$ threshold value

A couple of remarks are added with the aim to help the practical accomplishment of the χ^2 test. The test is applied in an invariable form only for the cases, when the domain of values of the random variable (the interval) is divided into $r \leqq 30$ subintervals. For greater values of r the normal density function is used instead of the density function $g_r(x)$ since for $r > 30$ the random variable $\rho = \sqrt{2\chi^2} - \sqrt{2r}$ has an approximately standard normal distribution $N(0,1)$ with a negligible error. Thus denoting the distribution function of the standardized normal distribution by Φ we have

$$1 - p = \int_c^{+\infty} g_r(x)\mathrm{d}x \approx \frac{1}{2}[1 - \Phi(\sqrt{2c} - \sqrt{2r})].$$

The decomposition should be carried out in such a way that the probabilities of the subsets be equal or almost equal, furthermore the value Np_j be greater than 20. In case of very good χ^2 tests (that is if the probability of the test is frequently less than 0.001) one should challenge the initial hypothesis, too. (For example applying the next test *a*.) For the sequence of numbers $u_i = i/m$ — provided $m \sim N$ — the value p will be very small, the hypothesis is accepted but the sequence is "too" uniform.)

We will consider some special cases, how can the uniform distribution of the generated numbers be tested by a χ^2 test.

(a) The forwardmost way of testing the uniformity is the direct application of Theorem 1. Split up the basic interval [0,1) into $r = 1000$ equal parts, the probability of a point's falling into a subinterval is 0.001, thus the number of points to be generated will be chosen to be $N = 20\,f000$ (greater values of N are admissible too, of course). Having generated N numbers u_i we evaluate the quantity k_N^2 from (2.21) and then we look up the value of $\chi^2(r-1,\beta)$ from a table (or evaluate it directly from $\int\limits_{x^2(r-1,\beta)}^{+\infty} g_{r-1}(x)\mathrm{d}x = 1-\beta$). Note that the value of $\chi^2(r-1,\beta)$ may be computed by using the inverse function $F_{r-1}(y)$ of the χ^2 distribution function with $r-1$ degree of freedom: $\chi^2(r-1,\beta) = F_{r-1}(\beta)$ and a subroutine evaluating this function generally can be found among the library subroutines of greater computers. Now if $k_N^2 < \chi^2(r-1,\beta)$ holds (or the probability of the test is less than β that is $\int\limits_0^{k_N^2} g_{r-1}(x)\mathrm{d}x < \beta$) then the initial hypothesis is accepted.

We have to be aware of the fact however that this test assesses the distribution, that is the uniformity of the first three digits only.

(b) The combination test examines by the χ^2 test the frequency of occurrence of any given quadruple (e.g. *aabb*, *aaab*, *aaaa*, etc.) among the quadruples produced by putting together four consecutive numbers $(u_i u_{i+1} u_{i+2} u_{i+3})$. Of course in practical executions u_i's are rounded to one or two places. Here the gap test is mentioned: the distribution of the length of the subsequence is tested, whose members are in a gap (between two given numbers), the frequency test: the frequency of any given digit is checked, the run-up and run-down test: the length of monotone increasing and monotone decreasing sequences of numbers is tested. In these cases the construction of the test is quite similar to the one described under (a).

(c) Some information can be obtained on the dependency of subsequent numbers, that is the distribution of the n-tuples if the uniformity test is applied to the n-tuples. More precisely one side of the n-dimensional unit cube $K = \{\mathbf{x} \mid 0 \leqq x_i \leqq 1,\ i = 1,\ldots,n\}$ is divided into k equal parts and the cube itself is divided into $r = k^n$ small cubes by parallel hyperplanes placed at equal distances apart. Construct the n-tuples $\mathbf{x}^{(i)} = (u_{(i-1)n}, u_{(i-1)n+1}, \ldots u_{in-1})$, the probability of $\mathbf{x}^{(i)}$'s falling into a small cube is $p_j = \frac{1}{k^n}$. After this preparation the χ^2-test can be stated in the usual way. Computer runs cannot be performed for great values of k^n because of the time required thus this test is not to be applied for parameter values $k \geqq 20, \quad n \geqq 5$ generally.

(d) A very important type of the statistical tests for randomness is based on the theory of ordered samples. By the aid of this type of tests we can examine the multidimensional behaviour of the generated random numbers in dimensions substantially greater than the value $n = 5$ mentioned in the previous paragraph.

Let $u_{(j-1)n}, \ldots, u_{jn-1}$ be uniformly distributed and introduce the notation $x_j = \max(u_{(j-1)n}, \ldots, u_{jn-1})$. According to a well-known result the distribution function of the random variable x_j is $G(x) = x^n, \quad x \in [0,1)$. On the other hand the random variable

$$u = G(x_j) = x_j^n$$

is uniformly distributed in the interval $[0,1)$, since $P\{u < z\} = P\{x_j < < \sqrt[n]{z}\} = z$ consequently the frequency test described under (a) can be applied to the random variable u. That is once more the interval $[0,1)$ is divided say into $r = 1000$ equal parts, and we will have $p_j = 0.001$. Generate now $N = 20\,000$ n-tuples, compute the corresponding values

$$y_j = [\max(u_{(j-1)n}, \ldots, u_{jn-1})]^n, \quad j = 1, \ldots, N$$

and evaluate the frequencies ν_j, the value k_N^2, etc.

A completely similar test can be built for the values of the minimum of n numbers. Let $x_j = \min(u_{(j-1)n}, \ldots, u_{jn-1})$, where $u_{(j-1)n}, \ldots, u_{jn-1})$ are uniformly distributed in $[0,1)$. The distribution function of the random variable x_j is $H_j(x) = 1 - (1 - x)^n$ and similarly to the previous reasonings we have now that the random variable

$$u = H(x_j) = 1 - (1 - x_j)^n$$

is uniformly distributed in $[0,1)$. The test basically follows the description of the test with the maximums, the uniformity test under (a) is executed for the numbers $y_j = 1 - (1 - \min(u_{(j-1)n}, \ldots, u_{jn-1}))^n$.

(e) Similarly to the test in (d), we can carry out examination of numbers with any non-uniform distribution. Consider the task of testing the generator that produces numbers x_i with distribution function $F(x)$. The numbers $F(x_i)$ are uniformly distributed in $[0,1)$ since $P\{F(x_i) \leqq z\} = = P\{x_i \leqq F^{-1}(z)\} = F(F^{-1}(z)) = z$, consequently the uniformity test in (a) can be applied to the numbers $F(x_i)$. Note that in actual computational practice statistical testing of non-uniform random numbers are rarely carried out, because most of the generators of non-uniform distributions used are theoretically accurate, that is if the initial values employed are uniformly distributed then the distribution of the transformed, non-uniform random number will be precisely the desired one.

(f) A statistical test basically differring from the previous ones will be described here. This ω^2 test helps us to examine the deviation of the empirical distribution determined by the generated samples from the theoretically desired one.

Let the distribution function of the random variable ξ be denoted by $F(x)$ and consider the realizations $x_1, \ldots, x_N$ of the random variable ξ. Denote the sample distribution function given by the samples $x_1, \ldots, x_N$ by $F_N^*(x)$, that is

$$F_N^*(x) = \frac{\#\{x_i \mid x_i < x\}}{N}$$

where # means the number of the elements in the subsequent set, that is now the number of the samples x_i being less than a fixed number x (see Fig. 10).

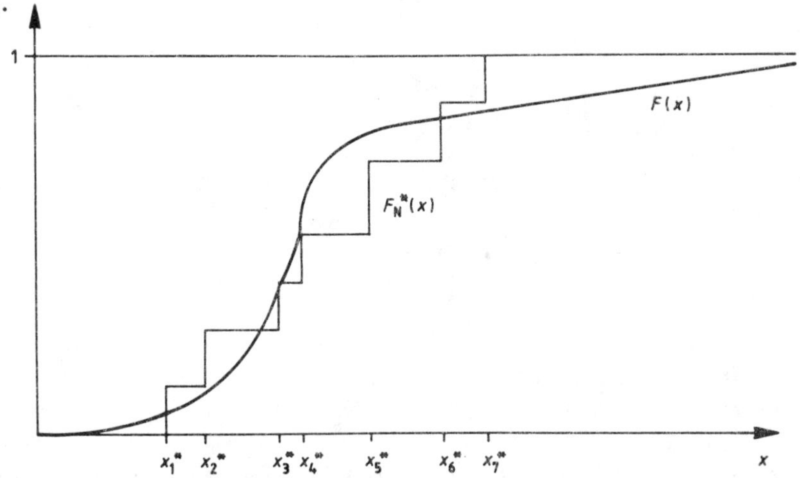

Fig. 11. The sample distribution function $F_N^*(x)$ for $N = 7$ (where $x_1^*, \ldots, x_7^*$ is the ordered sample from the samples $x_1, \ldots, x_7$)

To characterize the difference between $F(x)$ and $F_N^*(x)$ we will use the quantity

(2.22) $$\omega_N^2 = N \int_{-\infty}^{+\infty} [F(x) - F_N^*(x)]^2 \mathrm{d}F(x)$$

which is basically the weighted sum of the squared difference. The following theorem holds with respect to the quantity ω_N^2.

Theorem 2. For any random variable ξ with distribution function $F(x)$ we have

$$\lim_{N\to\infty} P\{\omega_N^2 < x\} = a_1(x), \quad x > 0$$

where the function $a_1(x)$ is independent of ξ (that is of $F(x)$).

Based on this theorem the ω^2 test can be constructed, similarly to the χ^2 test. For a given confidency level β the threshold value x_β can be determined, for which $a_1(x_\beta) = \beta$ (see Fig. 12).

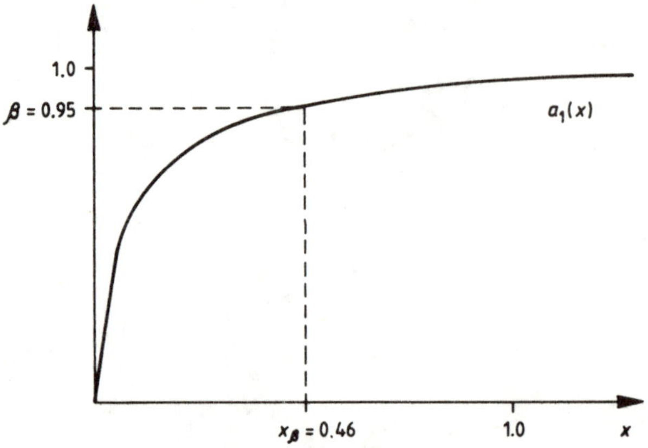

Fig. 12. The function $a_1(x)$ used in the ω^2 test

The actual realization of the test is: generation of samples $x_1, \ldots, x_N$, determination of the distribution function $F_N^*(x)$, evaluation of the value ω_N^2 according to formula (2.22). If now we have $\omega_N^2 < x_\beta$, then we accept the null hypothesis (that the samples $x_1, \ldots, x_N$ are distributed according to the distribution function $F(x)$), otherwise the null hypothesis is rejected.

The ω^2 test can be applied with a sample size of about $N = 50$. The main advantage of the test is that there is no need to partition the domain of the random variable's possible values. As a disadvantage of the test we have to point out that the samples x_i are to be ordered for the evaluation of the function $F_N^*(x)$, and determination of the ordered sequence $x_1^* \leqq x_2^* \leqq \ldots x_N^*$ requires a great amount of computation for large N. To facilitate the practical application we present a more suitable form of the expression in (2.22):

$$\omega_N^2 = \frac{1}{12N} + \sum_{i=1}^{N} \left[F(x_i^*) - \frac{j - 0.5}{N} \right]^2, \qquad (2.23)$$

which actually can be derived from (2.22).

Table 6

Some values of the function $a_1(x)$ in the ω^2 statistical test

x	$a_1(x)$
0.12	0.50
0.15	0.60
0.18	0.70
0.24	0.80
0.35	0.90
0.46	0.95
0.74	0.99
0.87	0.995
1.17	0.999
1.49	0.9998

(g) The following spectral test is to help the qualification of the congruential generator's multiplicators. The complete description of the spectral-test is rather lengthy, so is its algorithm, this is why only its essence will be outlined below. Define the frequency of occurrence of any given n-tuple as a function of this n-tuple. Then we define the discrete Fourier transform of this function (having integer valued function values dependent on integer numbers as arguments) and its spectrum. Based on the behaviour of the spectrum the randomness of the generated sequence can be concluded.

One detail of the spectrum-test is examined below: let our generator be based on $z_{i+1} \equiv bz_i + r (\text{mod } m)$ and assume that it has maximum period length. Consider the smallest wavelength in the spectrum, which is given by

$$\nu_n = \min \sqrt{s_1^2 + s_2^2 + \ldots + s_n^2}$$

where the minimum is to be taken for all those n-tuples of integers $s_1, s_2, \ldots, s_n$ that satisfy the following congruency

$$s_1 + s_2 b + s_3 b^2 + \ldots + s_n b^{n-1} \equiv 0 \pmod{m}.$$

As we have seen the n-tuples of the numbers produced by the generator lie on parallel hyperplanes (see the previous section). The distance of two parallel hyperplanes is $1/\nu_n$.

According to some other considerations the smallest wavelength is bounded from above as

$$\nu_n \leqq j_n m^{1/n},$$

where j_n is a constant $1 \leqq j_n \leqq 2$ for $n = 1, 2, \ldots, 8$, m is the modulus of the generator. Based on this inequality it seems to be a sound recommendation that congruential generators (without some modifications or composite use) are not to be used in higher dimensions.

2.9. Notes

2.2 In this topic we mostly followed the description given by Knuth. The two most interesting publications are by Kolmogorov (see the last unproven theorem from Section 2.2) and Chaitin. See also the related paper of Martin–Lof.

2.3 With regard to the description of earlier gadgets the book of Hammersley and Handscomb is recommended. The described generating device which uses gamma radiation was proposed, built and used by Miyatake and his associates.

2.4 In this section Zaremba's argumentation and Hammersley and Handscomb's book are followed. From amongst the latest publications in this area we mention Niederreiter's, Sós's and Braaten's papers.

2.5 The idea of the congruential generator is due to Lehmer. Here we followed mainly Dieter and Ahrens's book.

2.6 The linear recursive series and the use of primitive polynomials for producing sequences of numbers was well known for a long time in coding theory, but their use as random number generator was proposed by Tausworthe. We heavily relied in our description on the paper of Lewis and Payne, but the papers of Maritsas and Arvillias are also recommended. For a fast initialization procedure see a recent paper of Collings and Hembree.

2.7 The first known publication on the parallel hyperplanes and the upper bound on their number is due to Marsaglia (1968). The shuffling of the numbers described in this section was proposed by MacLaren and Marsaglia (1965).

2.8 The construction of statistical tests can be found in any statistics textbook (see for example Prékopa or Sobol). The tests under (d) were proposed by MacLaren and Marsaglia (1965). The full description of the spectral test can be found in Knuth's book or in Golder's papers.

CHAPTER 3

GENERAL METHODS FOR GENERATING NON-UNIFORM RANDOM NUMBERS

In simulational computations several probability distributions may occur and it becomes necessary to generate random numbers with these distributions. General methods will be discussed in this chapter by the aid of which a random variable uniformly distributed in the interval [0,1) could be transformed into one with the desired distribution. Theoretically it is possible to use a random variable with some other distribution (e.g. normal) as a starting point, but in practice the transformation of uniform random variables is the most common. The reasoning behind this practice is twofold: mathematically the uniform distribution can be described very simply and uniformly distributed random numbers can be generated in a fast and easy way. For generating a random variable with a given distribution we always choose the "most suitable" of several general methods, though to achieve a satisfactory numerical efficacy a combination of two or more methods is required in general. The term "most suitable" should always be interpreted in accordance with the actual computational task, so it could mean the fastest method, the one yielding the shortest program, the most convenient to be implemented, etc.

We assume that there is an infinite supply of the realizations $u_1, u_2, \ldots$ of the random variable u uniformly distributed in the interval [0,1) which are mutually independent. Our methods are exact in the sense that theoretically the transformed random variable is distributed with the required distribution; in practice the distribution of the transformed variable is influenced by the word length of the computer employed and the quality of the generated values $u_1, u_2, \ldots$ that were used as starting points. It must be noted here that, according to the accepted terminology, instead of the generation of random numbers we talk about taking random samples, since the generation of a number is actually drawing a sample from the given distribution.

3.1. Inversion method

The method produces a random number with desired distribution function by making use of the inverse of the distribution function. Sometimes the method is referred to as the *direct method.*

Consider the random variable ξ with density function $f(x), x \in [a, b]$. Its distribution function is

$$F(x) = \int_a^x f(y)\, dy.$$

Assume now that there exists the inverse function F^{-1} of the distribution function.

Theorem 1. Let u be a random variable uniformly distributed in $[0,1)$; then the distribution function of the random variable $F^{-1}(u)$ is $F(x)$.

Proof. It can readily be seen from the equality

$$P\{F^{-1}(u) \leqq x\} = P\{u \leqq F(x)\} = F(x).$$

In the case of distributions for which the inverse function of the distribution function can be given explicitly in analytical form, the above theorem gives an algorithm for generating random variables with distribution function $F(x)$ simultaneously. The method is illustrated in Fig. 13.

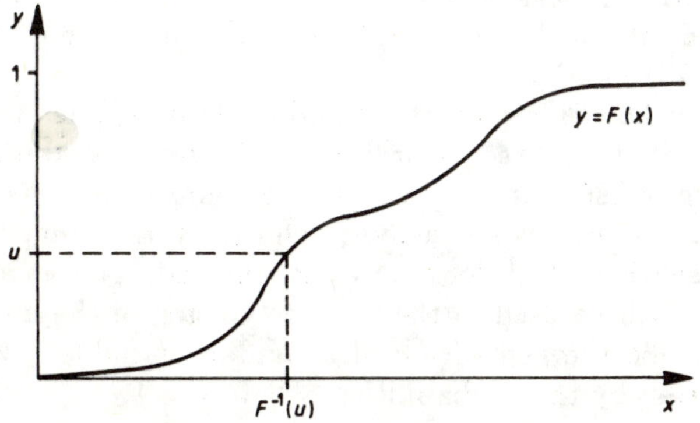

Fig. 13. Generation of a variable with distribution function $F(x)$ by the inversion method

C1 Continuous variable, by inversion

1. Generate u.
2. Deliver $x \leftarrow F^{-1}(u)$ as a value with distribution function $F(\cdot)$.

Example. Consider the standard exponential distribution with density function $f(x) = e^{-x}$, $x \geqq 0$. Its distribution function is

$$F(x) = \int_0^x e^{-y}\, dy = 1 - e^{-x}.$$

From the equation $y = 1 - e^{-x}$ the inverse function $x = -\ln(1 - y)$ is obtained, thus the random variable $-\log(1 - u)$ is exponentially distributed. (Throughout, all logarithms are natural unless otherwise indicated.) The random variable $(1 - u)$ is identically distributed with u, both being uniform $[0,1)$, thus could be interchanged preserving the distribution. Consequently the random variable $-\log(1 - u)$ has the same distribution as $-\log u$. Hence the following method has been obtained for generating exponentially distributed random variables.

E1 Exponential, by logarithm

1. Generate u.
2. Deliver $x \leftarrow -\log u$ as a sample from the standard exponential distribution.

If the distribution under consideration has a parameter $\lambda \neq 1$, that is, its distribution function is $F(x) = 1 - e^{-\lambda x}$, then similarly to the above reasoning it can be seen that the variable $-(1/\lambda)\log u$ will have distribution function $F(x)$.

Consider now how this general method could be applied for generating random variables distributed according to a discrete distribution function. As $F(x)$, being a distribution function, is monotone increasing, thus the equation $F(x) = u$ always has a root, defined by $x_0 = \inf\{x \mid u \leqq F(x)\}$ even in the case where F^{-1} does not exist. Instead of the inverse function this expression will be used in the discrete version of the method.

Let our task be the generation of a random variable ξ with discrete distribution given by the probabilities $p_i > 0$, $i = 1, \ldots, n$, that is

$$P\{\xi = x_i\} = p_i, \qquad i = 1, \ldots, n, \tag{3.1}$$

where $p_1 + \ldots + p_n = 1$, $x_1 \leqq x_2 \leqq \ldots \leqq x_n$. In this case the distribution function is $F(x) = P\{\xi < x\} = \sum_{i:x_i<x} p_i$, and using the function

$$G(y) = \inf\{x \mid y < F(x)\}$$

instead of the inverse function, the distribution of $G(u)$ has the required discrete distribution.

We introduce the notation $q_1 = p_1$, $q_2 = p_1+p_2, \ldots, q_n = p_1+\ldots+p_n$. With the aid of these total probabilities the determination of the function value $G(y)$ means the computation of such an index j for which the inequality

$$q_{j-1} \leqq y < q_j$$

is satisfied for the given y. So we have the corresponding algorithm.

D1 Discrete variable, by inversion

1. Generate u and set $j \leftarrow 1$.
2. If $u < q_j$ then go to Step 4.
3. Increase the index $j \leftarrow j + 1$ and go back to Step 2.
4. Deliver $x \leftarrow x_j$ as a sample from distribution (3.1).

Example. Let the discrete distribution be given by the following probabilities and values:

Table 7

$j =$	1	2	3	4	5
$p_j =$	0.1	0.2	0.2	0.1	0.4
$x_j =$	2	3	6	7	10

then the distribution function $F(x)$ and the corresponding function $G(y) = \inf\{x \mid y < F(x)\}$ are given by the following expressions:

$$F(x) = \begin{cases} 0, & \text{if} \quad x \leqq 2, \\ 0.1, & \text{if} \quad 2 < x \leqq 3, \\ 0.3, & \text{if} \quad 3 < x \leqq 6, \\ 0.5, & \text{if} \quad 6 < x \leqq 7, \\ 0.6, & \text{if} \quad 7 < x \leqq 10, \\ 1.0, & \text{if} \quad 10 < x, \end{cases}$$

$$G(y) = \begin{cases} 2, & \text{if } 0 \leqq y < 0.1, \\ 3, & \text{if } 0.1 \leqq y < 0.3, \\ 6, & \text{if } 0.3 \leqq y < 0.5, \\ 7, & \text{if } 0.5 \leqq y < 0.6, \\ 10, & \text{if } 0.6 \leqq y < 1. \end{cases}$$

The functions $F(x)$ and $G(y)$ are depicted in Fig. 14.

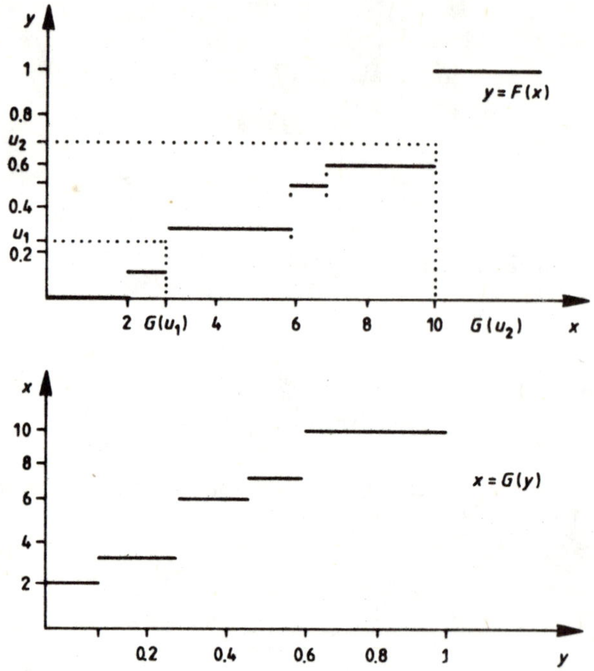

Fig. 14. A discrete distribution function and its "inverse" function

The algorithm is simple and can easily be coded. To generate one random number only $\overline{N} = 1$ uniformly distributed random number is required. Computational difficulties may arise in two cases: (i) if in the case of a continuous distribution there is not available an inverse function in an easily computable, closed form and (ii) if in the case of a discrete distribution the number n is great.

To use the inversion technique with an approximation to the inverse function is possible even if the inverse F^{-1} cannot be given analytically. Nevertheless, in this case the distribution of the generated numbers is only an approximation to the required distribution. The use of such approximating procedures — in our view — should be restricted to a very few exceptional cases. For example, if the existing, theoretically sound procedures have outstanding bad properties (slow run, large storage requirements, sophisticated program, etc.). In all other cases the use of the theoretically exact procedure is preferable since the computer realization of an algorithm distorts the distribution of the generated numbers all the same.

Consider the extent of the computational work of algorithm D1. The number of necessary comparisons in algorithms of this kind is at least $O(\log_2 n)$ since essentially we have here a search algorithm. An improvement of D1 approximating this limit is given in a subsequent section. Algorithm D1 delivers the value x_i with probability p_i after having completed i comparisons. Thus the expected number of comparisons is

$$S = \sum_{i=1}^{n} i p_i.$$

The computational work can be reduced by a simple trick. Before the generation procedures let the probabilities p_i be arranged in a decreasing order and let them be denoted by p_i^*, that is, $p_1^* \geqq p_2^* \geqq \cdots \geqq p_n^*$, and rearrange the values x_i similarly. If a random sample is generated by algorithm D1 from this distribution, which is identical to the distribution given in (3.1), then the expected number of comparisons can be worked out as

$$S^* = \sum_{i=1}^{n} i p_i^*.$$

It can easily be seen from this that $S \geqq S^*$ and the equality holds only in the case if $p_i = p_i^*$, $i = 1, \ldots, n$ (that is, the case of uniform distribution $p_i = 1/n$, $i = 1, \ldots, n$), having $S = S^* = (n-1)/2$. In the case of the previous, numerically given example distribution we have $S = 3.5$ and $S^* = 2$ and this decrease has been obtained by changing the first and the last probabilities only.

It must be noted that from the description of the inversion method it follows immediately that the job of producing non-uniform random numbers is primarily a numerical problem. Theoretically it would be possible to follow the next algorithm for generating numbers with distribution function $F(x)$. Produce all the real numbers u between 0 and 1 that could be represented on the computer at hand and compute and store the corresponding values $F^{-1}(u)$. After this preparation the generation procedure would consist of producing one u and looking up the corresponding $F^{-1}(u)$ only. In practice this method cannot be followed since the storage space required by the necessary data would be of astronomical size, not realizable even on today's computers. (The use of disks would slow down the generation procedure immensely.)

3.2. Acceptance-rejection method

Consider the random variable ξ with density function $f(x)$, $x \in [a, b]$. The gist of the acceptance-rejection method is that we do not produce the number from the distribution with density $f(x)$ directly, but indirectly. First we generate numbers from another distribution, which is handled more easily in practice, then they are accepted or rejected according to a specific rule. The method will first be demonstrated in a simple case.

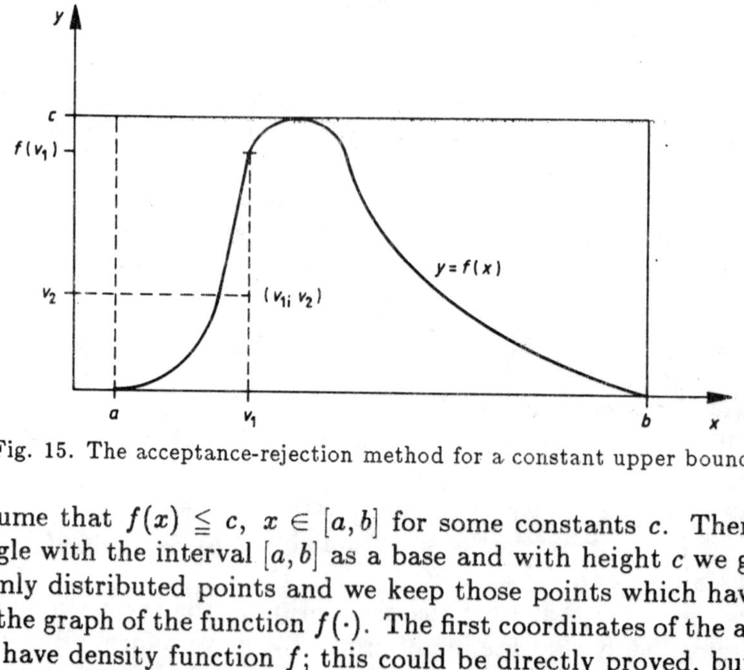

Fig. 15. The acceptance-rejection method for a constant upper bound c

Assume that $f(x) \leqq c$, $x \in [a, b]$ for some constants c. Then in the rectangle with the interval $[a, b]$ as a base and with height c we generate uniformly distributed points and we keep those points which have fallen under the graph of the function $f(\cdot)$. The first coordinates of the accepted points have density function f; this could be directly proved, but is also an immediate consequence of the more general Theorem 1 to be proved below. The algorithm sketched above is stated more formally as follows.

C2 Continuous variable, by rejection — constant upper bound

1. Generate u_1, u_2 independently. Let $v_1 \leftarrow a + (b - a)u_1$ and $v_2 \leftarrow cu_2$. At this moment the point (v_1, v_2) is uniformly distributed in the rectangle $\{(x, y) \mid a \leqq x \leqq b, 0 \leqq y \leqq c\}$.

2. If $v_2 < f(v_1)$ then deliver v_1 as a number with density function $f(x)$ (acceptance), otherwise repeat the whole procedure from the beginning (rejection).

The probability of acceptance which characterizes the procedure is called the efficiency of the acceptance-rejection method and is given by the following expression:

$$t = P\{v_2 < f(v_1)\} = \int_a^b \int_0^{f(v_1)} \frac{1}{c(b-a)}\, dv_2\, dv_1 = \frac{1}{c(b-a)},$$

since the random variables v_1 and v_2 are independent and their joint density function is $1/(c(b-a))$.

Obviously the method will be efficient if the constant c is small, thus the choice $c = \max_{x \in [a,b]} f(x)$ is recommendable. Geometrically, the efficiency is the ratio of the area under the graph of $f(x)$ to the area of the whole rectangle (a rough estimate is the number of accepted points divided by the number of all points).

Example 1. Let $f(x) = 2x$, $x \in [0,1]$ and choose the value $c = 2$. Then the computer implementation of procedure C2 for this particular density is:

EXAMP 1. Example for constant upper bound

1. Generate u_1, u_2. (The point in the rectangle would be $v_1 = u_1$, $v_2 = 2u_2$.)
2. If $u_2 < u_1$, then deliver u_1 as a sample with density $f(\cdot)$, otherwise go back to Step 1.

   The efficiency of the method is $t = 1/2$ for this distribution.

Example 2. Consider now the power density function $f(x) = nx^{n-1}$, $x \in [0,1]$. Numbers from this distribution can be produced by applying the inversion method since the distribution function is $F(x) = x^n$ and thus $u^{1/n}$ has density function $f(\cdot)$, if u is uniform. However, the computation of the nth root may be laborious, so there is justification for the following method. Let $c = n$ and then the algorithm based on the acceptance-rejection method is:

P1 Power distribution, from uniform

1. Generate u_1, u_2. (The uniform point would be $v_1 = u_1$, $v_2 = nu_2$.)
2. If $u_2 < u_1^{n-1}$ then deliver u_1, otherwise go back to Step 1.

The efficiency of the algorithm is $t = 1/n$, which shows the inadequacy of the method for large n.

Example 3. Consider the truncated exponential distribution with density function $f(x) = ce^{-x}$, $0 \leqq x \leqq 1$, where

$$1/c = \int_0^1 e^{-x}\, dx = \frac{e-1}{e} = 1/1.582.$$

An upper bound constant is $c = 1.582 = f(0)$, which can be used as a proportional to the density function $h(x) = 1$ of the uniform distribution (that is, $g(x) = c$).

The generating procedure for this exponential distribution is

E2 Exponential (truncated) by acceptance-rejection

1. Generate u and v.
2. If $u \leqq f(v)/g(v) = e^{-v}$, then deliver v, otherwise go back to Step 1.

Using a more general variant of the acceptance-rejection technique that is more suitable for practical purposes, we do not generate points in the base interval $[a, b]$ uniformly, but according to a density function $h(x)$ approximating the density function $f(x)$. Usually it should provide a better approximation to the density function $f(\cdot)$ than the constant function. Let $g(x)$ be a function possibly "mimicking" function $f(x)$ and being at the same time its upper bound in its interval of definition: $f(x) \leqq$ $\leqq g(x)$, $x \in [a, b]$. Let $c = \int_a^b g(x)\, dx$ be a finite value; then $h(x) = \frac{1}{c} g(x)$ is a density function. That is, we have

$$f(x) \leqq g(x) = ch(x), \qquad \text{where} \quad c \geqq 1. \tag{3.2}$$

Assume that we can generate random variables with density $h(x)$ easily or at least more easily than with density $f(x)$. The more general variant of the acceptance-rejection method can be described as follows.

C3 Continuous variable, by rejection — general upper bound

1. Generate u. Generate y according to the density function $h(x)$ in the interval $[a, b]$.
2. If $u \leqq f(y)/g(y)$ holds, then deliver $x \leftarrow y$ as a sample with density $f(\cdot)$ (acceptance), otherwise repeat the algorithm from Step 1 (rejection).

The next theorem proves the validity of the above algorithm.

Theorem 1. The numbers provided by Algorithm C3 are distributed according to the density function f.

Proof. The distribution function of the accepted values is

$$P\{x < p\} = P\{y < p \mid u \leqq f(y)/g(y)\}$$

where u, x and y are the quantities in Algorithm C3 and where p is a parameter. If here we state that u is uniformly distributed, y has density function h, y and u are independent variables, their joint density is $h(x)$, and furthermore we make use of the theorem of total probabilities, then we may write

$$(3.3)\quad P\{x < p\} = \frac{P\{y < p, u \leqq f(y)/g(y)\}}{P\{u \leqq f(y)/g(y)\}}$$

$$= \frac{\int_a^p \int_0^{f(r)/g(r)} h(r)\, du\, dr}{\int_a^b \int_0^{f(r)/g(r)} h(r)\, du\, dr} = \frac{\frac{1}{c}\int_a^p f(r)\, dr}{\frac{1}{c}\int_a^b f(r)\, dr} = \int_a^p f(r)\, dr.$$

Differentiating the last expression with respect to p we get the function $f(\cdot)$ as the density function of the delivered values.

The efficiency of Algorithm C3 is measured by the probability of acceptance, that is, the denominator of the first fraction in (3.3), which has the value $P\{u \leqq f(y)/g(y)\} = \frac{1}{c}$. This measure agrees with one's intuition of efficiency, in the sense that the closer $h(\cdot)$ approximates $f(\cdot)$ the more efficient is the method.

The acceptance-rejection technique can very easily be applied for the case of density functions taking on extreme shapes. (For applying the method it is enough to find an upper bound or an upper bounding function having finite integral.) On the other hand, in these cases its drawback becomes apparent: the algorithm becomes inefficient from a computational point of view, either because of the small value of $1/c$ or the evaluation of the ratio becomes difficult or time consuming.

Example 4. Let $f(x) = nx^{n-1}$, $x \in [0,1]$, $n > 1$ and choose the function $g(x) = nx$ as the enveloping function. The density function proportional to $g(\cdot)$ is $h(x) = 2x$. Application of the general technique C3 for this example yields the following algorithm.

P2 Power distribution, by linear upper bound

1. Generate u_1 and u_2.
2. Let $y \leftarrow \sqrt{u_2}$ (its density function is $h(\cdot)$).
3. [Comparison and trial.] If $u_1 \leqq y^{n-2}$, then deliver the value y, otherwise go back to Step 1.

In a computer implementation of the above algorithm, Step 2 will naturally be omitted and the inequality will be of the form $u_1^2 \leqq u_2^{n-2}$ (or for even n, $u_1 \leqq u_2^{(n-2)/2}$). The efficiency of the method is $2/n$ which has doubled the efficiency in Example 2 by a more suitable choice of upper bounding function. Of course, for large n neither of these methods can be recommended.

Theorem 2. In Algorithm C3 for generating one sample with density f the expected number of uniformly distributed random numbers required is $N = c(1+m)$ where m denotes the number of uniform samples needed to generate a random number with density h.

Proof. In one iteration (i.e. Step 1 and Step 2 only once in C3) a generated value is accepted with probability $1/c$. The probability of the event that the procedure will be finished in exactly the ith iteration is

$$(1 - 1/c)^{i-1}(1/c)$$

since it occurs if we reject the generated values in the first $(i-1)$ iterations and accept it in the last one. Thus the expected value k_0 of the number χ of the iterations is

$$k_0 = E\{\chi\} = \sum_{i=1}^{\infty} i(1 - 1/c)^{i-1}(1/c) = \frac{1}{c}\sum i(1 - 1/c)^{i-1}$$

Having differentiated the infinite series $\sum_{i=1}^{\infty} q^i = \frac{1}{1-q}$, $0 < q < 1$, we obtain $\sum_{i=1}^{\infty} iq^{i-1} = \frac{1}{1-q}$, where $q = 1 - 1/c$ is substituted, whereby we get $\sum_{i=1}^{\infty} i(1 - 1/c)^{i-1} = c^2$. Thus, for the expected value k_0 we have

$$k_0 = c.$$

In one iteration m uniform numbers are needed for generating one sample with density $h(\cdot)$ by assumption plus one additional sample for the decision variable. Thus, the average number of uniform variables required is

$$\overline{N} = c(1 + m).$$

In the frequently encountered case $m = 1$ the expected number of uniform variables becomes $\overline{N} = 2c$.

3.3. Decomposition method

In the decomposition or mixture method one requires the density function of the target distribution to be a mixture of other densities. The generation of the random variable is executed in two steps. First, a density function is chosen from the mixture; afterwards a variable is generated according to this density function.

Let our task be to generate random variables with density function $f(x)$, $x \in [a, b]$, where f can be written in the following form:

$$f(x) = \int_c^d g_y(x)\, \mathrm{d}H(y). \tag{3.4}$$

The term $g_y(x)$ denotes here a family of densities dependent on a parameter y, which in turn is a random variable in the interval $[c, d]$ with distribution function $H(\cdot)$. Assume that we may produce random numbers both with distribution function $H(\cdot)$ and with density function g_y. Then the procedure for generating random variables with density function $f(x)$ is by reason of (3.4):

C4 Continuous variable, by continuous decomposition

1. Generate y according to the distribution function H.
2. Generate a value x with density function g_y and deliver x as a sample from the distribution with density f.

Correctness of this procedure is established by the following theorem.

Theorem. The values generated by Algorithm C4 form a distribution with density $f(\cdot)$.

Proof. The distribution function of the values generated is

$$P\{x < r\} = \int_c^d P\{x < r \mid \eta = y\}\, \mathrm{d}H(y) = \int_c^d \int_{-\infty}^r g_y(z)\, \mathrm{d}z\, \mathrm{d}H(y).$$

Here we change the order of integration and make use of equation (3.4)

$$\int_{-\infty}^{r} \int_{c}^{d} g_y(z)\, dH(y)\, dz = \int_{-\infty}^{r} f(z)\, dz,$$

which after differentiation shows the density function to be f.

We may give another interpretation of the decomposition method. Let the joint density function of a random point (ξ, η) be $p(x, y)$, the density function of η be $h(\cdot)$, and the density function of ξ be $f(\cdot)$. Now we can write

$$f(x) = \int_{-\infty}^{+\infty} p(x, y)\, dy = \int_{-\infty}^{+\infty} g(x \mid y) h(y)\, dy, \tag{3.5}$$

where $g(x \mid y)$ is the conditional density function of the first variable. Thus (3.5) takes on the form of equation (3.4). Hence the first coordinate of the point (ξ, η) may be generated by first producing the second coordinate with density h and then a value of the variable ξ is generated according to the conditional density function $g(x \mid y)$.

Example 1. Assume that density function $f(x)$ can be written in the form

$$f(x) = n \int_{1}^{\infty} y^{-n} e^{-xy}\, dy, \qquad 0 \leqq x.$$

The density function of the parameter y is

$$h(y) = \int_{0}^{\infty} n\, y^{-n} e^{-xy}\, dx = n\, y^{-(n+1)},$$

and having it integrated we obtain its distribution function as

$$H(y) = \int_{1}^{y} n\, z^{-(n+1)}\, dz = 1 - y^{-n}.$$

From the equation $u = 1 - y^{-n}$ by the inversion method we have $y = (1-u)^{-1/n}$ being distributed with distribution function H if u is uniformly distributed in $[0,1)$. The conditional density function and the distribution function respectively may be worked out along the lines of equation (3.5) as

$$g_y(x) = y\, e^{-xy}, \qquad 0 \leqq x,$$

$$G_y(x) = \int_{0}^{x} y\, e^{-zy}\, dz = 1 - e^{-xy}.$$

From equation $1 - e^{-xy} = u$ we derive the random variable $x = -(\log u)/y$ obeying the distribution function G_y, where u is uniform. Summing up the above details, the next procedure generates random numbers with density function $f(\cdot)$.

EXAMP 2. Example for continuous decomposition

1. Generate u_1 and u_2. Set $y \leftarrow u_1^{1/n}$.
2. Deliver the value $x \leftarrow -(\log u_2)/y$.

Considering computational aspects, we note that efficient procedures could be constructed in the case where the distribution given by H is taken to be discrete, after which we need to build up the equation

$$f(x) = p_1 g_1(x) + p_2 g_2(x) + \ldots + p_n g_n(x) \tag{3.6}$$

with some probabilities p_i, $p_i \geqq 0$, $i = 1, \ldots, k$, $p_1 + \ldots + p_k = 1$, and densities $g_i(\cdot)$ instead of the rather elusive equation (3.4). The generation procedure based on decomposition of (3.6) is

C5 Continuous variable, by discrete decomposition

1. Generate a random index i from the set of numbers $1, 2, \ldots, k$ with probability p_i.
2. Generate the value x with density function g_i and deliver it as a sample with density function f.

The necessary procedure for establishing decomposition (3.6) of a given density function f can be illustrated geometrically in the following way. For each index i we have $f(x) \geqq p_i g_i(x)$, so that the graph of the function $p_i g_i(x)$ is always below the graph of $f(x)$. Consequently we consider the graphs of those simple functions $p_i g_i(\cdot)$ lying under the graph of $f(\cdot)$, whose corresponding density functions $g_i(\cdot)$ are prone to easy generation procedures. The area under the curve of $p_i g_i(\cdot)$ is p_i since $g_i(\cdot)$ is a density function. The better the sum of the inscribed curves approximates the function f, the nearer the sum of the related p_i's to 1. What one does in practice is that after taking some simple (constant or linear) functions g_i the last term of the mixture is made equal to the remaining part of the function; that is, we take the last density function $g_k(\cdot)$, the one determined by

$$p_k g_k(x) = f(x) - \sum_{i=1}^{k-1} p_i g_i(x).$$

Example. Consider the density function $f(x) = \dfrac{e}{e-1} e^{-x}$, $x \in [0, 1]$, a truncated version of the exponential density function. Split the interval

[0,1] into five equal sub-intervals and choose the constant densities of the uniform distribution in each sub-interval as the density functions in the mixture; that is, let $g_i(x) = 5$, $x \in [(i-1)/5, i/5]$, $i = 1, \dots, 5$. Values of probabilities p_i will be determined so that the function $p_i g_i(x)$ will be closest to the function $f(x)$. Then we arrive at the following equations, from which the values p_i can be computed:

$$\begin{aligned}
p_1 g_1(x) &= f(0.2) = 1.295\,, & p_1 &= 0.259\,,\\
p_2 g_2(x) &= f(0.4) = 1.060\,, & p_2 &= 0.212\,,\\
p_3 g_3(x) &= f(0.6) = 0.868\,, & p_3 &= 0.174\,,\\
p_4 g_4(x) &= f(0.8) = 0.711\,, & p_4 &= 0.142\,,\\
p_5 g_5(x) &= f(1.0) = 0.582\,, & p_5 &= 0.116\,.
\end{aligned}$$

Since we have $p_1 + \dots + p_5 = 0.903$, consequently $p_6 = 0.097$ emerges. The function g_6 will be determined to render the sum of the functions in the mixture to be exactly equal to $f(x)$:

$$p_6 g_6(x) = f(x) - \sum_{i=1}^{5} p_i g_i(x).$$

The function g_6 attains its maximum at $x = 0$, $g_6(0) = \dfrac{0.287}{0.097} = 2.959$. If for sampling with density function g_6 the acceptance-rejection technique is applied with a constant upper boundary function $c = g_6(0)$, then its efficiency would be roughly 1/3 which is considered to be a satisfactory rate of acceptance since generation with density function g_6 is only required in every tenth case on the average. The decomposition is illustrated in Figs. 16 and 17.

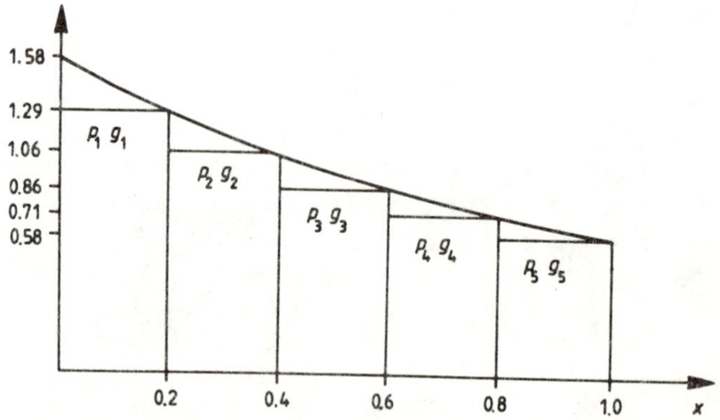

Fig. 16. The decomposition of the density function $f(x) = \dfrac{e}{e-1} e^{-x}$, $x \in [0, 1]$

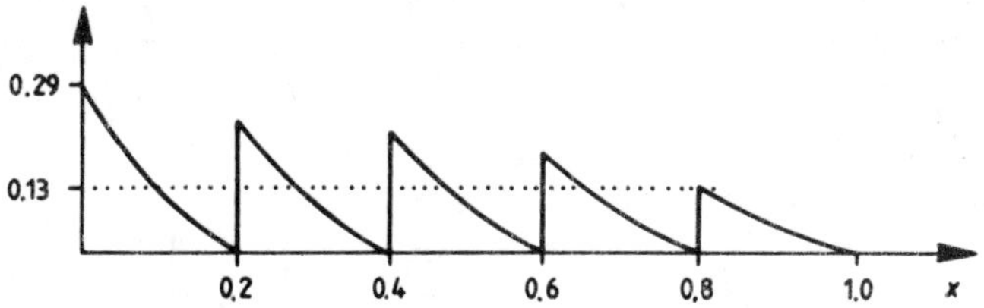

Fig. 17. The remainder function $p_6 g_6(x)$, $x \in [0,1]$

Summarizing what has been written above, random numbers with density function $f(x) = \frac{e}{e-1} e^{-x}$ are produced on the basis of the decomposition

$$f(x) = p_1 g_1(x) + \ldots + p_6 g_6(x)$$

by the following formal algorithm:

EXAMP 3. Example for discrete decomposition of exponential.

1. Generate an index i from the discrete distribution given by $P\{n = j\} = p_j$, $j = 1, \ldots, 6$. If $i = 6$ then go to Step 3.
2. [Generate a uniformly distributed number in the ith sub-interval.] Generate u and set $x \leftarrow 0.2(u + i - 1)$. Go to Step 4.
3. [Sample according to the "remainder" density function.] Generate u_1 and u_2, set $v \leftarrow 2.959\, u_2$. If now $v < f(u_1) - \sum_{i=1}^{5} p_i g_i(u_1)$ then set $x \leftarrow u_1$, otherwise repeat this step from the beginning.
4. [Finish.] Deliver the number x as a sample with density f.

The reason why the method of decomposition is deemed favourable from a computational point of view is that we can gear up the generators by its application. It is exemplified in the following simple case. Assume that generation of random variables with density function f can be accomplished rather slowly, e.g. one sample can be produced in 5 msec. If we have a decomposition

$$f(x) = 0.95\, g_1(x) + 0.05\, g_2(x),$$

where $g_1(x)$ is an "easy" function (generation of one sample takes 1 msec), $g_2(x)$ is a "hard" function to deal with (one sample is generated in 10 msec), then the corresponding decomposition-based generation procedure requires $1 * 0.95 + 10 * 0.05 = 1.45$ msec for producing one number with density f on the average (disregarding the time necessary to make a choice between g_1 and g_2). Therefore we have to find a decomposition according to which a fast generator will be used with great probability and even the generator corresponding to the remaining term with small probability is considerably slower, the resulting generator of the original largest density will be still faster than our original choice.

If the decomposition consists of two densities

$$f(x) = pg_1(x) + (1-p)g_2(x)$$

and one of them has probability very close to 1 (let, say, $p - \varepsilon$), then the generator of $f(\cdot)$ would generate samples using density $g_2(\cdot)$ almost all the time and only occasionally one with density $g_1(\cdot)$. This idea is incorporated into the following algorithm, which produces a number of random variables at once.

Let η be a random variable distributed geometrically with parameter p, let its distribution be denoted by GE(p). The probabilities of the distribution are

$$p_i = P\{\eta = i\} = p(1-p)^i, \qquad i = 0, 1, 2, \dots,$$

then we can formulate the above mentioned idea into the algorithm.

C6 Continuous variable, by modified discrete decomposition

1. Generate m from the distribution GE(p).
2. Generate m samples $x_1, x_2, \dots, x_m$ with density function $g_2(\cdot)$.
3. Generate one random number x_{m+1} with density $g_1(\cdot)$ and deliver the numbers $x_1, x_2, \dots, x_m, x_{m+1}$ as samples with density $f(\cdot)$.

Theorem. The random numbers produced by Algorithm C6 have density function $f(\cdot)$.

Proof. When we repeat Algorithm C6 the ith time, the number of generated samples will be m_i, the subtotals of the overall number of generated samples will be denoted by n_i, that is, $n_i = \sum_{j=1}^{i}(m_j + 1)$, $i = 1, 2, \dots, n_0 = 0$. The sequence of obtained random numbers will be referred to as $\{t_j\}$. First observe that the distribution of the subtotals n_i is negative binomial, being the sum of geometrically distributed random numbers. Thus we have

$$P\{t_j \text{ is distributed with density } g_1\} = P\{n_k = j \text{ for some } k\}$$
$$= \sum_{k=1}^{j} P\{n_k = j\} = p \sum_{k=1}^{j} \binom{j-1}{k-1} p^{k-1}(1-p)^{j-k} = p$$

which shows that the numbers with density g_1 have the proper frequency in the outcome $\{t_j\}$. Moreover, the random samples $\{t_j\}$ are independent, since the numbers generated in one bundle

$$t_{n_i+1}, t_{n_i+2}, \ldots, t_{n_i+m_{i+1}}, t_{n_i+1}$$

are produced independently and no interference is allowed between samples of different bundles.

This modification saves us the generation of the uniform number used to select a member of the mixture in $m_i - 1$ cases, though the added sophistication of the generation of a geometrically distributed random number decreases its value. Algorithm C6 should be used only when $1 - p$ is large (p is close to zero), that is, the size of bundle m_i is big.

Returning to the general decomposition algorithm, here are some suggestions regarding its applications. The simplest way of using a mixture in order to speed up a generator is to decompose the density function of the target distribution into a weighted sum of the densities of uniform distributions and a remainder term as we have done in the numerical example. This works for any density function. If the base interval is divided into sufficiently small sub-intervals, then in the overwhelming majority of cases the algorithm chooses only an interval and produces a number uniformly distributed in it. Theoretically it could be done for any distribution but in practice such direct application of the decomposition technique could have two major stumbling blocks: (i) a sample is required from a discrete distribution with a large number of possible outcomes, and (ii) the data describing the numerous sub-intervals have to be stored. The first difficulty can be surmounted at a cost of more sophisticated algorithms and some additional storage requirements (see Section 3.6 or Section 3.10). The second difficulty can be surmounted only in a certain sense, more precisely we have to say that "time could be bought for storage" and vica versa naturally, meaning that the more storage space we use the faster generator can be facilitated.

3.4. Comparison method

This method of random number generation is somewhat similar to the acceptance-rejection technique. It generates an initial sample and a chain of random numbers, then the initial sample is accepted or rejected according to the properties of the chain.

First, properties of a series of random variables will be examined. The method itself will be outlined later. Let t be a number from interval $[0,1)$ and generate the independent numbers $u_1, u_2, \ldots, u_k$ uniformly from interval $[0,1)$ until the chain of inequalities $t > u_1 > u_2 > \ldots > u_{k-1} \leqq u_k$ is satisfied, that is, until a u_1 is greater than its predecessor. If for the first number u_1 generated we have $t \leqq u_1$, then set $k = 1$. For the sequence of numbers produced in the above way the following theorem holds.

Theorem 1. For a given $t \in [0,1)$ define the function

$$h(t) = P\{t > u_1 > \ldots > u_{k-1} \leqq u_k, k \text{ odd}\},$$

then we have $h(t) = e^{-t}$.

Proof. By considering only the generation procedure of the sequence $u_1, \ldots, u_k$ one can observe immediately that

$$P\{t > u_1 > \ldots > u_{k-1} \leqq u_k\} = P\{t > u_1 > \ldots > u_{k-1}\} - P\{t > u_1 > \ldots > u_{k-1} > u_k\}. \tag{3.7}$$

To evaluate the probability $P\{t > u_1 > \ldots > u_k\}$ it suffices to note that u_i is less than t with probability t and thus the probability of the event that all the k numbers $u_1, \ldots, u_k$ will be less than t is t^k. The numbers $u_1, \ldots, u_k$ have $k!$ different permutations and the k inequalities in the brackets will be satisfied in just one case, hence

$$P\{t > u_1 > \ldots > u_k\} = \frac{t^k}{k!}.$$

Thus equation (3.7) can be reformulated as

$$P\{t > u_1 > \ldots > u_{k-1} \leqq u_k\} = \frac{t^{k-1}}{(k-1)!} - \frac{t^k}{k!}.$$

The function at issue now becomes

$$h(t) = P\{t > u_1 > \ldots > u_{k-1} \leqq u_k, k \text{ odd }\} = \sum_{\substack{k=1 \\ k \text{ odd}}}^{\infty} \left(\frac{t^{k-1}}{(k-1)!} - \frac{t^k}{k!} \right) = \sum_{k=0}^{\infty} (-1)^k \frac{k}{k!} = e^{-t}.$$

Proof (second). Let $h(t)$ be the function previously defined, A be the event, that for a given chain k odd. Let B be the event that k is even, that is $u_1 < t$ and after u_1 we have an odd number of u_i's. For the probability of events A and B one easily obtains

$$P(A) = h(t),$$

$$P(B) = \int_0^t P\{u_1 > u_2 > \ldots > u_{k-1} < u_k, k-1 \text{ odd }\}\, du_1$$

$$= \int_0^t h(u_1)\, du_1.$$

Since A and B are mutually exclusive events, thus

$$1 = h(t) + \int_0^t h(u_1)\, du_1.$$

Differentiating this equation with respect to t one has

$$h'(t) = -h(t),$$

whereby $h(t) = \mathrm{e}^{-t} + c$ with a constant c. Having the condition $h(0) = 1$ it implies $c = 0$, $h(t) = \mathrm{e}^{-t}$.

Now we are ready to discuss the method of comparisons. Assume that we want to generate random variables with density function $f(x) = ce^{-G(x)}$, $x \in [a,b]$ and furthermore we know that $0 \leqq G(x) \leqq 1$, $x \in [a,b]$ holds (here c is a constant $1/c = \int_a^b \mathrm{e}^{-G(s)} ds$). Utilization of the above investigated sequences of random variables leads us to the following algorithm:

C7 Continuous variable, by comparison

1. [A uniformly distributed point is chosen in $[a,b]$.] Generate u, set $x \leftarrow a + (b-a)u$, $k \leftarrow 0$, $t \leftarrow G(x)$, $u_0 \leftarrow t$.
2. [Generating a chain of variables.] Increase $k \leftarrow k+1$ and generate u_k. If $u_{k-1} \leqq u_k$, then go to Step 3 (finish the generation of a chain) otherwise repeat this step from the beginning.
3. [Rejection.] If k even, go back to Step 1 (rejection of a chain).
4. If k odd, deliver the number x, as one with density function $f(\cdot)$.

Theorem 2. Algorithm C7 produces numbers with density function $f(x) = c\exp(-G(x))$.

Proof. For a given value $t = G(x)$ the probability of accepting x — according to Theorem 1 — is $\exp(-G(x))$. Since x is uniformly generated in $[a,b]$, thus

$$P\{x < r \mid x \text{ is accepted}\} = \frac{P\{x < r, x \text{ accepted}\}}{P\{x \text{ accepted}\}}$$

$$= \frac{\int_a^r \mathrm{e}^{-G(s)} \frac{1}{b-a} ds}{\int_a^b \mathrm{e}^{-G(s)} \frac{1}{b-a} ds} = \frac{1}{\int_a^b \mathrm{e}^{-G(s)} ds} \int_a^r \mathrm{e}^{-G(s)} ds$$

$$= c \int_a^r \mathrm{e}^{-G(s)} ds$$

This completes our proof.

In order to render our method to be even more generally applicable, we would like to get rid of the restriction $0 \leqq G(x) \leqq 1$. It could be done in the following way. Assume for sake of simplicity $G(x)$ is monotone increasing and let the density function

$$f(x) = c\,\mathrm{e}^{-G(x)}, \quad x \in [a, +\infty) \tag{3.8}$$

be given, where contrary to the assumption of the previous theorem we do not assume that $0 \leqq G(x) \leqq 1$ holds. The density function will be decomposed into a mixture. Let a partitioning of the halfline $[a, +\infty)$ be given by the points $b_1, b_2, \ldots,\ a = b_1 < b_2 < \ldots < b_k < \ldots$ with the property that the greatest change of the function $G(x)$ in one sub-interval $[b_i, b_{i+1})$ is at most 1, that is

$$0 \leqq G(x) - G(b_i) \leqq 1, \quad \text{if} \quad x \in [b_i, b_{i+1}), \quad i = 1, 2, \ldots .$$

Density functions $g_i(x)$ are defined in the following way:

$$p_i g_i(x) = \begin{cases} f(x), & x \in [b_i, b_{i+1}), \\ 0, & \text{otherwise.} \end{cases}$$

The value of the constant p_i may be determined from the equation $\int_{b_i}^{b_{i+1}} f(x)\, g_i \mathrm{d}x = p_i$, thus we may write

$$f(x) = \sum_{i=1}^{\infty} p_i g_i(x). \tag{3.9}$$

Now it is evident that in the function

$$g_i(x) = \frac{c}{p_i} \mathrm{e}^{-G(x)} = \frac{c}{p_i}\, \mathrm{e}^{-G(b_i)}\, \mathrm{e}^{-[G(x) - G(b_i)]}, \quad x \in [b_i, b_{i+1})$$

the exponent $G(x) - G(b_i)$ has changed in value at most 1, thus samples with density g_i could be generated by the method of comparison.

Summing up our investigations, generation of random variables with density $f(x)$ in (3.8) can be carried out in two steps when using the comparison method and the mixture in (3.9).

C8 Continuous variable, by decomposition and comparison

1. Sample a density function g_i with probability p_i.
2. Generate a random number with density function g_i and deliver it as a sample with density $f(x)$.

Since any density function can be written in the form

$$h(x) = e^{-(-\log h(x))},$$

that is, in the form given by (3.8), it follows that samples from any distribution could be generated by the above variant C8 of the comparison method.

Evaluate now the number $\overline{N}$ of uniformly distributed random samples needed in Algorithm C7 of the comparison method to obtain one number with the desired distribution. The probability of the event that for a given $t = G(x)$ the generation of chain $t > u_1 > \ldots > u_{k-1} \leqq u_k$ terminates after having generated exactly k uniform numbers is

$$\frac{t^{k-1}}{(k-1)!} - \frac{t^k}{k!}$$

as it has been shown in the proof of Theorem 1. Then the expected value k_0 of the index k determined by the sequence is the following:

$$k_0 = \sum_{k=1}^{\infty} k\left[\frac{t^{k-1}}{(k-1)!} - \frac{t^k}{k!}\right] = \sum_{k=0}^{\infty}(k+1)\frac{t^k}{k!} - \sum_{k=1}^{\infty} k\frac{t^k}{k!}$$
$$= \sum_{k=0}^{\infty}[(k+1) - k]\frac{t^k}{k!} = e^t. \qquad (3.10)$$

Denote the probability of index k determined by the sequence being odd by p_1 (obviously $p_1 = e^{-t}$). The probability that we reject $(n-1)$ sequences and accept the nth is $(1-p_1)^{n-1}p_1$. In each of the n sequences the expected value of the required random numbers equals k_0. Thus the expected value of the variables required in the sequences is

$$m = p_1 k_0 + 2(1-p_1)p_1 k_0 + 3(1-p_1)^2 p_1 k_0 + \ldots +$$
$$+ n(1-p_1)^{n-1} p_1 k_0 + \ldots$$
$$= p_1 k_0 \sum_{k=0}^{\infty} (k+1)(1-p_1)^k = p_1 k_0 \frac{1}{p_1^2} = \frac{k_0}{p_1},$$

since the last infinite series is the derivative of the geometrical series $\frac{1}{1-q} = 1+q+q^2+\ldots$ if we substitute $q = 1-p_1$. In the final generation procedure for each sequence an additional variable is also required apart from k_0, namely for generating a variable x in the interval $[a,b]$. Thus finally we arrive at

Theorem 3. The expected number of the uniform random variables required to generate one sample with density function $f(x) = c\,e^{-G(x)}$, $x \in [a,b]$ by Algorithm C6 is

$$\overline{N} = \frac{1 + \frac{1}{b-a} \int_a^b e^{G(x)}\,dx}{\frac{1}{b-a} \int_a^b e^{-G(x)}\,dx}.$$

Proof. Change k_0 in the expression of m for $k_0 + 1$ in the light of the last remark and consider that $k_0 = e^t$ and $p_1 = e^{-t}$ holds only for fixed t. Taking into account the fact that t is uniform in $[a,b]$ the replacements are obvious.

Example. Consider the standard exponential density function $f(x) = e^{-x}$, $x \geqq 0$. Split up the halfline $[0,+\infty)$ into sub-intervals $[i, i+1)$, $i = 0,1,2,\ldots$ The probability of selecting the interval $[i,i+1)$ is

$$p_i = \int_i^{i+1} e^{-x}\,dx = e^{-1} - e^{-(i+1)} = e^{-1}(1-e^{-i})$$

and in the sub-interval random variables with density function $g_i(x) = \frac{e^{-x}}{p_i}$, $x \in [i, i+1)$ will be generated. For these densities the equation $f(x) = \sum_{i-0}^{\infty} p_i g_i(x)$ holds and the function x also changes at most 1 in the sub-interval $[i, i+1]$, consequently the second variant (Algorithm C8) of the comparison method is applicable.

A remark would be proper here before the detailed description of the algorithm. The probability of turning out an odd index in the sequence $t > u_1 > \ldots \leqq u_k$ is e^{-t}, and for the present density function $f(x) = e^{-x}$ the value t is chosen uniformly in $[0,1)$ thus

$$P\{t \text{ uniform}, t > u_1 > \ldots > u_{k-1} \leqq u_k,\ k \text{ even}\}$$
$$= \int_0^1 (1 - e^{-t})dt = e^{-1}.$$

Therefore the probability p_i of selecting sub-interval $[i, i+1)$ is equal to the probability of the event that in i sequences the index turned out to be even and in the $(i+1)$st sequence the index becomes odd. This fact will be economized in Step 5 of the following algorithm:

E3 Exponential, by comparison

1. Initialize $i \leftarrow 0$.
2. Generate u and set $t \leftarrow u$.
3. Generate u_1 and if $u < u_1$ holds, then go to Step 6.
4. Generate u, if $u_1 > u$, then go back to Step 3.
5. [The sequence is rejected since the index is even.] Increase the index of the sub-interval $i \leftarrow i+1$ and go back to Step 2 for starting a new sequence.
6. Deliver the number $x \leftarrow i + t$ as exponentially distributed.

Since the algorithm starts the generation of a new sequence in Step 2, if the index in the previous sequence was even then a value in the sub-interval $[i, i+1)$ is produced exactly with probability p_i according to the remark made prior to the algorithm.

The number of uniformly distributed samples required can be worked out by expression (3.10) as

$$\overline{N} = \frac{1+m}{p_i} = \frac{1 + \int_0^1 e^x \, dx}{\int_0^1 e^{-x} \, dx} = \frac{e}{1 - e^{-1}} \approx 4.3$$

which could be considered rather high.

3.5. Economical method

We modify the already described method of acceptance-rejection for random number generation to yield a technique rejecting no generated sample. Let our task be to generate random numbers with density function $f(x)$, $x \in [a,b]$, and let $h(x)$, $x \in [a,b]$ be another density function. Introduce the following notation:

$$I^+ = \{x \mid h(x) > f(x)\}$$
$$I^- = \{x \mid h(x) \leqq f(x)\}$$

and let T be a monotone increasing transformation of the set I^+ into I^-, that is, $T : y = T(x)$, $x \in I^+$, $y \in I^-$. Actually the reasoning will be similar for T being monotone decreasing but we will stick to the monotone increasing version. Furthermore, to keep the presentation as simple as possible, we also impose some other restrictions. Suppose $h(x) > f(x)$ if $x \in [a,c)$, $f(x) \geqq h(x)$ if $x \in [c,b]$, $f(c) = h(c)$ for some $c \in [a,b]$. Finally, assume that with a transformation T for any $x \in I^+ = [a,c]$ the following equilibrium equation holds:

$$\int_a^x [h(y) - f(y)]\, dy = \int_c^{T(x)} [f(y) - h(y)]\, dy.$$

Under the previous assumptions, samples with density function $f(x)$ can be produced by the following random number generator.

C9 Continuous variable, by economical

1. Generate x with density $h(\cdot)$.
2. If $x \in I^-$, then deliver x.
3. [Test.] Generate u. If $f(x)/h(x) \leqq u$, then transform this value by T, that is, deliver the value $T(x)$, otherwise deliver x.

The validity of the algorithm is rendered by the following theorem.

Theorem. Algorithm C9 produces random numbers distributed with density function $f(x)$.

Proof. Denote the value produced as the outcome of the algorithm by y and let u be uniform in [0,1), x distributed with density h. Corresponding to the three possible cases that may occur, we present the probability $P\{y \leqq d\}$, $d \in [a,b]$ as a sum of three other probabilities according to the theorem of total probabilities

$$\begin{aligned} P\{y \leqq d\} = {} & P\{x \leqq d \mid x \in I^-\} \cdot P\{x \in I^-\} \\ & + P\{x \leqq d \mid x \in I^+, f(x)/h(x) > u\} \cdot P\{x \in I^+, f(x)/h(x) > u\} \\ & + P\{T(x) \leqq d \mid x \in I^+, f(x)/h(x) \leqq u\} \cdot P\{x \in I^+, f(x)/h(x) \leqq u\}. \end{aligned}$$

The first case (Step 2) is the simplest one

$$P\{x \leqq d \mid x \in I^-\} \cdot P\{x \in I^-\} = P\{x \leqq d,\ x \in I^-\} = \int_{\substack{a\\ z\in I^-}}^{d} h(z)\, \mathrm{d}z.$$

In the second case, when the generated value is accepted, then we have

$$P\{x \leqq d \mid x \in I^+,\ f(x)/h(c) > u\} P\{x \in I^+,\ f(x)/h(x) > u\}$$

$$= P\{x \leqq d, x \in I^+, f(x)/h(x) > u\} = \int_{\substack{a\\ x\in I^+}}^{d} \int_{0}^{f(x)/h(x)} h(x) \mathrm{d}u\, \mathrm{d}x = \int_{\substack{a\\ x\in I^+}}^{d} f(x) \mathrm{d}x.$$

Finally, the last part deals with the case where transformation T is applied:

$$P_3 = P\{T(x) \leq d,\ x \in I^+,\ f(x)/h(x) \leqq u\}$$

$$= \int_{\substack{a\\ x\in I^+}}^{T^{-1}(d)} \int_{f(x)/h(x)}^{1} h(x)\, \mathrm{d}u\, \mathrm{d}x$$

$$= \int_{\substack{a\\ x\in I^+}}^{T^{-1}(d)} [h(x) - f(x)]\, \mathrm{d}x.$$

Making use of the equilibrium equation, the last integral can be rewritten as

$$P_3 = \int_{\substack{c\\ x\in I^-}}^{d} [f(x) - h(x)]\, \mathrm{d}x.$$

Summing up the probabilities we have

$$P\{y \leqq d\} = \int_{\substack{a\\ x\in I^-}}^{d} h(x)\, \mathrm{d}x + \int_{\substack{a\\ x\in I^+}}^{d} f(x)\, \mathrm{d}x + \int_{\substack{a\\ x\in I^-}}^{d} [f(x) - h(x)]\, \mathrm{d}x$$

$$= \int_{a}^{d} f(x) \mathrm{d}x$$

which gives the required result.

Assume that for generating one variable with density function h only one uniformly distributed sample is required. Then in order to obtain one random number with density function f by the economical algorithm C9, we need

$$\overline{N} = \int_{\substack{a\\ x\in I^-}}^{b} h(x)\, \mathrm{d}x + 2 \int_{\substack{a\\ x\in I^+}}^{b} h(x)\, \mathrm{d}x = 1 + \int_{\substack{a\\ x\in I^+}}^{b} h(x)\, \mathrm{d}x$$

uniformly distributed numbers on the average, and it is at most 2.

Essentially the economical method performs transformation of values from interval I^+ to interval I^-, if necessary. The transformation T transfers some of the values in interval I^+ generated in an amount greater than desired to the interval I^-, where originally less than the desired number of samples are produced ($f(x) \geqq h(x)$ if $x \in I^-$). The equilibrium equation expresses that for a shortage cropping up to a given point x we can determine the same amount of surplus by transformation T.

The method is graphically illustrated in Fig. 18. More general cases can be covered by subdividing the support interval into sub-intervals, to which the previously described procedure could be separately applied and then by the decomposition technique we put together the parts.

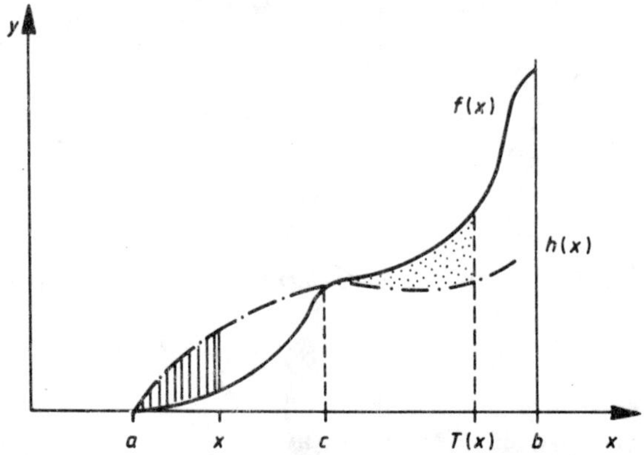

Fig. 18. Illustration to the economical method; the shaded surplus area is equal to the dotted shortage area, because of the equilibrium equation

The equilibrium equation can be written in two different ways, depending on the monotone increasing or decreasing property of the transformation T; the transformation will be denoted by $U(x)$ and $V(x)$ respectively, and the two equations are

$$\text{(3.13)} \qquad \int_a^x [h(y) - f(y)]\,\mathrm{d}y = \int_c^{U(x)} [f(y) - h(y)]\,\mathrm{d}y$$

$$\int_x^c [h(y) - f(y)]\,\mathrm{d}y = \int_c^{V(x)} [f(y) - h(y)]\,\mathrm{d}y.$$

From these equations, using the notation $K(x) = F(x) - H(x)$ one may obtain

$$K(U(x)) = K(c) - K(x) \quad (3.14)$$
$$K(V(x)) = K(x).$$

Thus the practical applicability greatly depends on the availability and ease of inversion of the function K, that is, whether $U(x), V(x)$ can be expressed explicitly from (3.13). For finding special transformations of type T other ad hoc methods should be considered also.

Example. Let $f(x) = 2x$, $x \in [0,1]$, and $h(x) = 1$ $x \in [0,1]$ uniform density function. Substituting $F(x) = x^2$, $H(x) = x$, $K(x) = x^2 - x$ into equations (3.14) the transformations

$$U(x) = 0.5 + \sqrt{x - x^2}$$
$$V(x) = 1 - x$$

can be derived. Since $V(x)$ is simpler, the algorithm employing $V(x)$ will be presented as an example of sampling random numbers with density $f(x)$ by the economical method.

L1 Linear density, by economical

1. Generate u.
2. If $u > 0.5$, then deliver u.
3. Generate v. If $v < 2u$, then deliver u, otherwise deliver $1 - u$.

To generate a random number with density $f(x)$, $N = 1.5$ uniformly distributed number is required on the average by this algorithm.

At this point it is thought to be appropriate to put in a few words on approximating densities. In many cases we are only given an empirical density function or we can approximate the density function by evaluating relative frequencies. If the approximation is composed of constant functions, then it is clearly a mixture of uniform distributions. If the approximating function is piecewise linear, then the economical method can be applied. Denote one term in the mixture of linear densities by $g(x) = mx + c$, $x \in [a,b]$ and assume $m > 0$. We can directly apply the previous Algorithm L1, or we may — which would be more preferable, especially for relatively large value of $ma + c$ — further decompose the function $g(x) = g_1(x) + g_1(x)$, where $g_1(x) = d_1(ma + c)$ is constant and $g_2(x) = d_2 m(x - a)$, where d_1 and d_2 are two suitable constants asserting $\int g_1(x)\,\mathrm{d}x = \int g_2(x)\,\mathrm{d}x = 1$.

3.6. Ratio-of-uniforms method

In this method a point is generated uniformly over a certain domain in the plane and the random number with the desired distribution is obtained by taking the ratio of the two coordinates of the uniformly distributed point.

In order to describe the method, consider a density function $f(x)$ and a domain

$$D = \{(x,y) \mid 0 \leqq x \leqq x\sqrt{f(y/x)}\}.$$

The algorithm for producing random numbers with density function $f(x)$ is the following.

C10 Continuous variable, by ratio-of-uniforms

1. Generate a sample point (u,v) uniformly distributed over domain D.
2. Deliver $x \leftarrow v/u$ as a random number with density function $f(x)$.

The validity of the algorithm is stated and shown in the following theorem.

Theorem. Algorithm C10 produces numbers with density function $f(x)$.

Proof. We are going to make use of the transformation theorem regarding multivariate density functions, presented in Section 1.3. The joint density function of the point (u,v) is $g(p_1,p_2) = c$, if $(p_1,p_2) \in D$ (i.e. $0 \leqq p_1 \leqq \sqrt{f(p_2/p_1)}$, where c is a constant determined by the equation $l/c = \iint\limits_D \mathrm{d}p_1\, \mathrm{d}p_2$. In the algorithm we used the transformation $T : (p_1,p_2) \to (t_1,t_2)$ where $t_1 = p_2/p_1,\ t_2 = p_1$, its inverse is $T^{-1} : (t_1,t_2) \to (p_1,p_2)$, where $p_1 = t_2,\ p_2 = t_1t_2$. The Jacobian of this latter is $\dfrac{\partial T^{-1}}{\partial \mathbf{t}} = -t_2$ whence the density function of the transformed point $T(u,v) = (x,y)$ follows

$$h(t_1,t_2) = g(T(p_1,p_2))|_{\mathbf{t}} \cdot \left|\frac{\partial T^{-1}}{\partial \mathbf{t}}\right| = \begin{cases} ct_2, & \text{if } \ 0 \leqq t_2 \leqq \sqrt{f(t_1)} \\ 0, & \text{otherwise.} \end{cases}$$

The marginal density function of the first coordinate (what we actually need) and the constant c can be evaluated from the equation:

$$\int\limits_0^{\sqrt{f(t_1)}} h(t_1,t_2)\, \mathrm{d}t_2 = \int\limits_0^{\sqrt{f(t_1)}} ct_2\, \mathrm{d}t_2 = \frac{c}{2} f(t_1).$$

If we substitute here $c = 2$, then the result is just as required: the density function of the first coordinate is $f(\cdot)$ and the area of the domain D is $1/c = 1/2$.

Example 1 (Normal distribution). Consider the positive half of the standard normal distribution (samples from the standard normal distribution can be obtained by attaching a random sign to these "positive" random numbers). Its density function is $f(x) = \sqrt{\frac{2}{\pi}} \cdot e^{-\frac{x^2}{2}}$, the corresponding domain is

$$D = \{(x,y)|0 \leqq x \leqq \sqrt{\sqrt{\frac{2}{\pi}} \exp\{-y^2/2x^2\}}\}.$$

After streamlining the inequalities we get

$$x^2 \leqq \sqrt{\frac{2}{\pi}} \exp\{-y^2/2x^2\},$$

$$\log\left(\sqrt{\frac{\pi}{2}} x^2\right) \leqq -y^2/2x^2,$$

$$y^2 \leqq 4x^2 \log\left(\sqrt{\frac{\pi}{2}} \cdot x^2\right).$$

Example 2 (Exponential distribution). Consider the standard exponential density function $f(x) = e^{-x}$, $x \geqq 0$, then we have $D = \{(x,y) \mid 0 \leqq x \leqq \exp\{-y/2x\}\}$, that is $y \leqq -2x \log x$.

The method can easily be extended for any function $k(x)$ with finite integral K, $K = \int k(x)\,dx$. Denote $D_k = \{(x,y) \mid 0 \leqq x \leqq \sqrt{k(y/x)}\}$, then it could be shown in a step-by-step way similar to the reasoning in the theorem, that if (u,v) is uniformly distributed in D_k, then v/u has density function $f(\cdot)$ proportional to $k(\cdot)$, that is $f(x) = \frac{1}{K} k(x)$ and the are of D_k is $K/2$.

The domain D_k is bounded in the x direction by the graph of $x = \sqrt{k(y/x)}$, if $k(\cdot)$ is bounded, and in many cases this equation can be solved explicitly for y in terms of x. Furthermore D_k is bounded in the y direction if $x^2 k(x)$ is bounded. Generally for producing points uniformly distributed in domain D_k (or D), not very simple ones but bounded, one has to resort to acceptance-rejection techniques using simple enveloping domains (rectangles, circles) as will be described in the next chapter.

At this point we may wonder if there are not other possible transformations of uniform random numbers instead of the ratio-of-uniform.

Actually there have been some other special cases presented. Using the same notation as for the function $k(x)$ with finite integral K, define the domains

$$D_k^* = \{(x, y) \mid 0 \leqq x \leqq k(x + y)\},$$
$$D_k^{**} = \{(x, y) \mid 0 \leqq x \leqq [k(y/\sqrt{u})]^{2/3}\}.$$

If (u, v) is uniformly distributed in D_k^* (or in D_k^{**}), then it can be proved that the random variable $u + v$ (or $v/\sqrt{u}$) is distributed with density function $f(x) = \dfrac{1}{K}k(x)$.

The greatest appeal of the method is its simplicity. Possible points of hidden difficulties may be encountered in generating a point uniformly in D, that might have been done by a very low ratio of acceptance.

3.7. Variants of the acceptance-rejection method

The acceptance-rejection technique is the most flexible way of handling the generation of random numbers from a given distribution. Its simplicity, universality and occasional weaknesses (which actually seem to be correlated properties of the method) have inspired many researchers to develop variants that are maybe more specific, but at the same time are certainly more efficient. Some of these variants will be presented here.

The basic notation and the algorithm of acceptance-rejection method are repeated here for convenience. Random variables with density function $f(\cdot)$ are required and we are given an upper bounding function $g(x)$, for which $f(x) \leqq g(x) = ch(x)$, $c = \int g(x)\,dx$, thus $h(x)$ is a density. The formal algorithm is

C3 Continuous variable, by rejection

1. Generate u. Generate y with density $h(\cdot)$.
2. If $u \leqq f(y)/g(y)$ then deliver $x \leftarrow y$, otherwise go back to Step 1.

An idea for improving the method by getting rid of the comparison of Step 2 (which could be very time consuming for complicated functions f and/or g) in most cases has been incorporated in the squeeze method. Let $s(\cdot)$ and $g(\cdot)$ be lower bounding and upper bounding functions with respect to density function $f(\cdot)$, respectively (that is, $s(x) \leqq f(x) \leqq g(x)$), where the functions $s(\cdot)$ and $g(\cdot)$ and the ratio $s(x)/g(x)$ can be evaluated at small costs, they should be constant, linear or other simple functions.

The term *squeeze method* comes from this squeezing the function $f(\cdot)$ between two functions $s(\cdot)$ and $g(\cdot)$. Preserving the notation $ch(x) = g(x)$, $c = \int g(x)\,dx$ the squeeze method's formal algorithm can be stated as:

C11 Continuous variable, by acceptance-rejection + squeeze modification

1. Generate u. Generate y with density h.
2. [Fast acceptance.] If $u \leqq s(y)/g(y)$, then deliver $x \leftarrow y$.
3. [Slow acceptance.] If $u \leqq f(y)/g(y)$, then deliver $x \leftarrow y$, otherwise go back to Step 1.

Example. Consider the truncated exponential distribution with density function $f(x) = \dfrac{e}{e-1}e^{-x}$, $0 \leqq x \leqq 1$, investigated in Example 3 of Section 3.2 as a demonstration of the acceptance-rejection technique (algorithm E2). The upper bounding function was $g(x) = \dfrac{e}{e-1} \approx 1.582$ and the corresponding density function is $h(x) = 1$. Here the modified version of the algorithm will be presented with a lower bounding constant function $s(x) = f(1) = \dfrac{1}{e-1} \approx 0.582$, as the idea of the squeeze method is being exploited.

EXAMP 4. Example for squeezing the truncated exponential

1. Generate u and v.
2. If $u \leqq s(v)/g(v) \approx 0.368$, then deliver v.
3. If $u \leqq f(v)/g(v) = e^{-v}$, then deliver v, otherwise go back to Step 1.

Thus the computation of the exponential function can be avoided in one-third of all the cases because of the immediate fast acceptance.

Similarly, one can construct a simple upper bounding function $t(x)$ for which $f(x) \leqq t(x) \leqq g(x)$, whereby a procedure with fast rejection can be obtained. The point of searching and constructing another upper bounding function $t(\cdot)$ besides the function $g(\cdot)$ is that it serves a different purpose. The function $t(\cdot)$ should be simple to evaluate whereas the function $g(\cdot)$ should be a density function multiplied by a constant according to which random numbers should easily be generated.

The acceptance-complement method is a technique where the mixture method has been combined with the acceptance-rejection method. Decompose the target density $f(x)$ into two subdensities $g_1(\cdot)$, $g_2(\cdot)$, where $g_1(x) \geqq 0$, $g_2(x) \geqq 0$, as

$$f(x) = g_1(x) + g_2(x) \tag{3.16}$$

and denote by p_1 and p_2 the following probabilities:

$$p_1 = \int g_1(x)\,dx, \qquad p_2 = \int g_2(x)\,dx.$$

Choose a density function $h_1(\cdot)$ dominating the subdensity $g_1(\cdot)$, that is, $h_1(x) \geqq g_1(x)$, and denote the density function proportional to $g_2(\cdot)$ by $f_2(\cdot)$, that is $f_2(x) = \frac{1}{p_2} g_2(x)$, then the algorithm can be stated as

C12 Continuous variable, by acceptance-complement

1. Generate u and generate x with density h_1.
2. If $u \leqq g_1(x)/h_1(x)$, then deliver x (acceptance, finish), otherwise proceed to the next step (this is actually rejection of x).
3. Generate y with density f_2 and deliver y.

Rewriting the decomposition of $f(x)$ in (3.16) as

$$f(x) = p_1 \left[\frac{1}{p_1} g_1(x)\right] + p_2 \left[\frac{1}{p_2} g_2(x)\right] = p_1 f_1(x) + p_2 f_2(x) \tag{3.17}$$

where the functions in the brackets (f_1 and f_2) are densities, one can see that Algorithm C11 is of decomposition type, where the generation with density $f_1(\cdot) = \frac{1}{p_1} g_1(\cdot)$ is performed by the acceptance-rejection technique with majorizing function $\frac{1}{p_1} h_1(x)$. The main point here is that choosing the first member of the mixture of f has the same probability p_1 as the probability of acceptance in Step 2, thus the acceptance-rejection has been exploited for two ends. Similarly, one can see that rejection occurs with the same probability $1 - p_1 = p_2$ as the choice of the second density function $\frac{1}{p_2} g_2(\cdot)$.

The method can be recommended for mixtures with p_1 close to 1 and $h_1(\cdot)$ well approximating $g_1(\cdot)$.

The inversion-rejection method has been developed for cases where we do not have the inverse F^{-1} of the target distribution but explicit forms of the density function $f(\cdot)$ and the distribution function $F(\cdot)$ are available. The basic idea of the method is simple. After having decomposed the support of the density function into sub-intervals: (i) the choice of a member of the decomposition is required and then (ii) the acceptance-rejection technique may be employed in the selected sub-interval. The term *inversion* in the name of the method refers to step (i) where selection of the sub-interval (generation of a sample from a discrete distribution) may be done by inversion (searching).

Effective use of the method can be conjectured only in cases where some additional knowledge of the behaviour of the density function is guaranteed, for example if we know that $f(\cdot)$ is monotone, or unimodal (that can be split up into two monotone parts) or convex over its support, etc. We describe the method for the monotone decreasing density function $f(x)$, $x \in [0,1]$. The sequential search will not be done on previously established sub-intervals but we create them by halving the length of the interval being investigated.

C13 Continuous variable, by inversion-rejection

1. [Inversion.] Generate u and initialize the interval (a, b) by $a \leftarrow 0.5$, $b \leftarrow 1$.
2. If $u < F(a)$, then shift the interval to the left and halve the length of it by setting $b \leftarrow a$, $a \leftarrow 0.5a$ and repeat this step from the beginning, otherwise proceed to the next step (at this point we know that the root x_0 of the equation $F(x_0) = u$ is in interval (a, b)).
3. [Acceptance-rejection.] Generate v and w. Initialize $y \leftarrow a+v(b-a)$; y is uniformly distributed in interval (a, b).
4. If $w \leqq f(y)/f(a)$, then deliver v, otherwise go back to Step 3.

The inversion part can be replaced with something more precise if need be, for example a back-and-forth searching until the length of the interval will be less than a prescribed precision. Having more details on the behaviour of $f(\cdot)$ the crude bound $f(a)$ could be changed into something better, e.g. a linear upper bounding function, or we can use the squeeze method here. The algorithm can be accelerated by storing the function values $f\left(\frac{1}{2^k}\right)$, $F\left(\frac{1}{2^k}\right)$ for $k = 0, 1, 2, \ldots$ also.

Finally, we present an application of the acceptance-rejection technique, where the target density f is very complicated, but can be approximated from above and below by a convergent series of simpler functions whereby the method is called the *series* method of random number generation. In particular, assume that there exist sequences of functions $f_{2n}(\cdot)$ and $f_{2n}(\cdot)$, $n = 1, 2, \ldots$ for which $f_{2n}(x) \leqq f(x) \leqq f_{2n}(x)$ and

$\lim_{n\to\infty} f_{2n}(x) = \lim_{n\to\infty} f_{un}(x) = f(x)$. Furthermore, there exists an upper bounding function $ch(x) \geqq f(x)$, where $h(\cdot)$ is a density function. The algorithm can be stated as follows.

C14 Continuous variable, by series

1. Generate u. Generate x with density $h(\cdot)$ and initialize $n \leftarrow 0$, $t \leftarrow uh(x)$.
2. Increase $n \leftarrow n+1$. If $t \leqq f_{2n}(x)$, then deliver x.
3. If $t > f_{un}(x)$ then go back to Step 1 (rejection).
4. [Proceed to the next function.] Go back to Step 2.

Several special cases of this general method deserve mentioning, but we give here only one example.

Assume that $f(x)$ can be written as an alternating sequence of the type

$$f(x) = g(x)(1 - a_1(x) + a_2(x) - a_3(x)\ldots), \tag{3.18}$$

where $g(x) = ch(x)$, $c = \int g(x)\,dx$ and $a_n(x)$ is a sequence of functions for which $a_n(x) \geqq a_{n+1}(x)$, for all x, $n = 1, 2, \ldots$ and $\lim_{n\to\infty} a_n(x) = 0$.

The algorithm for generating random numbers with density function $f(x)$ is

C15 Continuous variable, by series + alternating sequence

1. Generate x with density $h(\cdot)$. Generate u and set $n \leftarrow 0$, $s \leftarrow 1$.
2. Increase $n \leftarrow n+1$ and set $s \leftarrow s - a_n(x)$. If $u \leqq s$ then deliver x.
3. Increase $n \leftarrow n+1$ and $s \leftarrow s + a_n(x)$, if $u > s$, then repeat the whole procedure from Step 1.
4. Go back to Step 2.

Since for any odd k we have the inequalities

$$1 + \sum_{j=1}^{k} (-1)^j a_j(x) \leqq f(x)/g(x) \leqq 1 + \sum_{j=1}^{k+1} (-1)^j a_j(x), \tag{3.19}$$

thus the criterion of acceptance would be $u \leqq f(x)/g(x)$. Replacing it with the left hand side in the spirit of the squeeze method we have the relation $u \leqq 1 + \sum_{j=1}^{k} (-1)^j a_j(x)$ to be checked. Changing the sign we have $1-u \geqq \sum_{j=1}^{k}(-1)^{j+1}a_j(x)$ and $1-u$ can be replaced by its statistical equivalent u. According to these remarks a modified algorithm could be

C16 Continuous variable, by series-alternating squeeze

1. Generate u. Generate x with density $h(\cdot)$, set $n \leftarrow 0$, $s \leftarrow 0$.
2. Set $n \leftarrow n+1$, $s \leftarrow s + a_n(x)$, if $u \geqq s$, then deliver x.
3. Set $n \leftarrow n+1$, $s \leftarrow s - a_n(x)$, if $u < s$, then return to Step 1, otherwise go to Step 2.

These last two algorithms (C14 and C15) create the possibility for approximating a density function by an infinite series, the individual terms of which could be much easier to evaluate, and having a generator without ever evaluating the target density. Practical usefulness depends on the form of series and how far we have to go in the series (the size of n) before deciding upon acceptance or rejection, that is, how fast is the convergence of the series to the target density.

3.8. Generating discrete variables by binary search

At the inversion method we already have presented an algorithm for generating samples of the random variable ξ given by the probabilities

(3.20) $$P\{\xi = x_i\} = p_i, \quad p_i > 0, \quad i \geqq 1, \dots, k.$$

The model corresponding to Algorithm D1 is the following. We have in an urn balls numbered from 1 to k and we have to choose the ith ball with probability p_i. If the ith ball was chosen, the quantity x_i is the delivered (generated) value. In adapting the model to a computer, difficulty arises if k (the number of balls) is great and the choice is hard. Had we two balls in the urn, only one comparison would be necessary.

In what follows a procedure is described where in the sense of the previous remark the number of balls in one urn is decreased by combining probabilities, thus reducing the number of comparisons as compared with Algorithm D1.

Define the numbers $p(j,i)$ as follows. Let

(3.21) $$p(j,0) = p_j, \qquad j = 1, \dots, k.$$

These are called the probabilities of the 0th level. The probabilities $p(j,1)$ of the 1st level are thought of as the sum of two consecutive 0th level probabilities

$$p(j,1) = p(2j-1,0) + p(2j,0), \qquad j = 1,\ldots,k_1, \tag{3.22}$$

where $k_1 = [(k+1)/2]$, and the brackets [ ] mean the (lower) integer part of the number in brackets ([3.2]=[3.7]=3). Of course in the case of k odd the last sum will consist of one term only:

$$p(k_1,1) = p(2k_1-1,0) = p(k,0).$$

Similarly, the probabilities $p(j,2)$, $j = 1,\ldots,k_2$, of the 2nd, $\ldots, p(j,2)$, $j = 1,\ldots,k_l$ of the lth level are constructed as the sum of two consecutive probabilities of the previous level. Thus probabilities of higher and higher levels are manufactured until there are no more probabilities, that is, the last mth level is reached when we have only one probability

$$p(1,m) = p(1,m-1) + p(2,m-1) = 1,$$

that is, $k_m = [(k_{m-1}+1)/2] = 1$. Furthermore, compute the total probabilities

$$\begin{aligned} \overline{p}(1,i) &= p(1,i), \\ \overline{p}(j,i) &= p(1,i) + p(2,i) + \ldots + p(j,i) \\ &= \overline{p}(j-1,i) + p(j,i), \; j = 2,\ldots,k, \; i = 1,\ldots,m-1. \end{aligned} \tag{3.23}$$

The value $\overline{p}(j,i)$ is the joint probability of the probabilities of level i with index j or less.

The procedure of generation will advance on the binary tree just produced in a top–down manner; the branch to proceed along is chosen according to the probabilities $\overline{p}(j,i)$ and on all levels only one comparison is required to proceed. The preparation and the generator itself are formally described in what follows.

D2 Discrete variable, by binary search

Preparation

1. Initialize $i \leftarrow 0$, $k_0 \leftarrow k$, $p(j,0) = p_j$, $j = 1,\ldots,k$.
2. Increase index $i \leftarrow i+1$ and set $j \leftarrow 1$, $k_1 \leftarrow [(k_{i-1}+1)/2]$.
3. [Determine the probabilities of level i.] Let

$$p(j,i) \leftarrow p(2_j,i-1) + p(2_j,i-1), \; j = 1,\ldots,k_i-1.$$

4. If $k_{i-1} = 2k_i$, then set

$$p(k_i, i) \leftarrow p(k_{i-1} - 1, i - 1) + p(k_{i-1}, i - 1),$$

otherwise set $p(k_i, i) \leftarrow p(k_{i-1}, i - 1)$.
5. If $k_i \neq 1$, then go back to Step 2.
6. Initialize $m \leftarrow i$, $i \leftarrow 0$.
7. [Compute the total probabilities.] Increase index $i \leftarrow i + 1$ and set $\overline{p}(1, i) \leftarrow p(1, i)$. Evaluate the sums

$$\overline{p}(j, i) \leftarrow \overline{p}(j - 1, i) + p(j, i), \qquad j = 2, \ldots, k_i.$$

8. [Finish.] If $i = m - 1$, then finish the preparation otherwise go back to Step 7.

Generation

1. Generate u and initialize $j \leftarrow 1$, $i \leftarrow m - 1$.
2. [Select an index j on the level i.] If $u < \overline{p}(j, i)$, then set $j \leftarrow 2j - 1$, otherwise set $j \leftarrow 2j + 1$.
3. [Reduce the counter of level.] Decrease $i \leftarrow i - 1$, if $i = 0$, then go to Step 4, otherwise go back to Step 2.
4. If $u < \overline{p}(j, 0)$, then deliver the value $x \leftarrow x_j$, otherwise deliver $x \leftarrow x_{j+1}$.

Example. The working of the algorithm is illustrated in the following discrete distribution with $n = 13$ elements. The probabilities $p_j = p(j, 0)$, $j = 1, \ldots, 13$ are

$$0.02, \quad 0.03, \quad 0.03, \quad 0.03, \quad 0.04, \quad 0.05, \quad 0.05,$$

$$0.05, \quad 0.10, \quad 0.10, \quad 0.10, \quad 0.20, \quad 0.20$$

consecutively. From these probabilities the values $p(j, i)$, $\overline{p}(j, i)$ are computed by the preparation part of Algorithm D2 forming altogether $m = 4$ levels (total probabilities are at the bottom of each box):

Assume we have generated the value $u = 0.79$. Algorithm D2 first examines the inequality $u < \overline{p}(1, 3) = 0.3$. Since it is not satisfied the values $j \leftarrow 3$ and $i \leftarrow 2$ are set and the condition $u < \overline{p}(3, 2) = 0.8$ is examined. It is fulfilled, thus $j \leftarrow 5$ and $i \leftarrow 1$ turn out, then from the inequality $u < \overline{p}(5, 1) = 0.5$ we get $j \leftarrow 11$, $i \leftarrow 0$. Finally, because $u < \overline{p}(11, 0) = 0.6$ is not satisfied, the algorithm delivers the 12 term of the distribution. (Some other examples: for $u = 0.85$, $j \leftarrow 13$; for $u = 0.32$, $j \leftarrow 9$ and for $u = 0.17$ we have $j \leftarrow 6$.)

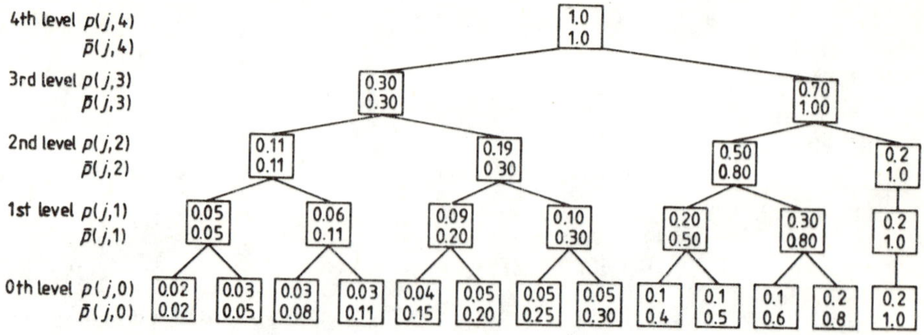

Scheme 1. The probabilities $p(j,i)$ and $\bar{p}(j,i)$ of a discrete distribution to Algorithm D2

Algorithm D2 is based on the following model of urns: in one urn there are two balls only, each one representing a subsequent urn. We have to choose between two balls and then in the urn indicated by the ball chosen there are similarly two balls, etc. For generating one sample by Algorithm D2, $[\log_2 n] + 1$ comparisons are required. To store the total probabilities $\bar{p}(j,i)$, $i = 1, 2, \ldots, m$, about n words are required and this, together with the original probabilities $p_j = p(j,0)$, means we need $2n$ words in the computer storage (the probabilities $p(j,i)$, $j = 1, \ldots, m$, $i = 1, \ldots, k_i$ are needed only during the preparation).

Algorithms similar to D2 could be constructed in such a way that probabilities $p(j,1)$ on the first level are produced not as a sum of 2, but 3, 4 or 10 consecutive probabilities of level 0. Probabilities on level 2 are formed as the sum of 3, 4 or 10 probabilities on level 1, respectively, etc. In these cases the searching procedure requires more comparison to make the transition from level i to level $i-1$ than in the case of Algorithm D2, but the number of levels is smaller. Practical usefulness advocates the choices of octal, decimal and hexadecimal search techniques, but basically some other methods (of decomposition type, described in Section 3.10) should be preferred.

3.9. Generating discrete variables by table method

The method for producing realizations of a random variable with discrete distribution to be described here could be substantially faster than the searching procedures presented previously. Before giving a formal description the main steps of the algorithm's operation are illustrated by a numerical example. Let the discrete distribution given by the following probabilities and values:

Table 8

Probability	Value
0.321	a
0.034	b
0.212	c
0.433	d

Most probably the fastest method for generating samples with such distribution would be the next one. Consider a vector with 1000 elements, store the value "a" in 321 elements, value "b" in 34 elements, value "c" in 212 elements and value "d" in 433 elements. After this preparation to produce one random value generate a random index over the set of integers $1, 2, \ldots, 1000$, that is, generate the random number $u = 0.d_1d_2 \ldots, d_z$ uniformly in the interval [0,1), where d_j denotes the jth decimal digit of u and let the index i be equal to the number $d_1d_2d_3 + 1$. Then i is uniformly distributed over the set $1, 2, \ldots, 1000$ of integer numbers. Deliver the ith element of the stored vector as a sample from the desired distribution. The soundness of the procedure can easily be shown by an urn model. Consider an urn having 1000 balls. We have the inscription "a" on 321 balls, inscription "b" on 34 balls, inscription "c" on 212 balls and inscription "d" on all the remaining 433 balls. One of them is randomly (uniformly) chosen. The computer algorithm is a natural implementation of this urn model. Since the procedure requires the generation of one number and the picking up of the element of the table with the index just obtained, it is very fast. A drawback of the method is that in most practical cases the number of the values to be stored may be quite large. Nevertheless, the memory requirement could be substantially decreased by a simple modification while not really affecting the speed of the algorithm.

The modified procedure is based on the following urn model. Consider three urns that correspond to the three decimal digits appearing in the statement of the probabilities. The first urn will be sampled with a probability corresponding to the sum of the first digits (decimals), the second urn will be chosen with a probability corresponding to the sum of the second digits (hundredths) and the third urn will be chosen with probability of the third digits. The contents of the urns is determined in the following way. In the urn of the decimals we store as many balls with inscription "a" as the decimal in the probability of the value "a", as many balls with inscription "b" as the decimals in the probability of "b", etc. The probability of choosing a specific urn and its contents is given in Table 9.

Table 9

The probability of selecting the urn	The contents of the urn	
	Number of balls	Inscription
1st urn 0.9 (decimals)	3	a
	2	c
	4	d
2nd urn 0.09 (hundredths)	2	a
	3	b
	1	c
	3	d
3rd urn 0.01 (thousandth)	1	a
	4	b
	2	c
	3	d
Total	28	

The generation procedure has been combined from two steps: first, an urn is chosen according to the corresponding probability and then one ball (that is, the value it is representing) is picked up from the balls in the urn — but this is chosen with uniform distribution.

Contrary to the one thousand constants of the previous version, after the modification storage for only 28 values is needed in the computer implementation of the numerical example. The contents of separate urns are filled into different "levels" of a vector, the levels are separated by starting addresses, and index modifiers are used to help in the location of different levels inside the vector. Obviously the algorithm can be constructed not only in decimal but arbitrary system. In the example discussed as demonstration the probabilities and the procedure were given in decimal system, however, the formal description of the general algorithm is detailed in binary system.

Let the probabilities $p_i = 0.b_{i1}b_{i2}\dots b_{im}$, $i = 1, 2, \dots, k$ be given, where m denotes the number of binary bits of a real number on the computer, b_{ij} is the jth bit in the ith probability. The discrete distribution of concern is determined by the equations $P\{\xi = x_i\} = p_i$, $i = 1, 2, \dots, k$.

D3 Discrete variable, by table method

Preparation

1. Compute the column sum s_j of the jth bits $s_j \leftarrow \sum_{i=1}^{k} b_{ij}$, $j = 1, 2, \dots, m$.
2. [Determine the probabilities t_j of the levels.] Set $t_j^* \leftarrow s_j/2^j$, $j = 1, 2, \dots, m$ and compute the total probability of level $t_1 \leftarrow t_1^*$, $t_j \leftarrow t_{j-1} + t_j$, $j = 2, 3, \dots, m$.
3. [The starting address of the level i is c_i.] Set $c_1 \leftarrow 0$, then determine $c_{i+1} \leftarrow c_i + s_i$, $i = 1, 2, \dots, m-1$. Compute the index $d_1 \leftarrow 0$, $d_j \leftarrow t_{j-1}2^j - c_j$, $j = 2, 3, \dots, m$.
4. [Store the values $x_i = 1, 2, \dots, k$ into the elements of the vector $\mathbf{A}$.] Initialize $q \leftarrow 0$ and $j \leftarrow 0$.
5. [Increase the index of the currently investigated level.] Increase $q \leftarrow q + 1$.
6. This step has to be performed for each index $i = 1, 2, \dots, k$: consider the bit b_{iq}, if $b_{iq} = 1$, then the next element of the vector $\mathbf{A}$ is filled up with the ith value of the distribution $A(j) \leftarrow x_i$, $j \leftarrow j + 1$.
7. If $q = m$, then the preparation is finished, otherwise go back to Step 5.

Generation

1. Generate u uniformly in [0,1), where the binary form of u is $u = 0.\ u_1u_2\dots u_m$ and set $j \leftarrow 1$ for the index of the level to be examined at the beginning.
2. [Registering the level.] If $u < t_j$, then deliver the number $x \leftarrow A(u_1\dots u_j - d_j)$. (The number $u_1\dots u_j$ is to be considered as written in base 2 system.)
3. Increase $j \leftarrow j + 1$ and go back to Step 2.

From the different quantities determined in the course of preparation the only values to be preserved are the vector $\mathbf{A}$, the total probabilities of levels t_j and the index modifiers d_j, $j = 1, 2, \ldots, m$. Note that if during preparation it turns out that $s_j = 0$ for some indices $j = 1, \ldots, r$, then obviously we do not need the corresponding values t_j and d_j; furthermore, in the Generation part of Algorithm D3, not the indices $1, 2, \ldots, m$, but indices of levels $r+1, r+2, \ldots, m$, that is, their re-indexed versions $(1, 2, \ldots, m-r)$, are investigated.

Example. Let the discrete distribution be given by the following values and probabilities ($P\{\xi = x_i\} = p_i$, $i = 1, \ldots, 4$), (see Table 10):

Table 10

	p_i (in binary form)	x_i
1	0.1010_2	0
2	0.0001_2	3
3	0.0011_2	6
4	0.0010_2	7

Using the Preparation part of Algorithm D3 one can obtain the values s_j : $1, 0, 2, 2$ consecutively, the total probabilities of levels t_j are 0.5, 0.5, 0.875, 1.0, the starting adresses are 0,1,1,4, the index modifiers 0,1,3 and 10 and finally the vector $\mathbf{A}$ is (0,0,6,7,3,6). If the value $u = 0.1101_2 = 0.812510_{10}$ has been generated, then D3 produces 7 from the discrete distribution, if $u = 0.10102 = 0.625_{10}$ then 0 is produced.

Similar algorithms can be built in different base systems; the octal and the hexadecimal systems could be proposed for practical reasons. From the possible variants the above described binary based algorithm requires the least memory and the greatest number of comparisons on the average (the necessary storage is basically $s_1 + \ldots + s_m$ words).

An advantage of the method is that only one uniformly distributed number is required, and the number of comparisons is equal to or less than the number of levels (digits in the representation of probabilities) not of order $\log_2 k$. A drawback of the algorithm is the feature that full efficiency can be attained by machine coded version only, which could be troublesome for users who do not want to get involved in technicalities. Furthermore, the insertion of a machine coded routine into a high level language program, and in some cases the great number of constants to be stored, may also cause difficulties. It must be noted here that in cases where the probabilities of a distribution are given with many digits it could be wortwhile to fabricate a hybrid algorithm: for a couple of first digits we apply the table method, while the other digits are covered with a simple search technique — according to the decomposition method this could result in a fast and relatively short algorithm.

3.10. Generating discrete variables by decomposition

In this section we describe a version of the decomposition method adapted for generating discrete random variables which is highly efficient and simpler than the table method.

Consider the discrete probability distribution given by the probabilities

$$P\{\xi = x_i\} = p_i, \qquad p_i > 0, \qquad i = 1, \ldots, k. \tag{3.24}$$

Its density function — that is, the probabilities p_i — will be produced as the sum of k densities, the weights of them will all be equal to $1/k$ (their choice is equally probable), and each density represents a discrete distribution with two possible values only (two-point distributions). Denote these distributions by the probabilities (q_{j1}, q_{j2}), $j = 1, \ldots, k$ where

$$P\{\eta_j = x_{j1}\} = q_{j1}$$
$$P\{\eta_j = x_{j2}\} = q_{j2}$$

and assume, furthermore, that the equations

$$p_i = \frac{1}{k} \sum_{ml: x_i = x_{ml}} q_{ml}, \quad m = 1, \ldots, k, \quad i = 1, \ldots, k, \quad l = 1, 2, \tag{3.25}$$

hold, where summing up is executed for those indices ml which fulfil $x_i = x_{ml}$, l being 1 or 2. Given the above notation, discrete variables with distribution in (3.24) will be produced by the next procedure.

D4 Discrete variable, by decomposition

1. [Selecting the jth distribution.] Generate u and set $j \leftarrow [ku]+1$ (here j is uniformly distributed over the set of integer numbers $1, 2, \ldots, k$).
2. [Generating a sample from the jth distribution.] Generate v, if $v < q_{j1}$ then deliver $x \leftarrow x_{j1}$, otherwise deliver $x \leftarrow x_{j2}$.

The procedure could be refined easily. Instead of generating one number in the first and one in the second step, it is sufficient to generate only one since if u is uniform in the interval [0,1) then $j \leftarrow [ku] + 1$ is uniform over the integers $1, 2, \ldots, k$ and $v \leftarrow ku - j + 1$ remains uniform in $[0,1)$. Such kind of saving by splitting up the generated number — the random index is produced from the first few bits, while the remainder is reusable — is very common in actual practice of the random number generation. Problems emerge, however, if the index to be generated is greater than, say, 1000, since this means stripping away more than ten bits of the original uniform number and the remaining bits could not ensure a uniformity of satisfactory level from a computational point of view.

Note that we do not really need the values q_{j2} (anyway we know that $q_{j2} = 1 - q_{j1}$), thus it suffices to store the values q_{j1} in a vector $\mathbf{Q}$. Similarly, it suffices to store the values x_i, $i = 1, \ldots, k$ and the indices $j1, j2$ (instead of the values x_{j1}, x_{j2}; the previous ones will be stored in vector $\mathbf{X}$, the indices $j1$ of the values to be directly delivered in vector $\mathbf{K}$, while the indices $j2$ of the values to be delivered second (transformed values) in vector $\mathbf{T}$. With this notation we arrive at the following modified version of the algorithm.

D5 Discrete variable, by modified decomposition

1. Generate u. Let $j \leftarrow [ku] + 1$, $u \leftarrow ku - j + 1$, then j is uniform over the integers $1, 2, \ldots, k$, u is uniform in $[0,1)$ and independent of j.
2. [Generation of a value from the jth distribution.] If $u < Q(j)$, then deliver the value $x \leftarrow X(K(j))$, otherwise deliver $x \leftarrow X(T(j))$.

Though the validity of the procedure (i.e., D5 delivers samples with distribution given in (3.24)) is obvious, it still remains to be shown that (3.24) can be decomposed as a sum of k distributions, that are equally probable and each of them has only two possible values. To this end we need the following theorem.

Theorem 1. Any discrete distribution (3.24) with k possible values can be given as the weighted sum of two other distributions, one of them having two possible values and the other one being a $(k-1)$-point distribution.

Proof. Our proof will be constructive; by giving the two distributions we demonstrate the theorem. For the sake of simplicity, assume $p_1 \leqq 1/k$, $p_2 \geqq 1/k$. Two probabilities with such properties have to exist in the distribution and they can be reindexed to 1 and 2 respectively if need be. The random variable η has two possible values with probabilities q_1 and

q_2, for which

$$q_1 = P\{\eta = x_1\} = kp_1$$
$$q_2 = P\{\eta = x_2\} = 1 - kp_1.$$

The random variable ς attains $k - 1$ values with probabilities p_i, $i = 2, 3, \ldots, k$. The distribution is given by the equations

$$p'_2 = \frac{k(p_1 + p_2) - 1}{k - 1} = P\{\varsigma = x_2\}$$
$$p'_j = \frac{k}{k-1} p_j = P\{\varsigma = x_j\}, \qquad j = 3, 4, \ldots, k.$$

It can readily be seen that (q_1, q_2) and $(p'_2, \ldots, p'_k)$ form distributions. The equations $q_1 + q_2 = 1$ and $p'_2 + \ldots + p'_k = 1$ are evident, $q_1 \geqq 0$, $q_2 \geqq 0$ is valid because of the assumptions, also because of $p_2 \geqq 1/k$ we have $k(p_1 + p_2) \geqq 1$, thus $p'_2 \geqq 0$ and $(p_1 + p_2) \leqq 1$ gives us $p'_2 \leqq 1$. Taking into account $p_2 \geqq 1/k$ we have $p_j \leqq \frac{k-1}{k}$, $j = 3, 4, \ldots, k$ resulting in $p'_j \leqq 1$.

Finally, we show that the original distribution $(p_1, p_2 \ldots, p_k)$ is a weighted sum of (q_1, q_2) and $(p'_2, p'_3, \ldots, p'_k)$, the weights being $1/k$ and $(k-1)/k$, respectively, since we have

$$p_1 = \frac{1}{k} q_1$$
$$p_2 = \frac{1}{k} q_2 + \frac{k-1}{k} p'_2$$
$$p_j = \frac{k-1}{k} p'_j, \qquad j = 3, 4, \ldots, k,$$

which completes the proof.

And now we can show how a decomposition of the original distribution can be constructed.

Theorem 2. Any distribution $(p_1, p_2, \ldots p_n)$ can be presented as the weighted sum of k distributions, each taking on only two possible values and having weight $1/k$.

Proof. The first half of the assertion can be obtained immediately from the previous theorem, or rather from applying it k-fold consecutively. The only remaining detail to be considered is the weights of the distributions. When the original distribution $(p_1, p_2, \ldots p_k)$ is decomposed into the two-point distribution (q_{11}, q_{12}) and the $(k-1)$-point distribution $(p'_2, \ldots, p'_k)$ then their weights are $1/k$ and $(k-1)/k$ respectively. The second one is further broken down into a two-point distribution (q_{21}, q_{22}) with weight $1/(k-1)$ and a $(k-2)$-point distribution with weight $(k-2)/(k-1)$. Substituting this sum into the decomposition of the original distribution $(p_1, p_2, \ldots p_k)$ one obtains that the overall weight of the distribution (q_{21}, q_{22}) is $\frac{k-1}{k} \cdot \frac{1}{k-1} = \frac{1}{k}$, that is, exactly as we stated. Similarly, the soundness of the theorem can be proven for the weights of the subsequent two-point distributions (q_{j1}, q_{j2}), $j = 3, \ldots, k$.

Example. Consider the distribution determined by the values and probabilities in Table 11.

Table 11

No.	Probability	Value
1	0.10	0
2	0.15	1
3	0.45	5
4	0.30	6

Pick up the probabilities p_1 and p_3 (now $k = 1/4$), then we obtain the two-point distribution

$$q_{11} = 0.4, \qquad x_{11} = 0,$$
$$q_{12} = 0.6, \qquad x_{12} = 5,$$

by applying Theorem 1. The remaining distribution $p_2' = 0,2$, $p_3' = 0.4$, $p_4' = 0.4$. Repeatedly making use of Theorem 1, selecting probabilities p_2' and p_3' the two-point distribution is

$$q_{21} = 0.6, \qquad x_{21} = 1,$$
$$q_{22} = 0.4, \qquad x_{22} = 5,$$

the remainder is $p_3'' = 0.4$, $p_4'' = 0.6$. From this we arrive at

$$q_{31} = 0.8, \qquad x_{31} = 5,$$
$$q_{32} = 0.2, \qquad x_{32} = 6,$$

and the remainder $q_{41} = 1.0$, $x_{41} = 6$, $q_{42} = 0$, which is a degenerated distribution.

According to Theorems 1 and 2 the formal description of the preparation of Algorithm D5 — that is, determination of the vectors $\mathbf{Q}$, $\mathbf{K}$ and T in Algorithm D5 — can be stated as follows.

D6 Preparation to D5

1. [Initialize.] Determine the sets of indices U (surplus) and H (shortage) by investigating p_i for all $i = 1, \dots, k$: if $p_i \geqq 1/k$, then store index i in the set U, otherwise in the set H. Set $j \leftarrow 0$.
2. If the set H is not empty then increase $j \leftarrow j+1$ and proceed to Step 4.
3. [The remaining $k - j + 1$ two-point distributions are generated — $q_{l1} = 1$, $q_{l2} = 0$ for all $l \in U$.] If U is empty, finish the preparation. Let u be an index from U, increase $j \leftarrow j+1$, set $Q(j) = 1$, $K(j) = u$, $T(j) = 0$. Decrease the probability $p_u \leftarrow p_u - 1/k$. If $p_u = 0$, then cancel index u : set $U \leftarrow U - \{u\}$. If $p_u < 1/k$, then u is transferred to the set of shortage indices $H \leftarrow H + \{u\}$, $U \leftarrow U - \{u\}$. If $j = k$, then finish, otherwise go back to Step 2.
4. [Determine the jth two-point distribution.] Select an element from the set H, denote it by h and select an element u from the set U.
5. Set $Q(j) \leftarrow kp_h$, $K(j) \leftarrow h$, $T(j) \leftarrow u$.
6. [Modify the probabilities of the remainder distribution and the sets of indices.] Let $p_u \leftarrow p_u + p_h - 1/k$, cancel index $h : H \leftarrow H - \{h\}$. If $p_u = 0$, then cancel index $u : U \leftarrow U - \{u\}$. If $p_u < 1/k$, then u is to be transferred to the set $H : H \leftarrow H + \{u\}$, $U \leftarrow U - \{u\}$. If $j = k$, then finish, otherwise go back to Step 2.

Example. Applying this preparatory algorithm to the previous numerical example one obtains the following vectors:

$$\mathbf{Q} = (0.4 \quad 0.6 \quad 0.8 \quad 1)$$

$$\mathbf{K} = (1 \quad 2 \quad 3 \quad 4)$$

$$\mathbf{T} = (3 \quad 3 \quad 4 \quad 0)\,.$$

Actually this preparation could be modified to yield an algorithm handling the degenerated cases (Step 3) more elegantly, but we thought clarity of presentation to be a more important issue. Note that modification of probabilities is not done in the same way as the proof runs — the difference is only a constant multiplier.

This algorithm is the one to be recommended when someone wants to generate samples from a discrete distribution having a fairly large number of support points at reasonable speed but does not want to get entangled in sophisticated programming tricks.

3.11. Generating discrete variables by economical method

The economical method has been devised as a modification of the acceptance-rejection technique, where no generated value is rejected. In this section we show how this idea can be applied to any discrete distribution.

Every discrete distribution concentrated on n points can be represented as an equally weighted mixture of n two-point distributions as it was described in the previous section. These distributions will be used in the discrete version of the economical method.

Consider first a two point distribution of the random variable α, given $P\{\alpha = a_1\} = p$, $P\{\alpha = a_2\} = 1 - p$ (assume $p < \frac{1}{2}$). The uniform distribution with probabilities $\frac{1}{2}, \frac{1}{2}$ can be used as a starting distribution, that is uniformly distributed indices will be generated and transformed if necessary to yield realizations of α. Index 1 will be the surplus index (since $p < \frac{1}{2}$) and index 2 is the shortage index. The straightforward application of the economical method will yield the procedure.

D7 Discrete variable by economical (preliminary)

1. Generate u.
2. If $u > \frac{1}{2}$ (that is we have $i = 2$, which is a shortage index) then deliver a_2.
3. Else if $u < p$ then deliver a_1, otherwise deliver a_2 ($i = 1$ is transformed into $i = 2$).

Here the transformation of the surplus values into shortage values was done by the transformation $T(1) = 2$. We take advantage of this simplicity by decomposing the original distribution into a mixture of two-point distributions and applying the previous trick repeatedly.

To proceed formally consider the n-point distribution of ξ given by $P\{\xi = x_i\} = p_i$, $p_i > 0$, $i = 1, \dots, n$, $p_1 + p_2 + \dots + p_n = 1$ and n two-point distributions defined as $P\{\eta_i = y_{ji}\} = q_{ji}$, $j = 1, 2, q_{1i} + q_{2i} = 1$, $q_{1i}, q_{2i} > 0$, $i = 1, \dots, n$, where the distribution of ξ is given as the equally weighted sum of the distributions of the η_i. Assume without loss of generality $q_{1i} \leqq q_{2i}$ (degenerated distributions with $P\{\eta_i = y_{2i}\} = 1$ should be transformed into $P\{\eta_i = y_{1i}\} = q_{1i} = 1/2$, $P\{\eta_i = y_{2i}\} = q_{2i} = \frac{1}{2}$, where $y_{1i} = y_{2i}$).

Construct the vector $\mathbf{r} = (r_1, \ldots, r_{2n})$ the following way. Probabilities $q_{1s} = q_{2s} = \frac{1}{2}$ are placed in the middle of the vector $\mathbf{r}$ and not equal probabilities $q_{1i} < q_{2i}$ are filled in from the outside in. More precisely define r_k, z_k by the following procedure.

D8 Preparation to D9

1. Initialize $k \leftarrow 1$, $m \leftarrow 1$, $i \leftarrow 0$.
2. Increase $i \leftarrow i + 1$. If $i > n$, then go to Step 5.
3. If $q_{1i} = q_{2i}$ then go to Step 4, otherwise set $r_k \leftarrow q_{1i}$, $r_{2n-k+1} \leftarrow q_{2i}$, $z_k \leftarrow y_{1i}$, $z_{2n-k+1} \leftarrow y_{2i}$, $k \leftarrow k+1$ and go back to Step 2.
4. Set $z_{n-m+1} \leftarrow y_{1i}$, $z_{n+m} \leftarrow y_{2i}$, $m \leftarrow m+1$ and go back to Step 2.
5. Compute and save the quantities $n^+ \leftarrow k$, $n \leftarrow n + m$.

Note that the values $r_{n^+}, \ldots, r_{2n}$ will not be needed in the final generator. The indices from 1 to $n^+ - 1$ are surplus indices, the shortage indices are those from n^- to $2n$ while probabilities with indices between n^+ and $n^- - 1$ are equal to 1/2 and can be considered as shortage indices. The transformation T of a surplus value with index i is simple $T(i) = 2n - i + 1$.

The generation procedure thus consists of uniformly choosing an index k between 1 and $2n$, if $k \geqq n^+$, then z_k is delivered, otherwise a random test is in order: Generate u uniformly, if $u/2 < r_k$ then deliver z_k, otherwise deliver the transformed value z_{2n-k+1}. The expected number of uniform deviates required is $\overline{N} = 1.5$ (or even less if $n^+ < n$). Compare this to the situation in the alias method where one uniform number is needed for choosing a distribution, another one for sampling with it, that is $\overline{N} = 2$.

The formal economical algorithm of generation makes use of the standard procedure of splitting up a uniformly distributed number into an index and a uniform number, reducing thus $\overline{N}$ practically to 1. $[x]$ denotes the integer part of x.

D9 Discrete variable, by economical

1. Generate u uniformly.
2. Compute $k \leftarrow [2nu] + 1$.
3. If $k \geqq n^+$ then deliver z_k.
4. Else compute $u^* \leftarrow (2nu - k + 1)/2$ if now $u^* < r_k$ then deliver z_k, otherwise deliver z_{2n-k+1}.

It is of theoretical interest to note that the average number of uniform numbers required can asymptotically be reduced to 1 by repeated application of this idea. (Theoretical interest only, because space requirements increase exponentially and the standard procedure of splitting the uniform to get an index and a uniform already had the practical effect of only requiring one uniform.) The same idea can repeatedly be applied to the $2n$ component vector $\mathbf{r} = (r_1, \ldots, r_{2n})$ in the following sense. Replace $\mathbf{r}$ with the vector $\mathbf{s} = (s_1, \ldots, s_{4n})$, let $s_i = 2r_i$, $s_{4n-i+1} = 2(r_{2n-i} - 1/2)$, $i = 1, 2, \ldots, n^+ - 1$, where we assume $s_i < s_{4n-i+1}$, otherwise s_i and s_{4n-i+1} are exchanged. The remaining probabilities s^i, $i = n^+, n^+ + 1, \ldots, 4n - n^+$ are equal to $1/2$. Then s and the corresponding generator is similar to $\mathbf{r}$ and the generator D6, the difference is that here we have at least $2n$ uniform members out of the $4n$ probabilities. The original distribution $p_1, \ldots, p_n$ now can be represented as the equally weighted sum of the $2n$ distribution (s_i, s_{4n-i+1}). After the kth repeated application of the construction at least $[1 - 2/(2^k)]$th part of the probabilities will be equal to $1/2$, yielding an automatic procedure for obtaining a uniform approximation to the original distribution $p_1, \ldots, p_n$, and a generator where in limit only $\overline{N} = 1$ uniform numbers are needed. This is one way "to bargain time for storage space".

3.12. Notes

3.1 The inversion method, together with the acceptance-rejection technique and the continuous form of the decomposition method, first appeared in the paper of Butler (1956) but they seem to have been known previously. See, for example, von Neumann (1951), to whom the idea of the acceptance-rejection technique — among several other inspirations — is commonly attributed.

3.2 The acceptance-rejection technique is the most widely known, and possibly the most widely used, because it is based on a very simple, easily understandable idea. The reader is cautioned, however, since generally it is *not* the most effective method.

3.3 The discrete form of the decomposition method was published by Marsaglia (1961) and he was also the person who indicated its practical exploitability.

3.4 The comparison method for the exponential distribution was discovered by von Neumann in 1949. It is not clear to what extent he was aware of its implicit power. The currently known general form of the method has been obtained as a result of the effort of several researchers. Forsythe (1971) noticed that the method can be modified for density functions of type $\exp\{-G(x)\}$, Ahrens and Dieter (1973) made several improvements greatly enhancing the numerical efficiency of the method, while Atkinson and Pearce (1976) observed how one can apply the procedure to any density function.

3.5 The method was published by Deák (1982); for an application see Iván and Deák (1984).

3.6 The method was presented by Kinderman and Monahan (1977). The variants employing transformations of uniform variates other than the ratio were developed by Barbu (1983).

3.7 The squeeze method was implicitly implemented by Marsaglia (1967); the general method was described in Marsaglia (1977). The acceptance complement method was given in Krommal and Peterson (1981). It has been shown by Deák (1982) that this method is actually a decomposition type technique. The inversion-rejection method and the series method were invented by Devroye (1984 and 1981). It should also be noted that our presentations of the methods do not always follow the reasonings of the original papers. Some other special techniques for generating random numbers include Lux (1978), where derivative of the density function is shown, and Devroye (1981), where one is given the characteristic function of the distribution.

3.9 The table method was published by Marsaglia (1963). In his original paper another modification — further reducing the storage requirements — was also applied: the different levels were "slid" one into another, so that some values appearing in neighbouring levels were made to overlap.

3.10 The procedure first published by Walker (1977) under the name "alias" appeared there in a form different from our presentation. It was recognized by Krommal and Peterson (1979) that this method is actually a decomposition type procedure, and we mainly followed their way of reasoning. A more general form of Theorem 2 appears in Dieter (1982) stating that any k-point distribution can be expressed as r distributions with s possible points, where according to Peterson and Krommal $r = \lceil (k-1)/(s-1) \rceil$ (here $\lceil x \rceil$ is the smallest integer greater or equal to x), which for the case $s = 2$ treated in 3.10 gives the minimal value to be $r = k - 1$.

3.11 The discrete version of the economical method was published by Deák (1986).

CHAPTER 4

RANDOM NUMBER GENERATORS FOR COMMON DISTRIBUTIONS

4.1. Introduction

When creating a random number generator for a specific distribution several ways can be chosen. We can start with one of the general methods described in the previous chapter and apply the same for the distribution in consideration. (Of course sometimes this application can be done in different ways, for instance having chosen the acceptance-rejection technique we still have a free choice among the possible upper-bounding functions.) Also in case of a different procedure we can scrutinize the given distribution and making use of some specific feature of the distribution a generator can be built. We will see examples for both cases. Algorithm E1 in the previous chapter is application of the inversion technique for the exponential distribution, while algorithm N2 in the next section uses a special property of the normal distribution to produce normally distributed numbers.

Since several different algorithms can be built for one distribution a natural question to ask is which generator is better, which one is to be used? The generators can be compared with regard to several criteria and the choice of the generator is made according to the criterium which one is important for us. Such criteria may include (i) How many uniformly distributed numbers are necessary on the average to produce one sample (pseudorandom number) with the given density function, (ii) what is the sophistication degree of the algorithm, how easy is the computer implementation of the program, can it be written in a high-level language or it requires assembler coding, (iii) how long is the computer program, what is the memory requirement, (iv) how fast is the generator, how much time is necessary for generating one sample.

In case of a simple program requiring a couple of hundred or thousand samples the most important criterium generally is the convenience of programming. In case of a program run frequently and using a great number of random samples the speed may become the most important consideration or sometimes the memory requested. When discussing the individual random number generators in this chapter we will examine all four criteria to facilitate the proper choice.

Our methods apart from a few exceptions are theoretically precise. It means that if the input of the algorithm is a uniformly distributed random variable, then the outcome is a random variable with the desired density function. Approximate methods for generating random numbers will only be described if they are widespread, or in some respect they offer much better performance compared to the exact algorithm (faster, the programming is easier, memory required is less, etc.). In our opinion such approximate methods should be used with caution and only in exceptional cases, since for most distributions we have theoretically sound procedures.

4.2 Normal distribution

The generation of normally distributed random numbers is very frequently required in simulational computations; this is why the normal distribution is discussed in this and the next section in a little bit greater detail than the other distributions. Here some simpler algorithms are described while in the following section two methods are covered that make possible the manifacturing of faster generators by employing more sophisticated procedures.

The distinguished role of the normal distribution can be accounted for by the central limit theorems, asserting that under suitable mild conditions the sum of identically distributed random variables is normally distributed. The density function $f(x)$ of the one-dimensional normal distribution with mean m and standard deviation σ can be given as

$$f(x) = \frac{1}{\sqrt{2\pi}\sigma} \exp\left\{-\frac{(x-m)^2}{2\sigma^2}\right\}, \quad -\infty < x < +\infty, \tag{4.1}$$

and the corresponding distribution function is

$$F(x) = \int_{-\infty}^{x} f(y)\, dy.$$

This distribution is denoted by $N(m, \sigma)$ usually. A special case of this distribution $N(0, 1)$ with expected value 0 and variance 1, whose density and distribution functions are φ and Φ respectively (see Figs 19 and 20):

$$\varphi(x) = \frac{1}{\sqrt{2\pi}} e^{-x^2/2}, \qquad -\infty < x < +\infty$$

$$\Phi(x) = \int_{-\infty}^{x} \varphi(z) dz. \tag{4.2}$$

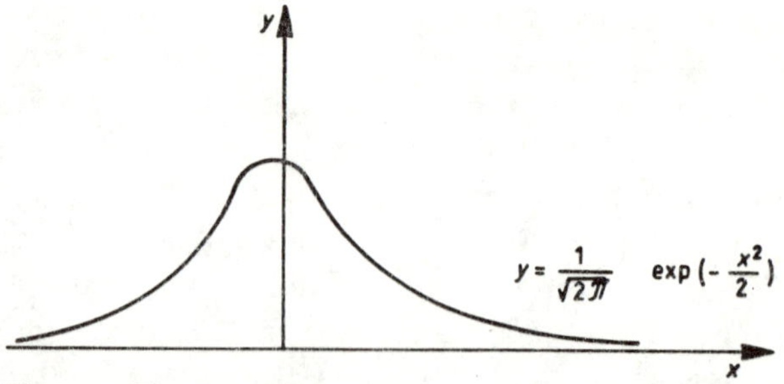

Fig. 19. The density function of the standard normal distribution $N(0,1)$

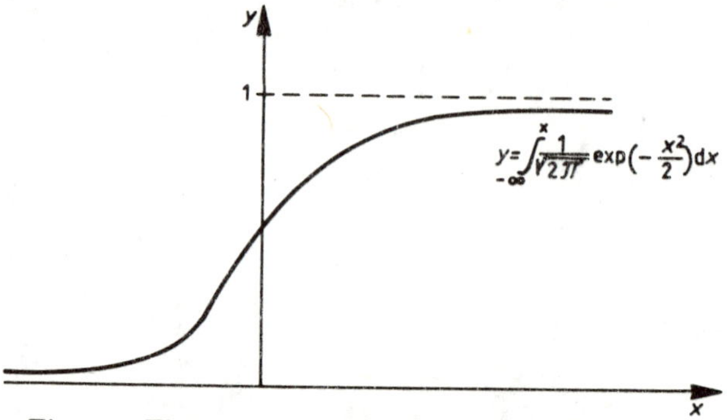

Fig. 20. The distribution function of the standard normal distribution $N(0,1)$

Using these notations one of the central limit theorems can be formulated as follows — we do not intend to state the most general result:

Theorem 1. If $\xi_1, \xi_2, \ldots, \xi_n, \ldots$ are independent, identically distributed random variables with finite variance, with $E(\xi_n) = m$, $D^2(\xi_n) = \sigma^2$, $n = 1, 2, \ldots$ then we have

$$\lim_{n\to\infty} P\left\{\frac{\xi_1 + \xi_2 + \ldots + \xi_n - nm}{\sigma\sqrt{n}} < x\right\} =$$
$$= \frac{1}{\sqrt{2\pi}} \int_{-\infty}^{x} e^{-y^2/2}\, dy = \int_{-\infty}^{x} \varphi(y) dy.$$

Essentially this theorem states that the normalized arithmetical mean of identically distributed random variables is normally distributed in limit.

If the density function of the random variable ξ is φ then the random variable $\eta = \sigma\xi + m$ has density function f. It means that in the sequel we can restrict our treatment to generating standardized normal distribution with expected value 0 and dispersion 1; since if we obtain samples $x_1, x_2, \ldots,$ of ξ, then $y_1 = \sigma x_1 + m,\ y_2 = \sigma x_2 + m, \ldots$ will be samples with density function $f(x)$.

The following approximate procedure can immediately be suggested on the basis of Theorem 1 for generating samples with density function φ: take the arithmetic mean of $n = 12$ identically, uniformly distributed random numbers. The underlying formal algorithm is the following:

N1 Normal, approximate

1. Generate the uniformly distributed numbers $u_1, u_2, \ldots, u_{12}$ in interval $[0,1)$.
2. Set $y \leftarrow u_1 + \ldots + u_{12}$.
3. Deliver the value $x \leftarrow y - 6$ as a sample with density ϕ.

The reason why we have chosen the sum of 12 numbers is that in this case the divisor $\sigma\sqrt{n}$ is exactly 1. Generating less uniformly distributed numbers in the first step the algorithm becomes faster, though the deviation of the produced numbers' distribution will be greater from the normal distribution, thus it is not to be recommended.

The advantage of the method is its simplicity, can very easily be implemented on a computer using no external references to subroutines. A disadvantage of the algorithm is the need for a great number of uniform samples ($\overline{N} = 12$) which implies that the algorithm is slow and besides the distribution of the generated samples will only approximate the desired standardized normal. The algorithm can be speeded up a little if it is coded in assembler and the additions are carried out using the fixed-point integer numbers and not the floating point numbers in interval $[0,1)$, because the floating point instructions are always slower than the fixed point operations.

To present the next method some preparations are necessary; three theorems are stated.

Theorem 2. Let ξ and η be two independent normally distributed random variables with $N(0,1)$, then the random variables $\rho = \sqrt{\xi^2 + \eta^2}$ and $\psi =$ arc tan (ξ/η) are independent of each other, their distribution is χ with 2 degrees of freedom and uniform in interval $[0, 2\pi)$ respectively.

Proof. The joint distribution function of the random variables ξ and η is

$$f(x_1, x_2) = \frac{1}{\sqrt{2\pi}} e^{-x_1^2/2} \cdot \frac{1}{\sqrt{2\pi}} \cdot e^{-x_2^2/2} = \frac{1}{2\pi} e^{-(x_1^2+x_2^2)/2}.$$

Apply now the polar transformation $\xi = \rho \cos \psi$, $\eta = \rho \sin \psi$. Using the notation $x_1 = y_2 \cdot \cos y_1$, $x_2 = y_2 \cdot \sin y_1$ the Jacobian of the transformation is

$$\begin{vmatrix} \frac{\partial x_1}{\partial y_1} & \frac{\partial x_1}{\partial y_2} \\ \frac{\partial x_2}{\partial x_1} & \frac{\partial x_2}{\partial y_2} \end{vmatrix} = y_1,$$

which — according to Theorem 2 of Section 3 in Chapter 1 — implies that the joint density function of the random variables $\rho = \sqrt{\xi^2 + \eta^2}$ and $\psi =$ arc tan (ξ/η) is given by

$$h(y_1, y_2) = \frac{1}{2\pi} y_1 e^{-y_1^2/2}.$$

Since it can be decomposed into two terms

$$g_1(y_1) = y_1 \cdot e^{-y_1^2/2}, \qquad 0 \leqq y_1 \leqq \infty$$

$$g_2(y_2) = \frac{1}{2\pi}, \qquad 0 \leqq 2 \leqq 2\pi$$

thus ρ and ψ are independent indeed. The uniformity of the variable ψ follows from the form of the density function g_2 while the χ distribution of the variable ρ is evident from its definition. (The density function of the χ-distributed random variable χ with k degrees of freedom is $2x^{k-1} e^{-x^2/2}/(2^{n/2}\Gamma(n/2))$, $x \geqq 0$.)

Theorem 3. Let u be uniformly distributed in $[0, 1)$, then the random variable $\sqrt{-2 \ln u}$ is χ-distributed with 2 degrees of freedom.

Proof. All what we have to do is write down the distribution function's definition

$$P\{\sqrt{-2 \ln u} < x\} = P\{\ln u > -x^2/2\} = P\{u > e^{-x^2/2}\} = 1 - e^{-x^2/2},$$

which proves the theorem.

The following reversion of Theorem 2 is also valid, it will give rise to the next algorithm for generating normal numbers.

Theorem 4. Let u_1 and u_2 be independent uniformly distributed random numbers in $[0,1)$ then the random variables $\xi = \sqrt{-2 \ln u_1} \cos(2\pi u_2)$ and $\eta = \sqrt{-2 \ln u_1}(\sin 2\pi\, u_2)$ are independent and distributed according to the standardized normal distribution.

Proof. The joint density function of the variables u_1 and u_2 is

$$h(y_1, y_2) = 1,\ 0 \leqq y_1, y_2 \leqq 1.$$

The Jacobian of the inverse transformation $y_1 = e^{-(x_1^2+x_2^2)/2}$, $y_2 = -\operatorname{arc\,tan}(x_1/x_2)2\pi$ is the following

$$\begin{vmatrix} \frac{\partial y_1}{\partial x_1} & \frac{\partial y_1}{\partial x_2} \\ \frac{\partial y_2}{\partial x_1} & \frac{\partial y_2}{\partial x_2} \end{vmatrix} = \frac{1}{2\pi} \cdot e^{-(x_1^2+x_2^2)/2}.$$

Consequently the joint density function of the random variables ξ and η can be obtained as

$$f(x_1, x_2) = \frac{1}{\sqrt{2\pi}} \cdot e^{-x_1^2/2} \frac{1}{\sqrt{2\pi}} \cdot e^{-x_2^2/2},$$

which proves the assertion of the theorem: ξ and η are independent and their distribution is $N(0,1)$.

N2 Normal, polar transformation

Preparation

1. Initialize the index i as $i \leftarrow +1$.

Generation

1. Set $i \leftarrow -i$. If $i > 0$ the go to Step 3.
2. Generate u_1 and u_2. Set $r \leftarrow (-2 \ln u_1)^{1/2}$ and $z \leftarrow 2\pi\, u_2$. Compute $y \leftarrow r \cos z$ (this value y is saved until the next call to the algorithm).Deliver the value $x \leftarrow r \sin z$.
3. Deliver the value $x \leftarrow y$.

The proper working of this algorithm is stated and subsequently proved in Theorem 4. The index i is needed for distinguishing between two types of calls, a call when a generated value is already stored and a call when no previously generated value is present (the original algorithm prepares two normally distributed numbers at the same time, algorithm N2 delivers this prepared value in case of each second call). The algorithm is simple, as an advantage we have to notice that only one uniformly distributed number is required to produce one normal sample ($\overline{N} = 1$). On the other side a disadvantage of the method is that the relatively slowly computable trigonometric functions and the square root is involved. In the sequel it will be shown how the use of the function sine and cosine can be avoided at a cost of a little sophistication.

Since u_2 is uniform in $[0,1)$ thus the point $(\cos(2\pi u_2),\ \sin(2\pi u_2))$ is uniformly distributed on the unit circle $F_2 = \{\mathbf{x} \mid x_1^2 + x_2^2 = 1\}$. In the modified polar algorithm a different method is used for producing this point $(\cos(2\pi u_2),\ \sin(2\pi u_2))$. Consider the points (v_1, v_2) uniformly distributed in the square $K_2 = \{\mathbf{x} \mid -1 \leqq x_1, x_2 \leqq 1\}$. If only the points falling into the unit circle $S_2 = \{\mathbf{x} \mid x_1^2 + x_2^2 \leq 1\}$ are accepted then their distributions is clearly uniform in S_2, what implies that their direction is uniformly distributed on the circle F_2 (projecting the points in S_2 to the circle line F_2, the projections on F_2 are uniform).

Theorem 5. Assume that the point (v_1, v_2) is uniformly distributed in the unit circle S_2, then it follows that

(i) the square of the random variable given by the equation $r = \sqrt{v_1^2 + v_2^2}$ is uniformly distributed in the interval $[0,1)$.

(ii) r is independent of the random variables $z_1 = v_1/\sqrt{v_1^2 + v_2^2}$ and $z_2 = v_2/\sqrt{v_1^2 + v_2^2}$ both,

(iii) the point (z_1, z_2) is uniformly distributed on the circle F_2.

Proof. The point (z_1, z_2) is on the circle F_2, since $z_1^2 + z_2^2 = 1$, the uniformity of the distribution is obvious thus (iii) holds. In order to prove the correctness of the theorem's first two statements the joint density function of the random variables r and z_1 will be derived. Denote the density function of the point (v_1, v_2) by h, then we have

$$h(y_1, y_2) = \frac{1}{\pi}, \qquad y_1^2 + y_2^2 \leqq 1.$$

The random variables r and z_1 are produced from the pair of random variables v_1, v_2 by the transformation $x_1 = \sqrt{y_1^2 + y_2^2}$, $x_2 = y_1/\sqrt{y_1^2 + y_2^2}$.

The inverse of this transformation is

$$y_1 = x_1 x_2,$$

$$y_2 = x_1\sqrt{1 - x_2^2},$$

whose Jacobian can be obtained after some simple calculations as $x_1/\sqrt{1 - x_2^2}$. From this using the transformation theorem on densities we can obtain the joint density function of the random variables $r_1 z_1$ as

$$f(x_1, x_2) = \frac{1}{2\pi\sqrt{1 - x_2^2}} 2x_1, \qquad -1 \leqq x_1, x_2 \leqq 1.$$

The independence of r and z_1 follows from the fact, that f can be decomposed into a product of expressions containing only one of the variables. Since the density function of the random variable r is $g(x_1) = 2x_1$, thus the density function of r^2 can be obtained as $\overline{g}(x_1) = g(\sqrt{x_1}) \cdot \dfrac{1}{2\sqrt{x_1}} = 1$, that is r^2 is really uniformly distributed in the interval $[0, 1)$.

The following modified version of the polar method where no trigonometric functions are involved is based on the theorem now proved.

N3 Normal, modified polar method

1. Set $i \leftarrow +1$.

Generation

1. Set $i \leftarrow -i$. If now $i > 0$ then go to Step 5.
2. Generate u_1 and u_2, let $v_1 \leftarrow 2u_1 - 1$, $v_2 \leftarrow 2u_2 - 1$, (v_1, v_2) is now uniform in the square K_2. Evaluate the quantity $w \leftarrow v_1^2 + v_2^2$.
3. [Test.] If $w > 1$, then go back to Step 2 (rejection, point (v_1, v_2) is not in the circle).
4. [The point (v_1, v_2) is uniform in the unit circle here.] Let $r \leftarrow [(-2 \ln w)/w]^{1/2}$ and let $y \leftarrow rv_2$, this value is to be saved until the next call to the algorithm. Deliver the number $x \leftarrow rv_1$ as a normal sample.
5. Deliver the number $x \leftarrow y$.

In the algorithm to produce one normally distributed sample $\overline{N} = 1.27$ uniformly distributed random numbers are required, because the probability of acceptance in Step 3 is 3.14/4. The procedure is simple, easily programmable. A drawback of the method is the use of the logarithmic and the square root functions that slow down the algorithm and makes it computer- (implementation) dependent. The reasonable speed and ease of programming make this generator N3 expedient in most cases unless we have special expectations with regard to the algorithm's performance.

4.3 Two fast normal generators

Two algorithms are described in this section for generating normal samples that are more sophisticated procedures than the ones proposed so far, they require storage of constants but they are also faster. Both of them are based on the decomposition technique, for the densities appearing in the first decomposition we will apply the economical method while in the second decomposition the comparison method is used extensively.

Two details are discussed and taken care of first; auxiliary algorithms are developed for generating samples according to an almost linear density function efficiently and for generating samples from the tail of the normal distribution (that is how can we generate normally distributed numbers that are greater than a given value).

Suppose that our task is to generate random numbers with a monotone decreasing density function $g(x)$, $x \in [0, c]$, $g(c) = 0$, which is almost linear, and for the sake of simplicity let it be convex (Fig. 21).

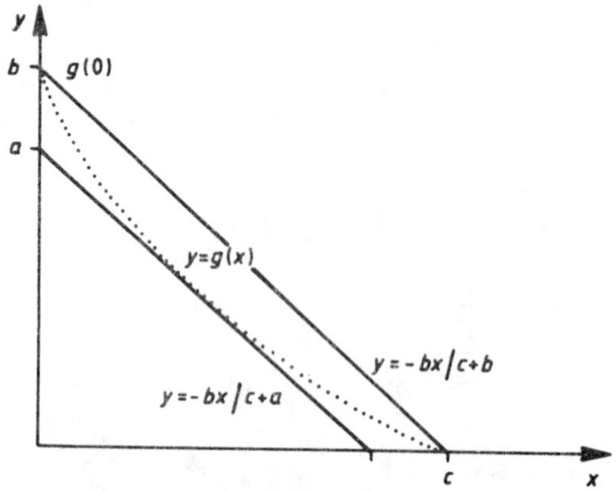

Fig. 21. Generating a variable with almost linear density $g(x)$, lying between upper and lower bounding lines

The acceptance-rejection technique is applied to this density function. Squeeze the graph of the density function $g(x)$ between two parallel lines, which using the notation $b = g(0)$ can be given as $y = -bx/c + a$ and

$y = -bx/c + b$ respectively, where a is a constant, $a < b$ and furthermore $-bx/c+a \leqq g(x) \leqq -bx/c+b$ for $x \in [0, c]$. The basic idea of the algorithm can be summarized the following way. Generate a point (x_0, y_0) uniformly in the triangle given by the upperbounding line $y = -bx/c + b$ and the coordinate axes. If this point is below the lower line $y = -bx/c + a$ (that is if the inequality $y_0 \leqq -bx_0/c + a$ holds) then it is certainly below the function $g(x)$, that is x_0 is accepted (fast acceptance). If the point (x_0, y_0) is between the two lines then we still have to check the inequality $g(x_0) \geqq y_0$; if it holds the value x_0 is accepted, too, and only in case of $g(x_0) < y_0$ will the point (x_0, y_0) be rejected. The main feature of the method is that evaluation of the function value $g(x_0)$ is required only in a small percentage of the cases.

Points uniformly distributed in the triangle can be generated the following way. Let u and v be two random variables, uniform in $[0, 1)$ and $M = \max(u, v)$, $m = \min(u, v)$. We show now that the point $(m, M - m)$ is uniformly distributed in the triangle given by the points $(0,0)$, $(0,1)$, $(1,0)$ as vertices. Since for the sum of coordinates $m + M - m \leqq 1$, the point $(m, M - m)$ is in the triangle. For the distribution function $F(x, y)$ of the point one can obtain

$$\begin{aligned} F(x,y) &= P\{m < x,\ M - m < y\} \\ &= P\{m < x,\ M - m < y,\ v < u\} + \\ &+ P\{m < x,\ M - m < y,\ u < v\}. \end{aligned}$$

The first probability on the right-hand side can be expressed as

$$\begin{aligned} P\{m < x,\ M - m < y,\ v < u\} &= P\{v < x,\ u - v < y\} = \\ &= \int_0^x \int_v^{y+v} du\, dv = xy, \end{aligned}$$

thus for the distribution function we have

$$F(x,y) = 2xy, \quad (x,y) \in \{(x,y) \mid 0 \leqq x, y \leqq 1,\ \ x + y \leqq 1\}$$

which means that the coordinates of the point $(m, M - m)$ are uniform. This means that the point $(cm,\ b(M - m))$ is uniformly distributed in the triangle given by its vertices $(0,0)$, $(c,0)$ and $(0,b)$ and this point falls below the line $y = -bx/c + a$, if $bM \leqq a$ holds. These preparations are summarized in the next auxiliary routine, by the aid of which random variables can efficiently generated with an almost linear function $g(x)$.

S1 Subalgorithm, almost linear density

1. Generate u and v uniformly in $[0,1)$, set $M \leftarrow \max(u,v)$, $m \leftarrow \min(u,v)$, $x_0 \leftarrow cm$, $y_0 \leftarrow bM$.
2. If $y_0 \leqq a$, then deliver x_0 (fast acceptance).
3. If $b(M-m) \leqq g(x_0)$, then deliver x_0 (slow acceptance), otherwise repeat the procedure from Step 1 (rejection).

Another method for generating points uniformly in the triangle $(0,0)$, $(0,1)$ $(1,0)$ can be worked out along the following lines: a random variable u with density function $f(x) = 2(1-x)$, $x \in [0,1)$ is generated (by algorithm L1, for instance), let v be uniform, then the point $(u,(1-u)v$, is uniform in the triangle.

Now we turn to the next detail: how to generate normally distributed random variables, subject to the condition of being greater than a fixed value b. In other words we are looking for a method to generate variables according to the density function

$$f(x) = c \cdot e^{-x^2/2}, \quad x \in [b, +\infty),$$

where the constant c is given by $\frac{1}{c} = \int_b^{\infty} \exp(-x^2/2)dx$. The function

$$g(x) = \alpha\, h(x) = \frac{c}{b} \cdot x \cdot e^{-x^2/2}, \quad x \geqq b$$

is an upper bound to $f(x)$. Our method for generating variables with density function $f(x)$ is an acceptance-rejection technique, where this $g(x)$ is the upper bound. Introduce the notations

$$G(x) = \int_6^x g(t)\, dt = \frac{c}{b}[e^{-b^2/2} - e^{-x^2/2}]$$

$$G(\infty) = \int_b^{\infty} g(t)\, dt = \frac{c}{b} \cdot e^{-b^2/2}$$

then the density function $h(x)$ proportional to the upper bounding function $g(x)$ can be given as

$$h(x) = \frac{g(x)}{G(\infty)} = x \cdot e^{-(x^2-b^2)/2}, \quad x \in [b, +\infty),$$

its distribution function

$$H(x) = \int_b^x h(t)\, dt = \frac{G(x)}{G(\infty)} = 1 - e^{-(x^2-b^2)/2}.$$

If we now express the variable x from the equation $1 - u = H(x)$ then the expression

$$x = (b^2 - 2 \ln u)^{1/2}$$

is obtained that can be used toward generating variables with density function h. The inequality to be tested in the acceptance-rejection algorithm is $f(x)/g(x) < u$, its left-hand side for the present functions f and g will take the form b/x, thus the condition could be rewritten as $b < ux$. According to these considerations the following algorithm A2 produces samples from the tail of the normal distribution.

S2 Subalgorithm, normal tail

1. Generate u and set $x \leftarrow (b^2 - 2 \ln u)^{1/2}$.
2. Generate v. If $b < vx$ then go back to Step 1 (rejection).
3. Deliver the number x as a normally distributed sample greater than b.

The probability of acceptance is given by the expression $1/G(\infty)$, which is more precisely

$$b \cdot e^{b^2/2} \int_b^{+\infty} e^{-x^2/2} dx, \tag{4.3}$$

which is large enough in case of $b \geqq 2$ (for instance we have 0.844 for $b = 2$ and 0.947 for $b = 3$).

And now we return to our original problem: we build a random number generator for producing samples from the standard normal distribution. Let the density function $\varphi(x)$ of the standard normal distribution be given in the form of a sum:

$$\varphi(x) = (2\pi)^{-1/2} \exp(-x^2/2) = q_1 f_1(x) + \ldots + q_5 f_5(x). \tag{4.4}$$

Table 12

Densities and probabilities of selection in decomposition (4.4)

i	Probability q_i	Interval $a_i \leq \lvert x \rvert \leqq b_i$		Density function $f_i(x)$
1	0.884	0.00	2.22	$c_1(2.22 - \lvert x \rvert)$
2	0.027	0.00	0.48	$c_2(\varphi(x) - q_1 f_1(x))$
3	0.047	0.48	1.59	$c_3(\varphi(x) - q_1 f_1(x))$
4	0.015	1.59	2.22	$c_4(\varphi(x) - q_1 f_1(x))$
5	0.027	2.22	$+\infty$	$c_5 \varphi(x)$

The probabilities q_i and density functions f_i are given in Table 1, the function f_i outside the given interval is assumed to be equal to zero, the constants c_i are determined by the equations $\int_{-\infty}^{+\infty} f_i(x)\,dx = 1$ that express the condition of being a density function. According to equation (4.4) the area below the density function $\varphi(x)$ is divided into five different parts: a great triangle given by $q_1 f_1(x)$, $x \in [0, b]$, three small almost triangle shaped remainders described by $q_2 f_2$, $q_3 f_3$ and $q_4 f_4$ and finally the distribution's tail $q_5 f_5 = q_5 c_5 \varphi(x)$. Decomposition (4.4) is graphically illustrated in Fig. 22.

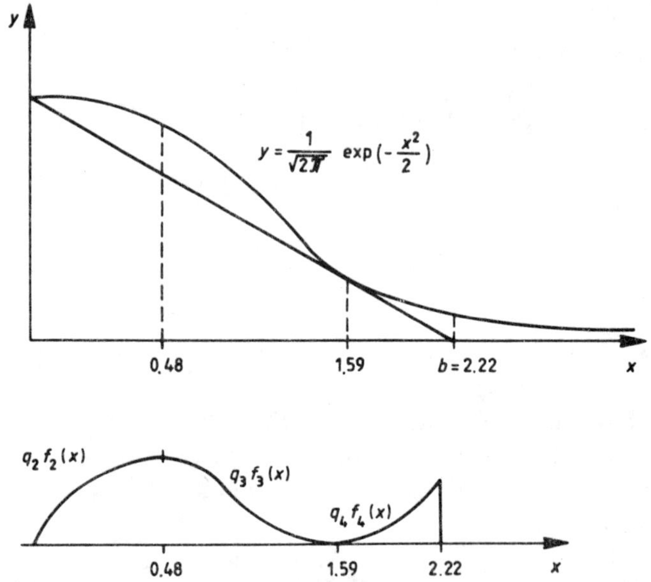

Fig. 22. Decomposition of the normal density function $\varphi(x)$

Generation of random numbers according to the linear density function f_1 is done using the economical method described in Section 3.5 (algorithm L1). Generating samples with almost linear densities f_2, f_3, f_4 can be done along the lines described in algorithm S1 and finally for sampling with density function f_5 algorithm S2 is used. In the final version of the algorithm the same computational trick is used as in algorithm D5 described in Section 3.8: the u variable which is used for selecting one of the densities f_i will be split into two parts, the known ("used") part is to be discarded, the remainder is employed as a uniformly distributed number.

In the formal description of the algorithm the notation

$$f(x) = \varphi(x) - q_1 f_1(x) = d_1 \exp\left(-x^2/2\right) - d_2(b - |x|)$$

is also used. The correct values of the constants $b, p_1, \ldots, c_{11}, \ldots$ are given in Table 2; some of them were evaluated from the following relations: $p_1 = p_2/2$, $p_2 = b/\sqrt{2\pi}$, $c_{11} = b/p_1$, $c_{12} = b/2$, $c_{21} = 1/p_2$, $c_{41} = 1/(p_3 - p_2)$, $c_{61} = 1/(p_4 - p_3)$, $c_{81} = 1/(p_5 - p_4)$, $c_{101} = 1/(1 - p_5)$, $c_{111} = b^2/2$, $d_1 = 1\sqrt{2\pi}$, $d_2 = d_1/b$.

Table 13

Constants appearing in normal algorithm N4

b	=	2.21603	58671	66471
p_1	=	0.44203	52011	49379
p_2	=	0.88407	04022	98758
p_3	=	0.93147	84468	00518
p_4	=	0.95872	08247	90463
p_5	=	0.97331	09541	73898
c_{11}	=	5.01325	65492	62003
c_{12}	=	1.10801	79335	83235
c_{21}	=	1.13113	16354	44180
c_{41}	=	21.09346	65310	41488
c_{51}	=	0.47972	74042	22441
c_{52}	=	1.10547	36610	22070
c_{53}	=	0.87283	49766	71790
c_{54}	=	0.04926	44963	73128
c_{61}	=	36.70751	50476	61921
c_{71}	=	0.47972	74042	22441
c_{72}	=	0.59550	71380	15940
c_{73}	=	0.80557	79244	23817
c_{74}	=	0.05337	75495	06886
c_{81}	=	68.53948	81511	71341
c_{91}	=	0.63083	48019	21960
c_{92}	=	0.75559	15316	67601
c_{93}	=	0.03424	05037	50111
c_{101}	=	37.46855	56956	85297
c_{111}	=	2.45540	74822	84127
d_1	=	0.39894	22804	01433
d_2	=	0.18002	51910	68563

N4 Normal, economical triangle

1. [Big triangle.] Generate u. If $u > p_1$, then go to Step 2, otherwise deliver the value $x \leftarrow c_{11}u - c_{12}$.
2. If $u > p_2$ then go to Step 4, otherwise set $x \leftarrow c_{21}u$.

3. Generate v_1 set $s \leftarrow v - 0.5$, $v \leftarrow |s|$, $t \leftarrow 1 - x$. If $v > t$, then exchange $x \leftarrow t$. If $s < 0$, then deliver the number $x \leftarrow -bx$, otherwise deliver $x \leftarrow bx$.
4. If $u > p_3$, then go to Step 6, otherwise let $x \leftarrow (u - p_2)c_{41}$.
5. [Second small triangle.] Generate v. Let $z \leftarrow u - v$, $x \leftarrow c_{51}+$ $+c_{52}\min(u,v)$. If $z < 0$, then change the sign $x \leftarrow -x$. If $\max(u,v) \leqq$ $\leqq c_{53}$, then deliver x. If $c_{54}|z| \leqq f(x)$, then deliver x, otherwise generate u and repeat this step from the very beginning.
6. If $u > p_4$ then go to Step 8, otherwise set $u \leftarrow (u - p_3)c_{61}$.
7. [First small triangle.] Generate v. Set $z \leftarrow u-v, x \leftarrow c_{71}-c_{72}\min(u,v)$. If $z < 0$, then change the sign $x \leftarrow -x$. If $\max(u,v) < c_{73}$ then deliver the value x. If $c_{74}|z| \leqq f(x)$ then deliver x, otherwise generate u and repeat this step from the very beginning.
8. If $u > p_5$ then go to Step 10, otherwise compute $u \leftarrow (u - p_4)c_{81}$.
9. [Third small triangle.] Generate v and set $z \leftarrow u - v$, $x \leftarrow b-$ $-c_{91}\min(u,c)$. If $z < 0$, then change the sign $x \leftarrow -x$. If $\max(u,v) <$ $< c_{92}$ then deliver the same x, otherwise generate u and repeat this step from the very beginning.
10. [The tail of the distribution.] Compute the number $u \leftarrow (u - p_5)c_{101}$.
11. Generate v, set $s \leftarrow v - 0.5$, $v = 2|s|$, $t \leftarrow c_{111} - \ln v$. If $u^2 t > c_{111}$ then generate u and repeat this step from the beginning otherwise set $x \leftarrow (2t)^{1/2}$ and deliver x if $s > 0$ or deliver $x \leftarrow -x$ if $s < 0$.

The expected value $\overline{N}$ of the necessary number of the uniformly distributed samples can be determined in the following way. To generate a variable with density function f_1 we need 1.5 uniformly distributed number. In case of the three small triangle shaped densities the joint acceptance's (the fast acceptance plus the acceptance in case of evaluating the function value together) probability is 0.827. For generating normally distributed random variables with absolute value greater than $b = 2.21$ the acceptance's probability can be evaluated from formula (4.3) as 0.849. These factors are to be weighted with the probabilities q_i of selection so we obtain

$$\overline{N} = 0.844 \cdot 1.5 + 0.0892\ (2/0.827) + 0.0267(2/0.849) = 1.604.$$

The advantage of the algorithm is the infrequent use of external subroutines (to evaluate the exponential function and the square root), and that it is faster than the modified polar method even in a high level language (in machine code it is considerably faster). As drawbacks of algorithm N4 we mention the need for storing 22 constants, we still need external routines (exp, sqrt) and the program is longer than that of previous algorithms. Finally the main feature of the algorithm is that its full efficiency is attained in machine coded version, but in this case its speed is almost the same as that of algorithm N5 to be described in the sequel.

The second fast method to be described in this section for generating normally distributed numbers is based on the decomposition method and the comparison method. First a preliminary version of the algorithm will be given, then some modifications speeding up the algorithm will be described and finally an implementation-ready version is presented.

In the main part of the algorithm — similarly to algorithm N4 — "positive" normal variables are generated, that is variables with density function $\psi(x) = \sqrt{\frac{2}{\pi}}\, e^{-x^2/2}$, $x \geqq 0$, and at the end of the procedure a random sign is attached to the generated sample. To apply the comparison method we decompose the halfline $[0, +\infty)$ into small subintervals $[a_i, a_i+1)$ by the aid of the splitting points $0 = a_1 < a_2 < a_3 < \ldots < a_k < \ldots$ in such a way that increase of the function $g(x) = x^2/2$ is less than 1 in any given subinterval. Instead of the function $g(x)$ after the appropriate factorization we will have the function $\hat{g}(x) = (x^2 - a_i^2)/2 = (w/2 + a_i)w$ in the exponent, where the notation $w = x - a_i$, $x \in [a_i, a_{i+1})$ was used. Denote the distribution function of the "positive" normal distribution by $\Psi(x)$, that is

$$\Psi(x) = \int_0^x \sqrt{\frac{2}{\pi}}\, e^{-z^2/2} dz.$$

The probability of selecting the ith subinterval $[a_i, a_{i+1})$ will be denoted by $p_i = \Psi(a_{i+1}) - \Psi(a_i)$. Similarly to algorithms C6 and C7 described in the previous sections the following procedure can be built for generating random variables with density function $\psi(x)$.

S3 Normal, by comparison, preliminary

1. Select one of the subintervals $[a_i, a_{i+1})$ with probability p_i.
2. Generate u and set $w \leftarrow u(a_{i+1} - a_i)$. Compute the value $t \leftarrow (w/2 + a_i)w$.
3. Generate u^*. If $u^* > t$, then go to Step 5 (this is the acceptance branch).
4. Generate u. If $u^* < u$ then go back to Step 2 (this is the case of the even index k, when the generated sequence $u, u^*, u, u^*, \ldots$ is rejected) otherwise set $t \leftarrow u$ and go back to Step 3 (the generation of the sequence continues).
5. Deliver the value $y \leftarrow w + a_i$.

This algorithm in the above described form requires at least three uniformly distributed random numbers: one in Step 1 for selecting one of the subinterval, one in Step 2 to generate a point uniformly in the subinterval, and finally at least one more in Step 3 or Step 4 for generating the chain of inequalities $u_1 > u_2 > \ldots < u_k$.

By a suitable selection of the splitting points a_i the probabilities $p_i = \Psi(a_{i+1}) - \Psi(a_i)$ are made more amenable to computer coding. This can be achieved for instance by dividing the halfline $[0, +\infty)$ into 32, equally probable parts, that is the points a_i be determined by

$$\frac{1}{32} = \Psi(a_{i+1}) - \Psi(a_i), \quad i = 1, \ldots, 31.$$

The last part $[a_{31}, +\infty)$ of the halfline $\left(\text{whose probability is also } \frac{1}{32}\right)$ is split up into decreasingly probable subintervals, the subsequent subintervals will have a probability 1/64, 1/128, 1/256,... etc. (The convenience of this decomposition will be evident from the following algorithm.) Having at hand a computer with e bit long words we will store and administrate $e - 5$ such intervals. Denote the length of these intervals by d_i, that is $d_i = \Psi^{-1}(1 - 2^i) - \Psi^{-1}(1 - 2^{i+1}), \ldots, \ i = 6, 7, \ldots, e - 5$, where Ψ^{-1} is the inverse of the distribution function Ψ. To generate samples from the tail of the distribution we might use algorithm S2 described above but for the sake of uniformity of treatment we use the same comparison method for producing normal samples with absolute value greater than $b = \Psi^{-1}(1 - 2^{-5}) \sim 2.16$, as in the case of the main body of the distribution. The selection of a tail interval with the corresponding probability can be done by examining the bits of a random number from the sixth bit to the eth one; if the ith bit examined is found to be different from zero then our choice is the ith tail interval. For these tail intervals the comparison method is applied without any change but in case of the first 31 intervals a couple of modifications are inserted into the original comparison technique.

Scrutinizing Step 3 in algorithm A3 we notice that evaluating the number w can be avoided in some cases. Denote the greatest value of the function $\hat{g}_i(x)$ in interval $[a_i, a_{i+1})$ by $t_{i\,\max}$, then the inequality $u^* > t_{i\,\max}$ implies the inequality $u^* > t = (w/2 + a_i)w$ no matter what value the variable w assumes in the interval $[0, \ a_{i+1} - a_i)$. This means that if the inequality $u^* > t_{i\,\max}$ holds then we do not need the value u produced in Step 2 for the test in Step 3. This situation can be exploited by exchanging the order of Step 2 and Step 3 and first of all the inequality $u^* > t_{i\,\max}$ is tested.

Table 14

Constants used in algorithm N5

i	a_i	t_i	h_i
1	0.00000000000000	0.00076738283767	0.03920617164634
2	0.03917608550309	0.00230687039764	0.03932704963665
3	0.07841241273311	0.00386061844387	0.03950999486086
4	0.11776987457909	0.00543845406707	0.03975702679515
5	0.15731068461017	0.00705069876857	0.04007092772490
6	0.19709908429430	0.00870839582019	0.04045532602655
7	0.23720210932878	0.01042356984914	0.04091480886081
8	0.27769043982157	0.01220953194966	0.04145507115859
9	0.31863936396437	0.01408124734637	0.04208311051344
10	0.36012989178957	0.01605578804548	0.04280748137995
11	0.40225006532172	0.01815290075142	0.04363862733472
12	0.44509652498551	0.02039573175398	0.04458931789605
13	0.48877641111466	0.02281176732513	0.04567522779560
14	0.53340970624127	0.02543407332319	0.04691571371696
15	0.57913216225555	0.02830295595118	0.04833486978119
16	0.62609901234641	0.03146822492920	0.04996298427702
17	0.67448975019607	0.03499233438388	0.05183858644724
18	0.72451438349236	0.03895482964836	0.05401138183398
19	0.77642176114792	0.04345878381672	0.05654656186515
20	0.83051087820539	0.04864034918076	0.05953130423884
21	0.88714655901887	0.05468333844273	0.06308488965373
22	0.94678175630104	0.06184222395816	0.06737503494905
23	1.00999016924958	0.07047982761667	0.07264543556657
24	1.07751556704027	0.08113194985866	0.07926471414968
25	1.15034938037600	0.09462443534524	0.08781922325338
26	1.22985875921658	0.11230007889456	0.09930398323927
27	1.31801089730353	0.13649799954975	0.11555994154118
28	1.41779713799625	0.17168856004707	0.14043438342816
29	1.53412054435253	0.22762405488269	0.18361418337460
30	1.67593972277344	0.33049802776911	0.27900163464163
31	1.86273186742164	0.58470309390507	0.70104742502766
32	2.15387469406144	–	–

Finally we make another observation allowing to decrease the number of required uniform samples. If $u^* > t_{i\,\max}$ holds, then u^* still remains uniformly distributed in the interval $[t_{i\,\max}, 1)$, that is the variable $u = (u^* - t_{i\,\max})/(1 - t_{i\,\max})$ is uniformly distributed in $[0, 1)$; this characteristic of re-usability is utilized in Step 3 of the forthcoming algorithm N5.

Table 14 (cont.)

Constants used in algorithm N5

i	d_i	i	d_i	i	d_i
1	0.67448975019607	17	0.15040938382813	33	0.10597677198479
2	0.47585963017993	18	0.14590257684509	34	0.10433484129317
3	0.38377116397654	19	0.14177003276856	35	0.10276601206127
4	0.32861132306910	20	0.13796317369537	36	0.10126505151402
5	0.29114282663980	21	0.13444176150074	37	0.09982723448906
6	0.26368432217502	22	0.13117215026483	38	0.09844828202068
7	0.24250845238097	23	0.12812596512583	39	0.09712430874765
8	0.22556744380930	24	0.12527909006226	40	0.09585177768776
9	0.21163416577204	25	0.12261088288608	41	0.09462746119186
10	0.19992426749317	26	0.12010355965651	42	0.09344840710526
11	0.18991075842246	27	0.11774170701949	43	0.09231190933664
12	0.18122518100691	28	0.11551189226063	44	0.09121548217294
13	0.17360140038056	29	0.11340234879117	45	0.09015683778986
14	0.16684190866667	30	0.11140272044119	46	0.08913386650005
15	0.16079672918053	31	0.10950385201710	47	0.08814461935364
16	0.15534971747692	32	0.10769761656476		

In the formal description of the algorithm the following notations are extensively used $t_i = \max_{x\in[a_i,a_i+1)} \hat{g}_i(x)$, $h_i = (a_{i+1} - a_i)/(1 - t_i)$, $d_i = \Psi^{-1}(1 - 2^i) - \Psi^{-1}(1 - 2^{-i+1})$, the quantities a_i, t_i, h_i, d_i are naturally stored in the computer prior to their use, the values of these constants are given in Table 14.

N5 Normal, comparison

1. Generate u, detach its first bit as a random sign s, that is set $s \leftarrow \text{sign}(2u - 1)$.
2. Consider the binary number given by the second, third ... sixth bits as an integer number and set the index i to this value (this corresponds to a uniform selection of the interval $[a_i, a_{i+1})$). If $i = 0$ then go to Step 8 (selecting the tail). Set $u^* \leftarrow 2^6u - [2^6u]$, $a \leftarrow a_i$, where u^* is uniformly distributed in $[0, 1)$ once more.
3. If $u^* > t_i$, then evaluate $w = (u^* - t_i)h_i$ and go immediately to Step 16.
4. Generate u. Compute $w \leftarrow u(a_{i+1} - a_i)$ and $t \leftarrow (w/2 + a_i)w$.
5. If $u^* > t$, then go to Step 16 (acceptance).
6. Generate u. If $u^* < u$, then generate u^* and then go back to Step 3 (the index k of the sequence is even, the sequence is rejected).

7. [Continue the generation of the sequence $u_1 > u_2 > \ldots$.] Set $t \leftarrow u$, generate u^* and go to Step 5.
8. [Generating a sample from the tail of the distribution.] Set $i \leftarrow 6$, $a \leftarrow a_{32}$.
9. Left shift the number u by one bit, $u \leftarrow 2u$. If $u > 1$, then go to Step 11.
10. Let $a \leftarrow a + d_i$, increase $i \leftarrow i + 1$ and go back to Step 9.
11. Set $u \leftarrow u - 1$.
12. Compute the values $w \leftarrow ud_i$ and $t \leftarrow (w/2 + a)w$.
13. [Generation of the sequence.] Generate u^*, if $u^* > t$, then go to Step 16 (acceptance).
14. Generate u. If $u^* < u$, then generate u and go back to Step 12.
15. Let $t \leftarrow u$ and go back to Step 13 (rejecting the generated sequence).
16. Set $x \leftarrow a + w$. If $s > 0$, then deliver the number x, otherwise deliver $x \leftarrow -x$ as a normally distributed number.

The expected value of the number of uniform random samples required in the algorithm is $\overline{N} = 1.23$. The expedient way to implement the algorithm is in assembler code-level language. Steps 1 and 2 can be realized by circular shifts (or left rotate) operations, this or similar instructions can be found on all computers. To achieve the greatest possible speed it is essential not to use the real numbers uniformly distributed in $[0, 1)$ but their product with the greatest possible integer number, which product in case of wordlength of e will be uniformly distributed integer numbers between 0 and $2^e - 1$. The essence of this modification is that fixed point operations are usually faster than floating point instructions.

This algorithm N5, is one of the fastest algorithms known. Its disadvantage is that this speed can be attained only at the cost of serious programming efforts (bit manipulation) and storing all the data about 150. Nevertheless if speed is the main consideration this is the algorithm to be used.

4.4 Exponential distribution

The density function of the general exponential distribution with parameter c is given as

$$g(x) = c \cdot e^{-cx}, \quad x \in [0, +\infty),$$

its distribution function accordingly has the form

$$G(x) = \int_0^x c \cdot e^{-cy} dy = 1 - e^{-cx}.$$

In case of the standard exponential distribution we have $c = 1$, that is its density function is (Fig. 23)

$$f(x) = e^{-x}, \quad x \geqq 0. \tag{4.5}$$

If ξ is a random variable with density function (4.5), then the random variable $\eta = \xi/c$ has an exponential distribution with parameter c as it can easily be shown by applying the transformation theorem from the first chapter. Further on thus it is sufficient to build a random number generator for the standardized exponential distribution with density function (4.5).

In the preceding chapter two methods were already discussed for generating exponentially distributed numbers: algorithm E1 by the aid of the inverting the distribution function, method E2 on the basis of the comparison method produce exponential samples. Here a refinement of method E2 is presented which is considerably faster than both E1 and E2.

Now we will have a decomposition of the halfline $[0, +\infty)$ different from that of in algorithm E2. Define the kth subinterval to be $[k \ln 2, (k+1) \ln 2)$, $k = 0, 1, 2, \ldots$, the probability of sampling from a subinterval thus described is

$$p_k = \int_{k \ln 2}^{(k+1)\ln 2} e^{-x} dx = 2^{-(k+1)}, \quad k = 0, 1, 2, \ldots \tag{4.6}$$

Define the density functions $g_k(x)$ of the decomposition so as to be proportional to the function e^{-x}, that is

$$g_k(x) = \frac{1}{p_k} e^{-x}, \quad x \in [k \ln 2, \ (k+1)\ln 2). \tag{4.7}$$

Using these notations the density function $f(x)$ can be presented in the form

$$f(x) = \sum_{k=0}^{\infty} p_k g_k(x). \tag{4.8}$$

Based on the decomposition technique random variables with density function $f(x)$ can be generated by selecting the kth interval with probability p_k, generating a sample in the interval $[0, \ln 2)$ with density function

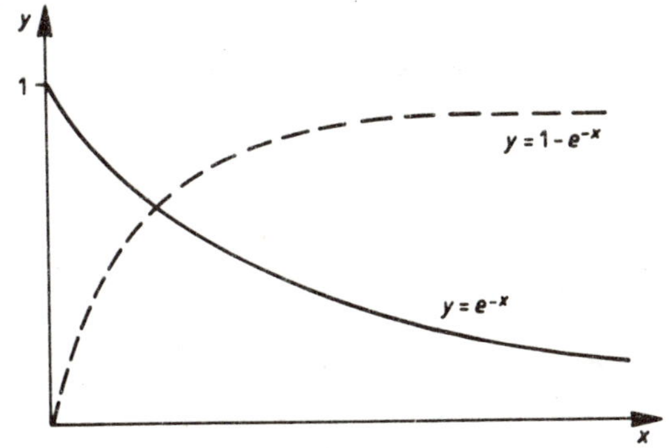

Fig. 23. The density and distribution function of the standard exponential distribution

$g_0(x) = 2e^{-x}$ and then adding to the same the number $k \ln 2$ (since the resulting number will be in the kth interval and its density function is proportional to e^{-x}). The first step, generating a sample from the discrete distribution given by the probabilities $p_0, p_1, \ldots, p_k, \ldots$ is an easy job because of the good selection of the splitting points $k \ln 2$; the number of the zeros after the decimal point in the binary form of a uniformly distributed number u has the same distribution $p_0, p_1, \ldots$. That is to generate a sample according to the probabilities in (4.6) we have to count only the number of the zeros in u up to the first nonzero bit (see Steps 2 and 3 in the next algorithm E3).

To produce a sample according to the density function $g_0(x) = 2e^{-x}$, $x \in [0, \ln 2)$ is little more difficult. We need the following

Theorem. Let $r_i = (\ln 2)^i/i!$, $i = 1, 2, \ldots$. If with probability r_i we generate the minimum of i numbers uniformly distributed in the interval $[0, \ln 2)$, then the density function of the resulting mixture is $g_0(x) = 2e^{-x}$, $x \in [0, \ln 2)$.

Proof. It is well known, that the distribution function of i random variable with distribution function $H(y)$ is $B_{1,i}(H(y))$, where $B_{1,i}$ is the distribution function of the beta distribution with parameters 1 and i, that is

$$B_{1,i}(z) = \frac{\Gamma(i+1)}{\Gamma(1)\Gamma(i)} \int_0^z (1-x)^{i-1} dx.$$

The distribution function of the random variable uniformly distributed in $[0, \ln 2)$ is $H(y) = y/\ln 2$ which implies that the minimum of i such variables has the distribution function

$$B_{1,i}(y/\ln 2) = \frac{\Gamma(i+1)}{\Gamma(1)\Gamma(i)} \int\limits_0^{y/\ln 2} (1+x)^{i-1} \mathrm{d}x = 1 - (1 - \frac{y}{\ln 2})^i.$$

Table 15

Constants in algorithm E3

q_1 =0.69314	71805	55945
q_2 =0.93337	36875	19046
q_3 =0.98887	77961	83868
q_4 =0.99849	59252	91496
q_5 =0.99982	92811	06139
q_6 =0.99998	33164	10073
q_7 =0.99999	85691	43877
q_8 =0.99999	98906	92556
q_9 =0.99999	99924	73416
q_{10}=0.99999	99995	28328
q_{11}=0.99999	99999	72881
q_{12}=0.99999	99999	98560
q_{13}=0.99999	99999	99929
q_{14}=0.99999	99999	99997
q_{15}=1.00000	00000	00000

Differentiating this last expression with respect to y the density function of the minimum of i numbers is obtained. Taking their sum with the weight r_i we arrive at

$$\sum_{i=1}^{\infty} r_i \frac{i}{\ln 2}(1 - \frac{y}{\ln 2})^{i-1} = \sum_{i=1}^{\infty} \frac{(\ln 2 - y)^{i-1}}{(i-1)!} = \mathrm{e}^{\ln 2 - y} = 2\mathrm{e}^{-y},$$

that proves the assertion of the theorem.

According to this theorem samples of the random variable with density function $2\mathrm{e}^{-x}$, $x \in [0, \ln 2)$ are produced by the following auxiliary algorithm.

S4 Subalgorithm, Exponential, truncated density

1. Generate u. If $u < \ln 2 = r_1$, then deliver the number u.
2. [Here we have $u \geqq \ln 2$.] Let $i \leftarrow 2$ and generate u^*. Set $u_{\min} \leftarrow u^*$.
3. Generate u^*. If $u^* < u_{\min}$ then replace $u_{\min} \leftarrow u^*$.
4. [Test: do we need more samples to compute the minimum?] If $u > r_1 + \ldots + r_i$, then increase the number of necessary samples $i \leftarrow i+1$ and go back to Step 3.

5. Deliver the number $x \leftarrow u_{\min} \ln 2$ as a sample with density $2e^{-x}$, $0 \leq x \leq \ln 2$.

Introduce the notation $q_i = r_1 + \ldots + r_i$, $i = 1, 2, \ldots$, the exact value of q_i is given in Table 13. Standard exponentially distributed random numbers are generated in the following algorithm E3 on the basis of decomposition (4.8) making use of the auxiliary routine S4 at the same time.

E4 Exponential, modified decomposition

1. Set $a \leftarrow 0$ and generate u.
2. If the first bit of u is 1, then go to Step 4.
3. [Determine the left-hand terminal point of the interval to be sampled. Note, that here the first bit of u is always 0.] Increase $a \leftarrow a + \ln 2$ and shift u to the left with one bit, that is let $u \leftarrow 2u$, this u is once again uniformly distributed in the interval $[0, 1)$, go back to Step 2.
4. [Begin the search for the minimum of i samples.] (Note that here the first bit of u is 1.) Shift left u with one bit, that is set $u \leftarrow 2u - 1$. If $u > \ln 2$, than go to Step 6.
5. [This corresponds to the case $i = 1$.] Deliver the number $x \leftarrow a + u$.
6. Set $i \leftarrow 2$ and generate u^*. Let $u_{\min} \leftarrow u^*$.
7. Generate a new u^* and if $u^* < u_{\min}$ then replace $u_{\min} \leftarrow u^*$.
8. If $u > q_i$, then increase the counter $i \leftarrow i + 1$ and go back to Step 7.
9. [Here $u < q_i$, that is i was selected with probability $r_i = q_i - q_{i-1}$.] Deliver the value $x \leftarrow a + u_{\min} \ln 2$ as a sample from the standard exponential.

The expected value $\overline{N}$ of number of necessary uniform deviates can be obtained by the following way. The procedure terminates in Step 5 with probability ln 2; in this case we needed just one uniform variable. If the algorithm delivers a value in Step 9 which happens with probability $1 - \ln 2$, then there were some additionally generated uniformly distributed numbers. The expected value of the number of these additionally generated uniform samples can be computed as

$$\sum_{i=2}^{\infty} i r_i = \sum_{i=2}^{\infty} (\ln 2)^i / (i-1)! = \ln 2 \sum_{i=1}^{\infty} (\ln 2)^i / i! = \ln 2.$$

Since one variable is always generated at the beginning thus we need one uniform number both in cases of Step 5 and Step 9, and in case of

Step 9 the number of the additional samples is ln 2 that is $\overline{N} = 1 + \ln 2 \sim$ ~ 1.69.

This algorithm is one of the fastest exponential algorithms. Its disadvantage is that the only economical way of performing the shift operations is via assembler coding that might be troublesome; the number of constants to be stored is not significant. Its advantage is the relatively simple program.

4.5 Gamma distribution

The density function of the standard gamma distribution with parameter b can be given as (see Fig. 24)

$$f(x) = \frac{1}{\Gamma(b)} e^{-x} x^{b-1}, \quad x \geqq 0, \; b > 0, \tag{4.9}$$

where $\Gamma(b)$ is the gamma function which is defined for integer values b as $\Gamma(b) = (b-1)!$ and in the general case

$$\Gamma(b) = \int_0^\infty e^{-x} x^{b-1} dx, \; b > 0.$$

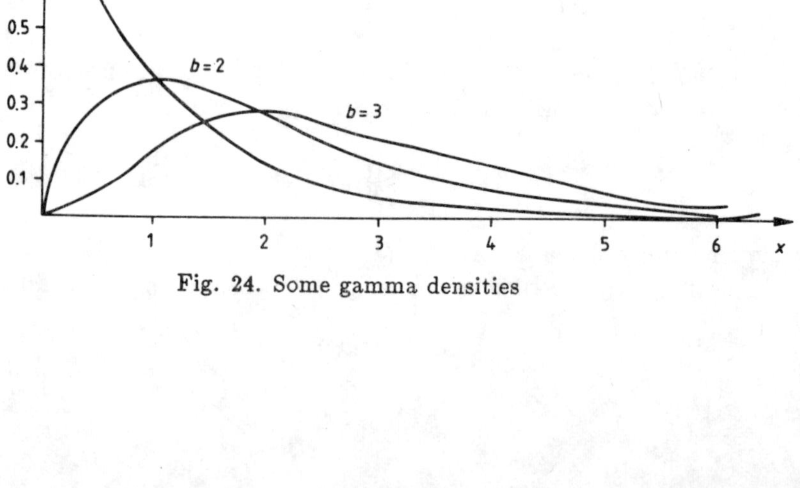

Fig. 24. Some gamma densities

We will consider the generation of standard gamma variables mostly. The nonstandard gamma distribution has two further parameters, denoting these by c and d respectively the nonstandard density function has the form

$$g_1(x) = \frac{1}{c^b\Gamma(b)} e^{-(x-d)/c}(x-d)^{b-1}. \tag{4.10}$$

This simplifies to the standard distribution's density by the choice $c = 1$, $d = 0$. In the sequel we restrict our discussion to the generation of random variables with density function (4.9), the more general nonstandard case can be derived from the standard case. Namely if ξ is a random variable with density function $f(x)$ then the density function of $c\xi + d$ is $g_1(x)$.

The density function of the generalized gamma distribution is

$$g_2(x) = \frac{g}{\Gamma(b)} e^{\{-x^a\}} x^{ab-1}, \quad x > 0.$$

Here if ξ has the standard gamma density function $f(x)$, then $\xi^{1/a}$ has density function $g_2(x)$, and if φ is a sample from the generalized gamma distribution, then φ^a is standard gamma. Of course $g_2(x)$ turns to be $f(x)$ by the choice $a = 1$.

From the family of densities given by (4.9) two particular cases deserve special attention. Choosing the parameter b to be 1, $f(x)$ turns to be the standard exponential density function; this case was discussed in the preceding section. The other case relevant to practice can be obtained substituting the parameter value $b = n/2$, where n is a natural number. Denote the variable determined by the density function thus obtained by ξ, then the distribution of the random variable 2ξ is said to be the χ^2 distribution with n degree of freedom; this special case is investigated at the end of this section.

We refer without a proof to the following well-known result from the probability theory.

Theorem 1. If ξ is a gamma distributed random variable with parameter α (that is denoted in the sequel as gamma (α) distributed) and η is a gamma (β) distributed, which is independent of ξ, then the random variable $\xi + \eta$ is gamma distributed with parameter $\alpha + \beta$.

According to this theorem any gamma distributed variable with a positive integer parameter n can be produced as the sum of n gamma variables with parameter $b = 1$ (i.e. standard exponential). If procedure E1 is used for generating exponentially distributed variables, then after streamlining the following algorithm emerges.

G1 Gamma, $n > 0$ integer parameter

1. Initialize the counters $i \leftarrow 1$, $p \leftarrow 1$.
2. Generate u and compute the product $p \leftarrow pu$.
3. If $i = n$ then deliver the value $x \leftarrow -\ln p$, otherwise increase the index $i \leftarrow i + 1$ and go back to Step 2.

If the parameter n of the distribution is not an integer, then a different algorithm is to be constructed. Naturally it is sufficient to take care of the parameter value $0 < b < 1$ because of Theorem 1 the integer part of the parameter can be dealt with algorithm G1. An acceptance-rejection technique is applied. Consider the majorizing function

$$g(x) = \begin{cases} x^{b-1}/\Gamma(b), & \text{if } 0 \leqq x \leqq 1 \\ e^{-x}/\Gamma(b), & \text{if } 1 < x, \end{cases}$$

which was built from two functions in order to achieve a satisfactorily high level of overall acceptance. For this function the inequality $f(x) \leqq g(x)$ really holds in case of $x \geqq 0$. The density function h which is proportional to the function $g(x)$ can be derived from $g(x)$ by dividing the same with the constant $\int\limits_0^\infty g(x)\mathrm{d}x = \dfrac{1}{b\Gamma(b)} + \dfrac{1}{\mathrm{e}\Gamma(b)}$, that is we have

$$h(x) = \begin{cases} \mathrm{e}bx^{b-1}/(\mathrm{e}+b), & 0 \leqq x < 1, \\ \mathrm{e}b\mathrm{e}^{-x}/(\mathrm{e}+b), & 1 \leqq x. \end{cases}$$

The related distribution function H can be determined by direct integration:

$$H(x) = \begin{cases} \dfrac{\mathrm{e}}{\mathrm{e}+b}x^b, & 0 \leqq x < 1, \\ \dfrac{\mathrm{e}}{\mathrm{e}+b} + \dfrac{b}{\mathrm{e}+b}[1 - \mathrm{e}^{1-x}], & 1 \leqq x. \end{cases}$$

Both parts of this latter can be inverted; the probability of the first part is $H(1) = \dfrac{e}{e+b}$, and the probability of the upper part of the distribution is $1 - \dfrac{\mathrm{e}}{\mathrm{e}+b} = \dfrac{b}{\mathrm{e}+b}$. The test for rejection is the scrutinization of the inequality $f(x)/g(x) < u$, which becomes $\mathrm{e}^{-x} < u$ for the case of $x < 1$, while if $x \geqq 1$ then the generated value is rejected in case of $x^{b-1} < u$. The above discussed details are incorporated into the following algorithm.

G2 Gamma, fractional part parameter $0 < b < 1$

1. [Select one of the two parts of the majorizing density function.] Generate u and let $p \leftarrow u(e+b)/e$. If $p > 1$ then go to Step 3.
2. [Generate a number in the interval $[0,1)$.] Set $x \leftarrow p^{1/b}$ and generate u^*. If $u^* > e^{-x}$, then go back to Step 1 (rejection of x), otherwise deliver the value x.
3. [Generate a value on the halfline $[1, +\infty)$.] Compute $x \leftarrow -\ln\left[((b+ +e)/e - p)/b\right]$ and generate u^*. If $u^* > x^{b-1}$ then go back to Step 1 (rejection), otherwise deliver the number x.

Algorithms G1 and G2 built together can generate a sample from any gamma distribution with an arbitrary parameter $b > 0$ as the next algorithm illustrates.

G3 Gamma, arbitrary parameter $b > 0$

1. Split up the parameter b into an integer and a fractional part $n \leftarrow [b]$, $f \leftarrow b - [b]$. Also initialize $y \leftarrow 0$, $z \leftarrow 0$.
2. If $n = 0$ then set $y \leftarrow 0$ and go to Step 3, otherwise generate a number y with parameter n by the aid of algorithm G1.
3. If $f = 0$ then set $z \leftarrow 0$ and go to Step 4, otherwise generate a sample z with parameter f according to algorithm G2.
4. Deliver the number $x \leftarrow y+z$ as a sample from the gamma distribution with parameter $b = n + f$.

The algorithm can easily be converted into a subroutine in a high level programming language. Computational difficulties, however emerge in cases of greater values of the parameter b ($b \geqq 10$). On the one hand the number of necessary uniform samples will be great (of course $N \geqq b$), on the other the product of more than ten numbers uniformly distributed in $[0,1)$ can easily underflow (the resulting product will be smaller than the smallest positive real number representable on the computer, causing thus the evaluation of the function ln be impossible, and by the same token distorting the tail of the distribution). These difficulties can be overcome only in a different procedure, which will be an acceptance-rejection method.

Consider the Cauchy distribution, which has the density function

$$g_0(x) = \frac{r}{\pi} \frac{1}{1 + r^2(x-s)^2}, \quad -\infty < x < +\infty, \ r > 0. \tag{4.11}$$

Its distribution function can be directly determined by integration, since we have

$$G_0(y) = \int_{-\infty}^{y} \frac{r}{\pi} \frac{1}{1 + r^2(x - s)^2} dx,$$

where substituting $z = r(x - s)$ we obtain

$$G_0(y) = \frac{1}{\pi} \arctan(r(y - s)) + \frac{1}{2}.$$

Applying now the inversion technique it can easily be seen that if u is uniformly distributed in $[0, 1)$ then the random variable $x = \frac{1}{r} \cdot \tan(\pi(u - \frac{1}{2})) + s$ has a density function $g(x)$.

Now an acceptance-rejection technique is described for generating gamma-distributed random numbers, the enveloping function will be proportional to a Cauchy density function. Consider the inequality

$$h(x) = \frac{1}{\Gamma(b)} \frac{e^{-(b-1)}(b-1)^{b-1}}{1 + \frac{1}{2b-1}(x - (b-1))^2} \geqq \frac{e^{-x} x^{b-1}}{\Gamma(b)} = f(x), \tag{4.12}$$

where the expression on the right-hand side is the density function of a gamma distribution with parameter b.

We show the inequality holds for values $x > 0$, $b > 1$. Rearranging (4.12) we obtain the inequality

$$e^{-(b-1)}(b-1)^{b-1} \geqq e^{-x} x^{b-1} \left[1 + \frac{1}{2b-1}(x - (b-1))^2\right] = \gamma(x),$$

which is to be proven. The derivative of the function $\gamma(x)$ on the right-hand side can be calculated as

$$\gamma'(x) = \frac{-1}{2b-1} e^{-x} x^{b-2} (x - (b-1))(x - b)^2.$$

Looking for the roots of $\gamma'(x) = 0$ we find that the only root for which the function $\gamma'(x)$ changes its sign is $x = b - 1$ (since we have $b > 1$). Thus the maximum of $\gamma(x)$ is attained at $x = b - 1$ and in this case (4.12) turns to be equality. This way we did not only show that $h(x) \geqq f(x)$ but the optimality of the upper-bounding nature of the function $h(x)$ in the sense that there is no constant $\alpha < 1$ for which $\alpha h(x) \geqq f(x)$ would hold.

The function $h(x)$ that we used for upper bound is proportional to a Cauchy distribution's density function, where the parameters are $r = \frac{1}{\sqrt{2b-1}}$ and $s = b - 1$. We have to determine the multiplier $a > 1$, which is the value of the ratio between the upper bounding function $h(x)$ and the Cauchy density function

$$(4.13)\qquad g(x) = \frac{1}{\sqrt{2b-1}\,\pi} \cdot \frac{1}{1 + \frac{1}{2b-1}(x-(b-1))^2}.$$

It can be derived by simple division

$$(4.14)\qquad \alpha = \frac{h(x)}{g(x)} = \frac{\pi\sqrt{2b-1}}{\Gamma(b)} e^{-(b-1)}(b-1)^{b-1}.$$

Random variables with density function $g(x)$ given in (4.13) can similarly be produced to the algorithm used for generating variables with density $g_0(x)$, that is employing the expression

$$x = \sqrt{2b-1}\tan\left(\pi\left(u - \frac{1}{2}\right)\right) + b - 1.$$

To complete the algorithm we need the acceptance's test which is the test of the inequality $f(x)/h(x) > u$. The left-hand side quotient in this inequality can be obtained after some computations

$$f(x)/h(x) = \left[1 + \frac{1}{2b-1}(x-(b-1))^2\right] e^{-(x(b-1))}x^{b-1}(b-1)^{-(b-1)},$$

what will be used in Step 3 of the next algorithm.

Now we summarize all the above discussed details in this formal description of a procedure for generating samples from a gamma (b) distribution with a parameter $b > 1$ according to the acceptance-rejection technique (to cover the complete range of parameter this algorithm G4 is to be augmented for instance by the algorithm G2 for parameter values $b \leqq 1$).

G4 Gamma, by Cauchy parameter $b > 1$

Preparation

1. Set $p \leftarrow b-1,\ q \leftarrow 2b-1,\ z \leftarrow \sqrt{2b-1}$.

Generation

1. Generate u. Set $t \leftarrow z\tan(\pi \cdot (u-0.5))$ and $x \leftarrow p + t$.
2. If $x < 0$ then go back to Step 1 (rejection, because for gamma distributed variables $x \geqq 0$ always).

3. Generate u^*. Compute the value $s \leftarrow (1+t^2/q)\exp\{p\ln(x/p)-t\}$, if now $s < u^*$, then go back to Step 2 (rejection on the basis of investigating the acceptance-rejection test), otherwise deliver the value x as gamma distributed with parameter b.

The expected value $\overline{N}$ of the number of uniformly distributed random variables required to generate one gamma (b) variable is

$$\overline{N} = \alpha + 1,$$

where α is the constant given in (4.14). The value of α is π in case of $b = 1$, and if $b \to \infty$ then $\alpha \to \sqrt{\pi}$, which implies that $\overline{N}$ varies between 4.14 and 2.77. The algorithm is easily implementable in high level languages, on the other side its drawback is the use of the exponential, tangent and logarithmic functions (or the exponential, tangent and power functions), furthermore the not very small $\overline{N}$ may not be a welcome feature.

The running time of a modified algorithm compared to that of the algorithm G4 can be halved or decreased to its third if we used a different function as upper bound and some decomposition ideas. We split the halfline $[0, +\infty)$ into two parts, in the first interval a normal enveloping function, in the remaining halfline an exponential upper bounding function is employed, and the algorithm is built according to the acceptance-rejection technique. The acceptance test is decomposed into fast and not so fast acceptance (similarly to algorithm S1 described in Section 4.3 for generating samples with almost linear density function), which results in accepting the generated sample in most of the cases after a very small amount of calculations.

We do not discuss in details the algorithm, only the formal description is given and the function $g(x)$, which is used as an upper bound of the gamma density function:

$$g(x) = \begin{cases} \exp\{-\dfrac{(x-(b-1))^2}{2(b-\sqrt{8b-3})}\}, & \text{if } 0 \leqq x \leqq \beta \\ (\dfrac{\beta}{b-1})^{b-1}\exp\{-(1-\dfrac{b-1}{\beta})x\}, & \text{if } \beta < x \end{cases}$$

where β is a constant, given by $\beta = b - 1 + \sqrt{6}\sqrt{b+\sqrt{8b/3}}$. The formal algorithm follows.

G5 Gamma, normal-exponential upper bound ($b > 2.533$)

1. Let $\mu \leftarrow b-1$, $V \leftarrow \sqrt{b}$, $\sigma^2 \leftarrow b + 1.63299316\ V$, $\sigma \leftarrow \sqrt{\sigma^2}$, $w \leftarrow \sigma^2/\mu$, $d \leftarrow 2.44948974\ \sigma$ and $\beta \leftarrow \mu + d$.
2. Generate u. If $u \leqq 0.0095722652$, then go to Step 8.
3. Generate s from the standard normal distribution and transform it $x \leftarrow \mu + \sigma s$. If $x < 0$ or $x > \beta$ then go back to Step 2 (the normal sample is not in the domain of the enveloping function g).
4. Generate u and set $S \leftarrow s^2/2$. If $s \geqq 0$ then go to Step 6.
5. If $u \leqq 1 - S((1-2s/v)W - 1)$, then deliver the value x, otherwise go to Step 7.
6. If $u \leqq 1 - S(W-1)$ then deliver x.
7. If $\ln u > \mu\ (1 + \ln\ (x/\mu)) - x + S$ then go back to Step 2 (rejection), otherwise deliver x.
8. Generate s from a standard exponential distribution and let $x \leftarrow \beta(1+$ $+s/d)$.
9. Generate u. If $\ln u > \mu(2 + \ln\ (x/\mu) - x/\beta) + 3.72032849 - \beta - 1n(\sigma d/\beta)$, then go back to Step 2 (rejection), otherwise deliver the value x.

On the average this algorithm uses 3–5 uniformly distributed uniform numbers and the time required to generate a sample is about one third of algorithm G4's running time that uses Cauchy upper bound.

For the generalized gamma density function $g_2(x)$ we develop another algorithm using acceptance-rejection technique. Consider a two parameter log logistic density function

$$h(x,\ \beta_1\ \beta_2) = \beta_1\beta_2\ x^{\beta_1}/(\beta_2 + x^{\beta_1})^2,\ x \geqq 0,$$

its distribution function being

$$H\ (x,\ \beta_1,\ \beta_2) = x^{\beta_1}/(\beta_2 + x^{\beta_1}),\quad x = 0,$$

which can be inverted, thus the random variable $\rho = [\beta_2\ u/\ (1-u)]^{1/\beta_2}$ has distribution function H if u is uniform. After some computations the best choices for β_1, β_2 and the constant α (for which $g_2(x) \leqq \alpha\ h\ (x,\ \beta_1,\ \beta_2)$) can be determined as

$$\beta_1^* = a\ \sqrt{2b-1},$$

$$\beta_2^* = b^{\sqrt{2b-1}},$$

$$\alpha^* = 4e^{-b}b^b/[\Gamma(b)\ \sqrt{2b-1}]$$

and the ratio

$$g_2(x)/\alpha^* \, h(x, \, \beta_1^*, \, \beta_2^*) = x^{ab-\beta_1^*} \, \mathrm{e}^{-x^q}(b^{\beta_1^*/a} + x^{\beta_1^*})^2/[4\beta_2^* \mathrm{e}^{-b} b^b].$$

Thus we arrive at the following formal description.

G6 Generalized gamma, log logistic upper bound, $b > 1$

Preparation

Compute $A \leftarrow b^{1/a}$, $B \leftarrow a\sqrt{2b-1}$.

Generation

1. Generate u_1 and u_2.
2. Compute $C \leftarrow \ln\,(u_1/\,(1-u_1))/B$.
3. If $\ln\,(u_1 u_2 (1-u_1)) \leqq b - 1.3862944 + abC + be^{aC}$ then deliver $x \leftarrow Ae^c$, otherwise repeat from Step 1.

To apply decomposition techniques for the gamma distribution is not an easy job because of the variable parameter b and what is induced by b the changing shape of $f(x)$. One may build a decomposition type algorithm in case of a fixed b, but it seems to be expedient only when a great amount of standard variables is required. Another more flexible way is discussed here.

Consider a flexible decomposition of the density function, dependent on b, governed by the points of inflections and the mode of the density function $f(x)$. The majorizing function is uniform over the main body of the distribution, linear (or triangular) over the shoulders and exponential in the tails. The whole region is split up into ten domains, in four of them uniform numbers are generated and accepted, in the six remainder domains acceptance-rejection is involved (two uniform, two triangular and two exponential). Employing squeeze techniques, too, the resulting algorithm seems to be the fastest one for $b > 1$.

A special case of the gamma distribution is the χ^2 distribution with parameter n (it is called the degree of freedom), whose density function is the following

$$k(x) = \frac{1}{2^{n/2}\Gamma\left(\dfrac{n}{2}\right)} \, \mathrm{e}^{-\frac{x}{2}} x^{\frac{n}{2}-1}, \quad x > 0. \tag{4.15}$$

If ξ is a gamma $\left(\dfrac{n}{2}\right)$ distributed random variable, then the variable $\eta = 2\xi$ has density function $k(x)$, since according to the definition we have η's distribution function,

$$P\{\eta < x\} = P\{\xi < \frac{x}{2}\} = \int_0^{x/2} \frac{1}{\Gamma\left(\frac{n}{2}\right)} e^{-y} y^{\frac{n}{2}-1} dy,$$

which after having differentiated will yield us on the right-hand side exactly $k(x)$.

G7 χ^2 distribution, integer parameter $n > 0$

1. Generate x according to the gamma $\left(\frac{n}{2}\right)$ density.
2. Deliver the value $x \leftarrow 2x$ as sample from the χ^2 distribution with n degree of freedom.

Naturally we can build an algorithm for generating variables with density $k(x)$ according to the general definition of the χ^2 distribution: we take the sum of the squares of n standard normal samples. This method can be used only for small parameter values n, for the computational work is linearly increasing with n, while in the previous algorithm (using in Step 1 procedure G4 or G5) the amount of computations basically is constant.

We can generate random variables from another related distribution, the Student distribution just by using its definition. This gives us

ST Student, parameter n

1. Generate x from the standard normal distribution and y from the χ^2 distribution with parameter n.
2. Deliver the value $x \leftarrow x\sqrt{n/y}$ as a sample from the Student (t) distribution.

4.6. Beta distribution

The density function of a beta distribution with parameters b and c is (see Fig. 25)

$$(4.16) \qquad f(x) = \frac{\Gamma(b+c)}{\Gamma(b)\Gamma(c)} x^{b-1}(1-x)^{c-1}, \quad 0 \leqq x \leqq 1, \quad b, c > 0.$$

This distribution is denoted sometimes as a beta (b, c) distribution. From the theory of order statistics the following result is well known.

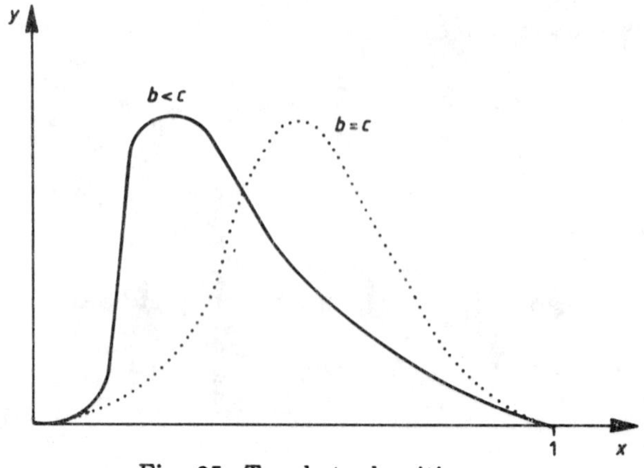

Fig. 25. Two beta densities

Theorem 1. Let the random variables $u_1, u_2, \ldots, u_n$ be uniformly distributed in the interval $[0,1)$ and consider the order statistics $u_1^*, u_2^*, \ldots, u_n^*$ obtained from those, where $u_1^* \leqq u_2^* \leqq \ldots \leqq u_n^*$. Then the kth variable in the sequence, that is u_k^* is a beta $(k, n-k+1)$ distributed random variable.

On the basis of this theorem samples can be obtained from a beta distribution with integer parameters by ordering.

B1 Beta, b, c integer

1. Generate the numbers $u_1, u_2, \ldots, u_{b+c-1}$.
2. Let u_b^* be the bth smallest one among them, that is $u_1^* \leqq \ldots \leqq u_b^* \leqq \ldots \leqq u_{b+c-1}^*$ and deliver the value u_b^* as a beta (b, c) distributed sample.

The method is very simple, the number of uniform variables required to produce one sample is $N = b + c - 1$. The ordering in Step 2 requires a fastly exploding amount of comparisons as $b + c$ increases. But if the sample is only partially ordered — only the bth smallest is determined — then the amount of calculations is only a linear function of $(b + c)$.

Another possibility for generating beta distributed random variables is offered through the next well-known theorem, which is given without a proof, too.

Theorem 2. If x and y are two, independent, gamma distributed random variables with parameters b and c respectively then the random variable $\frac{x}{x+y}$ is beta (b,c) distributed.

This result implies that any gamma generator can be used for producing beta samples. In general cases, when no specific requirements are present with regard to speed, programming convenience, etc. algorithm G4 augmented with G2 can be recommended.

B2 Beta, arbitrary parameters (b,c)

1. Generate x from the gamma (b) distribution and y from the gamma (c) distribution.
2. Deliver the number $x \leftarrow \frac{x}{x+y}$ as beta (b,c) distributed sample.

A separate procedure is described next for the special case $b = c$. An upper bound to the beta (b,b) density function is constructed which will be proportional to a normal density function and then the acceptance-rejection technique is applied. We give the formal algorithm without offering a proof on the validity of the procedure.

B3 Beta, symmetric density, $b = c > 1.5$

1. Set $A \leftarrow b-1$, $t \leftarrow (2A)^{1/2}$.
2. Generate the number s from the standard normal distribution and evaluate $x \leftarrow \frac{1}{2}(1+s/t)$.
3. [Rejection, if the value x is outside the domain of the beta function.] If $x < 0$ or $x > 1$ then go back to Step 2.
4. Generate u. If $u \leqq 1 - s^4/(8b-12)$ then deliver x.
5. If $u \geqq 1 - s^4/(8a-8) + \frac{1}{2}(s^4/(8a-8))^2$ the go back to Step 2.
6. If $\ln 4 > A \ln(4x(1-x)) + s^2/2$, then go back to Step 2.
7. Deliver x as a beta (b,b) distributed sample.

A straightforward decomposition type algorithm can be devised for beta distributions with parameters $b, c > 1$ along the following lines. (It is similar to the one mentioned at the gamma algorithms.) Compute the points of inflection and the mode of the density function $f(x)$, the decomposition is based on these points. The region of concern is split up into 10 domains. In four of them just uniform sampling is performed.

In the remaining six domains acceptance-rejection is used with squeeze modifications: in two domains uniform numbers are generated, in two domains random numbers with triangular densities are produced and in the two tail domains exponentially distributed numbers are sampled.

We underline the fact that all procedures described above for generating gamma and beta distributed variables are for variable parameters. If we have to generate a great number of samples from a distribution with a specific given parameter value, then of course we can manufacture a nicely tuned special algorithm. This can generally be achieved through the joint application of the decomposition technique and one or two other general methods. For example in the case of a beta distribution the interval $[0, 1]$ can be divided into 100 equal part and in each subinterval the density function of the uniform distribution is to be considered; the beta density function is to be approximated by their weighted sum, and as a 101-st member of the decomposition the toothlike remainder is taken (like the illustrative example described in Section 3.3). Of course this extra preparational work is worthwhile when there is a certain payoff through repeated use of the algorithm.

4.7. Binomial distribution

The random variable ξ has a binomial distribution with parameters (n, p) if

$$p_i = P\{\xi = i\} = \binom{n}{i} p^i (1-p)^{n-i}, \quad i = 0, 1, \ldots, n,$$

where n and $0 < p < 1$ are given. The probabilities p_i given above can be interpreted in the following way. Let A be an event which occurs with probability p (and of course it does not occur with probability $1 - p$). Taking now n independent experiments p_i is the probability of the event that A occurred exactly i times out of n. This justifies the following procedure.

BI1 Binomial, count off

1. Set $j \leftarrow 0$, $k \leftarrow 0$.
2. Generate u. If $u \leqq p$ then increase the counter $k \leftarrow k + 1$.
3. Let $j \leftarrow j + 1$. If $j < n$ the go back to Step 2 otherwise deliver the value k as a sample from the binomial with parameters (n, p).

The method is simple, the number of required uniform deviates and the computational works increase linearly dependent on the parameter n. This necessitates the construction of a procedure having favorable performance even for large values of n. The basic idea employed in this new procedure is that the value of the parameter n is decreased to a threshold number N (generally $N \sim 30$ can be taken) and then a sample is generated from the remainder, otherwise a symmetric beta distribution is used. The procedure is given without a proof.

BI2 Binomial, large parameter n

1. Initialize $m \leftarrow n$, $q \leftarrow p$, $k \leftarrow 0$, $y \leftarrow 0$, $h \leftarrow 1$.
2. If $m < N$, then generate a sample j by algorithm BI1 from the binomial distribution with parameters (m, q) and deliver the value $k \leftarrow k + j$.
3. If m is even, then decrease $m \leftarrow m - 1$ and generate u^*. If $u^* \leqq q$ then increase the index $k \leftarrow k + 1$.
4. Let $b \leftarrow (m+1)/2$ and generate a random number s from the symmetric beta (b, b) distribution. Compute the values $g \leftarrow hs$ and $z \leftarrow y + g$.
5. If $z \leqq p$ then modify $y \leftarrow z$, $h \leftarrow h - q$, $q \leftarrow (p - z)/h$ and $k \leftarrow k + a$, otherwise set $h \leftarrow g$, $q \leftarrow (p - y)h$.
6. Decrease the value of the parameter let $m \leftarrow b - 1$ and go back to Step 2 to repeat the iteration.

The algorithm for great values of n ($n \sim 1000$) is approximately 10–20 times faster than the previous procedure BI1.

4.8. Poisson distribution

A ξ discrete random variable is said to be from the Poisson distribution with parameter λ if its distribution is given by the probabilities

$$p_k = P\{\xi = k\} = \frac{\lambda^k}{k!} e^{-\lambda}, \quad k = 0, 1, 2, \ldots .$$

The Poisson distribution plays the same important role among the discrete distributions as the normal does among continuous distributions. The Poisson distribution can be derived through several ways, for example the Poisson distribution is obtained as the limiting distribution the binomial (n, p) if $n\,p = \lambda$ is a constant and $n \to \infty$. If we have a server where n customer arrive independently of each other and uniformly distributed in the time interval $[0, T]$ then the number of the customers arriving in the interval $[t_0, t_0 + t)$ is k with probability p_k in limit (assuming the rate of arrival $n/T = \lambda$ be constant, $n \to \infty$, $T \to \infty$). Poisson distribution is obtained in the case, too, if the time intervals between subsequent arrivals are assumed to be independent and exponentially distributed with parameter λ. This last way of obtaining Poisson distribution suggests the following algorithm where the sum of exponential random variables with density function $\lambda e^{-\lambda x}$ is observed, counting the number of variables until the sum becomes greater than 1. In other words we generate the numbers $u_1, u_2, \ldots, u_k, \ldots$ until the inequality $-\frac{\ln u_1}{\lambda} - \frac{\ln u_2}{\lambda} - \ldots - \frac{\ln u_k}{\lambda} \leqq 1$ is still satisfied, the number k of the variables generated indicates the number of events in $[0, 1]$.

P01 Poisson, count-down

1. Initialize $k \leftarrow 0$, $p \leftarrow 1$.
2. Generate u and set $p \leftarrow pu$.
3. If $p \geqq e^{-\lambda}$ then increase the index $k \leftarrow k + 1$ and go back to Step 2.
4. Deliver k as a sample from the Poisson (λ) distribution.

The procedure is simple, easily implementable. Its drawback is the linear increase in computational efforts in case of increasing λ. Because of this feature another, relatively simple procedure is presented next for producing Poisson (λ) variables for larger values of λ.

The algorithm to be described is based on the following reasonings. Consider the positive integer $n < \lambda$ (it is chosen to be a little smaller than λ itself, for example we can take $n = [7\lambda/8]$), and generate a sample x from the gamma (n) distribution. Then we have two cases

(i) if $x > \lambda$, then deliver a sample k from the binomial $(n - 1,\ \lambda/x)$ distribution,

(ii) if $x \leqq \lambda$ then generate a sample j from the Poisson $(\lambda - x)$ distribution and deliver the value $k \leftarrow n + j$.

Now we prove the validity of the procedure. In the first case the probability of the event that k is delivered can be evaluated as

$$p_k = \int_{\lambda}^{\infty} \binom{n-1}{k} \left(\frac{\lambda}{x}\right)^k \left(1 - \frac{\lambda}{x}\right)^{n-1-k} \frac{e^{-x} x^{n-1}}{(n-1)!} dx =$$

$$= \int_{\lambda}^{\infty} \frac{(n-1)!}{k!(n-1-k)!} \frac{\lambda^k (x-\lambda)^{n-1-k}}{x^{n-1}} \frac{e^{-x} x^{n-1}}{(n-1)!} dx =$$

$$= \frac{\lambda^k}{k!(n-1-k)!} \int_{\lambda}^{\infty} (x-\lambda)^{n-1-k} e^{-x} dx =$$

$$= \frac{\lambda^k e^{-\lambda}}{k!(n-1-k)!} \int_{0}^{\infty} t^{n-1-k} e^{-t} dt$$

where we substituted $t = x - \lambda$ in the last but one integral. The value of the last integral is $\Gamma(n-k) = (n-1-k)!$ which leaves us with $p_k = \lambda^k e^{-\lambda}/k!$ the desired result.

In the second case $0 \leqq x \leqq \lambda$ and the probability of generating k with the substitution $t = x/\lambda$ is

$$p_k = \int_{0}^{\infty} \frac{(\lambda - x)^{k-n} e^{-(\lambda - x)}}{(k-n)!} \frac{e^{-x} x^{n-1}}{\Gamma(n)} dx =$$

$$= \frac{e^{-\lambda}}{(k-n)!(n-1)!} \int_{0}^{\lambda} x^{n-1} (\lambda - x)^{k-n} dx =$$

$$= \frac{e^{-\lambda} \lambda^{n-1} \lambda^{k-n} \lambda}{(n-1)!(k-n)!} \int_{0}^{1} t^{n-1} (1-t)^{k-n} dx.$$

The last integral here is a beta integral, thus

$$p_k = \frac{e^{-\lambda} \lambda^k}{(n-1)!(1-n)!} \frac{\Gamma(n)\Gamma(k-n+1)}{\Gamma(k+1)} = \frac{e^{-\lambda} \lambda^k}{k!}.$$

Before giving a formal description of the algorithm we remark that the value $c = 20$ in Step 2 an arbitrary choice similarwise the value 7/8 in Step 6. This 7/8 choice makes possible to use a not very efficient procedure (procedure BI1) for generating variable from the binomial distribution in Steps 8–10, because we rarely need the generation of such variables. If the parameter λ is great ($\lambda > 1000$, say) then it is worthwhile to replace BI1 with a more efficient algorithm, for instance BI2.

P02 Poisson, from gamma

1. Initialize $k \leftarrow 0$, $w \leftarrow \lambda$.
2. If $w \geqq c$ (where $c = 20$) then go to Step 6.
3. [Case (ii).] Set $p \leftarrow 1$, $b \leftarrow e^{-w}$.
4. Generate u and set $p \leftarrow pu$. If $p < b$, then deliver the value k.
5. Increase the index $k \leftarrow k + 1$ and go back to Step 4.
6. Compute the integer part $n \leftarrow [7w/8]$ and generate a sample x from the gamma (n) distribution. If now $x > w$ then go to Step 8.
7. Let $k \leftarrow k + n$, $w \leftarrow w - x$ and go back to Step 2.
8. [Case (i).] Set $p \leftarrow w/x$.
9. Generate u. If $u \leqq p$ then increase the index $k \leftarrow k + 1$.
10. Let $n \leftarrow n - 1$. If $n > 1$, then return to Step 9, otherwise deliver the value k.

4.9 Miscellaneous distributions

Several other distributions' random number generator is discussed and presented in this section. Of these the first two is deemed to be worth of special attention because of their flexibility they can be used for generating approximate samples from other distributions.

Consider the Weibull distribution with three parameters, whose density function is

$$f(x) = \frac{c}{b}\left(\frac{x-a}{b}\right)^{c-1} \exp\left\{-\left(\frac{y-a}{b}\right)^{c}\right\},$$

from which the distribution function can immediately be derived as

$$F(x) = 1 - \exp\left\{-\left(\frac{x-a}{b}\right)^{c}\right\}.$$

Since this latter can be inverted, the random number generator is

WE Weibull

1. Generate u and compute $z \leftarrow -\ln u$.
2. Deliver the value $x \leftarrow bz^{1/c} + a$ as a sample with Weibull distribution.

Because of the three parameters of the distribution it can easily be fitted to other distributions. In practice the following routine is followed. Assume that ξ is a Weibull distributed random variable and introduce its shifted form $\eta = \xi - a$. The kth moment of this latter can be obtained

$$E(\eta^k) = b^k \Gamma(1 + k/c),$$

that is the expected value μ, variance σ^2 and the generalized third moment α_3 of the random variable ξ are given in the following equations

$$\mu = E(\xi) = a + b\Gamma(1 + 1/c),$$
$$\sigma^2 = E(\xi - \mu)^2 = b^2[\Gamma(1 + 2/c) - \Gamma(1 + 1/c)^2],$$
$$\alpha_3 = E(\frac{x - \mu}{\sigma})^3 =$$
$$= \frac{\Gamma(1 - 3/c) - 3\Gamma(1 + 2/c)\Gamma(1 + 1/c) + 2\Gamma(1 + 1/c)^3}{[\Gamma(1 + 2/c) - \Gamma(1 + 1/c)^2]^{3/2}}.$$

Assume that the distribution to be approximated has an expected value, standard deviation and generalized third moment m, s and β_3 respectively then the parameters of the approximating Weibull distribution can be computed from the system of equations $\mu = m$, $\sigma = s$, $\alpha_3 = \beta_3$. If these equations do not fully determine the parameters a, b, c then either we can arbitrarily fix them or a further requirement may be satisfied.

We can similarly proceed in case of Burr distribution that can be used for approximating other distributions too. Burr's distribution function is given as

$$G(x) = 1 - (1 + x^c)^{-k}, \; x \geqq 0, \; c, k > 0.$$

Its inverse exists so we have the following random number generator.

BU Burr

1. Generate u and set $z \leftarrow (1 - u)^{-1/k}$.
2. Deliver the value $x \leftarrow (z - 1)^{1/c}$ as a Burr-distributed sample.

The central moments of the distribution are

$$E(\xi^j) = \mu'_j = k\Gamma(k - j/c)\Gamma(j/c + 1)/\Gamma(k + 1), \; j < kc.$$

From these the first four standardized moments

$$\mu = \mu'_1, \qquad \sigma^2 = \mu'_2 - \mu^2,$$
$$\alpha_3 = (\mu'_3 - 3\mu'_2\mu + 2\mu^3)/\sigma^3$$
$$\alpha_4 = (\mu'_4 - 4\mu'_3\mu + 6\mu'_2\mu^2 - 3\mu^4)/\sigma^4.$$

If now the expected value, standard deviation, third and fourth standard moments of the distribution to be approximated are denoted by m, s, β_3, β_4 respectively, then first the equations $\alpha_3 = \beta_3$ and $\alpha_4 = \beta_4$ are solved for the unknown constants k and c. Using these k and c values a number x is generated by the BU procedure. The random number

$$x^* = \frac{x - (\mu - \sigma m/s)}{\sigma/s}$$

will have its first four moments be equal to that of the distribution to be approximated.

A random variable is called doubly exponential if its distribution function is of the form:

$$F(x) = \exp\{-\exp(-x)\}, \quad -\infty < x < +\infty.$$

Variables with this distribution can be generated by inverting the distribution function $F(x)$.

DE Doubly exponential

1. Generate u and set $y \leftarrow -\ln u$.
2. Deliver the value $x \leftarrow -\ln y$ as a doubly exponential distributed sample.

The exponential power distribution is given by its density function

$$f(x) = \frac{1}{2\Gamma\left(1 + \frac{1}{a}\right)} \exp\{-|x|^a\}, \qquad a \geqq 1.$$

This is a flexible family of symmetric distributions, including the uniform, normal and double exponential distributions.

A convenient algorithm can be obtained via the acceptance-rejection technique, using the double exponential density function

$$h(x) = \exp\{-|x|/b\}/2b,$$

where $b = \left(\frac{1}{a}\right)^{\frac{1}{a}}$. Multiplying $h(x)$ with the constant

$$c = \frac{1}{\Gamma\left(1 + \frac{1}{a}\right)} \cdot \left(\frac{1}{a}\right)^{\frac{1}{a}} \exp\left\{1 - \frac{1}{a}\right\}$$

the inequality $f(x) \leqq ch(x)$ holds, which gives rise to

PE Power exponential, via double exponential, $a > 1$

1. Generate a double exponential variable x with parameter b.
2. Generate u uniformly.

3. If $\ln u > [-|x|^a + |x|/b - 1 + \frac{1}{a}]$, then go to Step 2, otherwise deliver x.

The density function of a Rayleigh distributed random variable is

$$h(x) = \frac{1}{\sigma^2} x \exp\left\{-\frac{1}{2\sigma^2} y^2\right\}, \quad y > 0.$$

After integration and inverting the distribution function the following generator can be given.

RA Rayleigh

1. Generate u and compute the number $y \leftarrow \ln u$.
2. Deliver $x \leftarrow \sigma\sqrt{y}$ as Rayleigh distributed.

Considering the fact, that the random variable $-\ln u$ (similarly the random variable $-\ln(1-u)$) is exponentially distributed we can replace the first step in the algorithms WE, RE, FR, RA with a procedure that generates a sample from the standard exponential distribution.

A random variable is Frechet distributed if its distribution function is given in the form

$$G(x) = \exp(-x^{-\alpha}), \quad x \geqq 0, \quad \alpha < 0,$$

with a suitable constant α. Inverting the distribution function immediately yields the corresponding algorithm.

FR Frechet

1. Generate u and compute the value $y \leftarrow -\ln u$.
2. Deliver the number $x \leftarrow y^{-1/\alpha}$ as Frechet distributed.

Finally a discrete distribution's generator will be discussed. The probabilities p_i associated to the negative binomial distribution with parameter b are the following

$$p_i = \binom{b-1+i}{i} p^b (1-p)^i, \; b > 0, \; 0 \leqq p \leqq 1, \; i = 1, 2, \ldots$$

Note, that the parameter b is not necessarily an integer number. Using the usual search procedure (direct or inverse method) for this discrete distribution yields the following algorithm.

NB1 Negative binomial

1. Initialize the values $k \leftarrow 0$, $p \leftarrow p^b$, $s \leftarrow p$, $q \leftarrow 1-p$, $r \leftarrow (b-1)q$ and generate u.
2. If $u < s$ then deliver k.
3. Increase the index $k \leftarrow k+1$, compute $p \leftarrow p(q + r/k)$ $s \leftarrow s + p$ and go back to Step 2.

The algorithm is simple, easily implementable, but its drawback is that the computational work linearly dependent on the parameter b. In case of $b > 50$ this algorithm NB1 should be used only if a very few samples are required. To offer an alternative way of generating negative binomial numbers with parameter b we describe — without a proof — a different procedure whose computational work is virtually independent of b.

NB2 Negative binomial, from Poisson

1. Generate x from the gamma (b) distribution and compute $m \leftarrow x(1-$ $-p)/p$.
2. Generate the number k from the Poisson (m) distribution and deliver the same as a sample from the negative binomial distribution.

4.10. Notes

4.2 The polar method (N2) as a random number generator was presented by Mueller, the modified polar algorithm by Marsaglia (1964). A good summary of the normal and exponential generators was published by Ahrens and Dieter (1972). Some recent works are by Kinderman and Ramage (1976), Deák (1981).

4.3 The auxiliary algorithms S1 and S2 are due to Marsaglia (1964). The decomposition used in algorithm N4 was developed by Kinderman and Ramage, the presented form of the algorithm was published by the author (1981). The algorithm N5 was presented by Ahrens and Dieter under the name FL_5, being a member of an algorithms' family.

4.4 The procedure E3 that incorporates some ideas of Sibuya, too, was presented by Ahrens and Dieter (1974).

4.5 The algorithms G2, G3, G4, G5 were published by Ahrens and Dieter (1974). Algorithm G6 was published by Tadikamalla (1979). The decomposition type algorithm thought to be the fastest was developed by Schmeiser and Lal (1980) as algorithm G4PE. Other gamma generators can be found in the papers of Atkinson and Pearce (1976),

Wallace and Cheng (1977), Tadikamalla (1981), Cheng and Feast (1980). Best (1983) gave an improvement to G2.

4.6 The algorithm B3 was published in Ahrens and Dieter (1974), another method is in Cheng (1978). The decomposition type algorithm, mentioned at the end is due to Schmeiser and Babu (1980), it is probably the fastest method. For some other beta generators see the same paper.

4.7, 4.8, 4.9 The procedures BI2, P2, NB2 were published in the great overview of Ahrens and Dieter (1974). There are methods where time required for the generation of one binomial variable remains bounded even as $n \to \infty$, see Fishman (1979) and Devroye (1980), (1986). The double exponential generator was devised by Tadikamalla.

CHAPTER 5

GENERATION OF RANDOM VECTORS

5.1. General methods for generating random vectors

In case of drawing samples of one-dimensional random variables one only has to produce the "randomness" subject to the required regularity. Contrary to this when generating random vectors the dependence of the individual components on each other has to be taken into account as well. While the generation of "randomness" in case of random vectors is based on the same general methods described in the one-dimensional case, the correlation between the individual variables can be taken into consideration through the conditional distribution functions.

It is well known that if $\boldsymbol{\xi} = (\xi_n, \ldots, \xi_n)$ is a random vector with distribution function $G(x_1, x_2, \ldots, x_n) = P\{\xi_1 < x_1, \xi_2 < x_2, \ldots, \xi_n < x_n\}$, then it can be decomposed as a product of two distribution functions as

$$(5.1) \qquad G(x_1, x_2, \ldots, x_n) = G_1(x_1) G_{2\ldots n}(x_2, \ldots, x_n \mid x_1),$$

where G_1 is the unconditional distribution function of the first component, $G_{2\ldots n}$ is the conditional distribution function of the components $\xi_2, \ldots, \xi_n$ given the condition $\xi_1 = x_1$. Applying the same idea to the function $G_{2\ldots n}$ on the right-hand side of (5.1), $(n-1)$ times repeatedly we obtain the following decomposition:

$$(5.2) \qquad \begin{aligned} G(x_1, x_2, \ldots, x_n) =& G_1(x_1) G_2(x_2 \mid x_1) G_3(x_3 \mid x_1, x_2) \\ & \ldots G(x_n \mid x_1, \ldots, x_{n-1}), \end{aligned}$$

where $G_i(x_i \mid x_1, \ldots, x_{i-1})$ is the conditional distribution function of the random variable ξ_i given the values of the random variables $\xi_1, \ldots, \xi_{i-1}$. Making use of equation (5.2) the following procedure can be derived for generating random variables with distribution function $G(x_1, \ldots, x_n)$:

V1 Vector generation, conditional distribution function

1. Generate the value z_1 according to the distribution function $G_1(x_1)$ and set $i \longleftarrow 2$.
2. Generate the number z_i from the distribution given by the conditional distribution function $G_i(x_i \mid x_1 = z_1, \ldots, x_{i-1} = z_{i-1})$.

3. If $n > i$, then increase $i \longleftarrow i+1$ and go back to Step 2.
4. Deliver the vector $(z_1, \ldots, z_n)$ as a sample from the distribution with distribution function $G(x_1, \ldots, x_n)$.

Theorem 1. The vector $(z_1, \ldots, z_n)$ produced by Algorithm V1 has distribution function $G(x_1, \ldots, x_n)$ indeed.

Proof. It is known about the distribution functions of the components of the generated vector that because of the very procedure they are the conditional distribution function G_i's:

$$P\{z_i < x_i\} = G_i(x_i \mid x_n, \ldots, x_{i-1}) = \\ = P\{z_i < x_i \mid x_1 = z_1, \ldots, x_{i-1} = z_{i-1}\}.$$

Thus the distribution function of the generated vector can be derived as

$$P\{z_1 < x_1, \ldots, z_n < x_n\} = P\{z_1 < x_1\}P\{z_2 < x_2 \mid x_1 = z_1\} \ldots \\ \ldots P\{z_i < x_i \mid x_1 = z_1, \ldots, x_{i-1} = z_{i-1}\} \ldots \\ \ldots P\{z_n < x_n \mid x_1 = z_1, \ldots, x_{n-1} = z_{n-1}\} = \\ = G_1(x_1)G_2(x_2 \mid x_1) \ldots G_n(x_n \mid x_1, \ldots, x_{n-1}),$$

and this product equals to $G(x_1, \ldots, x_n)$ because of (5.2).

To enhance application of this general procedure to practical problems we derive the necessary equations in terms of density functions instead of distribution functions. Denote the joint density function by $g(x_1, x_2, \ldots, x_n)$ then the individual density functions $g_1, \ldots, g_n$ of the components to be generated can be evaluated using the following equations

$$g_1(x_1) = \int_{-\infty}^{+\infty} \cdots \int_{-\infty}^{+\infty} g(x_1, x_2, \ldots, x_n)\, \mathrm{d}x_2, \mathrm{d}x_3, \ldots, \mathrm{d}x_n,$$

$$g_2(x_2 \mid x_1) = \int_{-\infty}^{+\infty} \cdots \int_{-\infty}^{+\infty} \frac{g(x_1, x_2, \ldots, x_n)}{g_1(x_1)}\, \mathrm{d}x_3 \ldots \mathrm{d}x_n,$$

$$\vdots$$

$$g_{i+1}(x_{i+1} \mid x_1, \ldots, x_i) = \\ \int_{-\infty}^{+\infty} \cdots \int_{-\infty}^{+\infty} \frac{g(x_1, x_2, \ldots, x_n)}{g_1(x_1)g_2(x_2 \mid x_1) \ldots g_i(x_i \mid x_1, \ldots, x_{i-1})}\, \mathrm{d}x_{i+1} \ldots \mathrm{d}x_n,$$

$$\vdots$$

$$g_n(x_n \mid x_1, \ldots, x_{n-1}) = \int_{-\infty}^{+\infty} \cdots \int_{-\infty}^{+\infty} \frac{g(x_1, x_2, \ldots, x_n)}{(g_n(x_1) g_2(x_2 \mid x_1) \ldots g_{n-1}(x_{n-1} \mid x_1, \ldots, x_{n-2})},$$

where the right-hand side of the ith equation contains $n - i$ integral signs. When generating random vectors an additional difficulty is caused by the variety of the possible domains as compared to the case of the one-dimensional random numbers. In one dimension we have to deal with only two different types of domains, the interval and the halfline. In the multidimensional case there are infinitely many domains that can not be transformed into each other by simple transformations (translation, rotation, reduction or enlargement). This stumbling block appearing in multidimensional cases is one of the gravest problems of numerical mathematics; there is no reliable, reassuring solution until now. Useful procedures have been developed for domains that appear most frequently in practice, some of those will be presented in the forthcoming sections.

By the n-dimensional version of the acceptance-rejection technique uniformly distributed points can be generated in tangled domains, that analytically not or only in a cumbersome way can be dealt with.

Assume that we can produce uniformly distributed points in the simple bounded domain T, say a rectangle which entails the irregular, tangled domain H (Fig. 26). The enveloping domain T is generally chosen to be the smallest one of the possible rectangles.

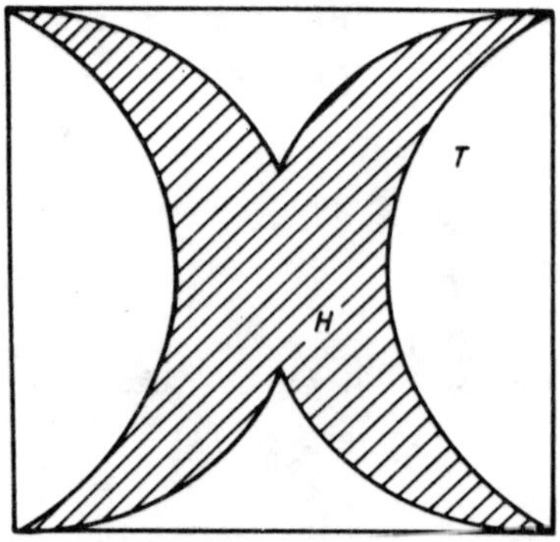

Fig. 26. Generating uniformly distributed points in a sophisticated domain H

V2 Vector generation, acceptance-rejection

1. Generate the point $\mathbf{x}$ uniformly in the rectangle T.
2. If $\mathbf{x} \in H$ is true, then deliver the point $\mathbf{x}$ as a sample from the uniform distribution in domain H (acceptance), otherwise go back to Step 1 (rejection).

Theorem 2. The points accepted in procedure V2 are uniformly distributed in domain H.

Proof. It immediately follows from the algorithm.

The efficiency of the algorithm can be measured by the ratio of the relevant domains' volumes, that is

$$p = V_H / V_T,$$

where V_H and V_T are the volumes of the domains H and T respectively. Generally the procedure is of relatively small efficiency because of the small value of p or in some cases because of the difficulties encountered in testing the relation $\mathbf{x} \in H$ in Step 2 of Algorithm V2. The efficiency can be increased if the domain T can be replaced by domain T' of volume $V_{T'}$, for which we have $T' \supset H$ and $V_T > V_{T'}$ and generation of uniformly distributed points in T' can be done with equal ease as in T. The test $\mathbf{x} \in H$ can be avoided in many cases similarly to the idea employed at the generation of random numbers with almost linear density function (Section 4.3), where most of the cases a simple fast test is carried out, instead of the exact test. In our case a domain T'' is looked for, which is in domain H, and the test $\mathbf{x} \in T''$ is easily executable. This modification means a considerable acceleration of the computational procedure if T'' is a good inner approximation of domain H, that is the ratio $V_{T''}/V_H$ is near to 1. Nevertheless one cannot expect very good performance from this acceptance-rejection technique, especially in great dimensions. The simplest illustration to this dimensional effect is the following: compute the ratio of a cube's volume with sides of length 0.95 and the unit cube's volume in $n = 100$ dimensions, which is

$$(0.95)^{100}/1 = 0.0059.$$

This number illustrates the principle: "volume is concentrated on the facets" in high dimensions, making difficult to obtain good acceptance ratios in case of irregular domain (see also the ratio of the volumes of a hypersphere and the entailing hypercube in Section 5.4).

All in all for generating randomly distributed points in multidimensional spaces one can employ (i) the general algorithms V1 and V2, (ii) the second transformation theory as exposed in Section 1.3 or (iii) the special properties of the individual distributions. One possible multidimensional application of the decomposition technique is shown for the multidimensional distribution function, but there is not a straightforward application for any distribution contrary what we have seen in case of univariate distributions. Partly this scarcity of the tools of generation accounts why the topic of generating random vectors is much less developed as compared to the producing samples of one-dimensional random variables.

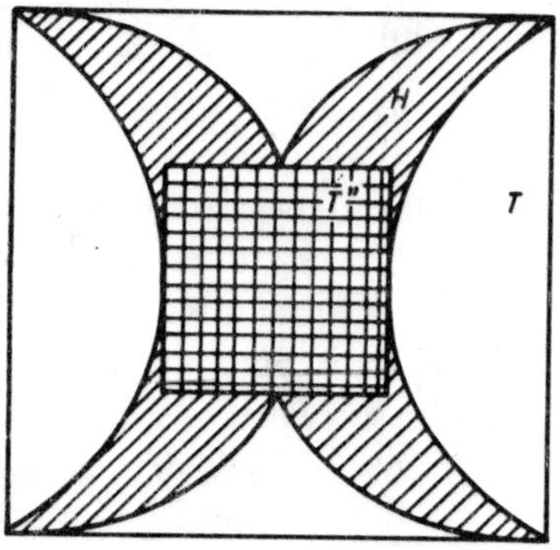

Fig. 27. Simplified acceptance test

5.2. Uniform distribution in rectangle

Consider the task of generating uniformly distributed points in a rectangle that has his edges parallel to the coordinate axes and one of its vertices is the origin. The general algorithm V1 is applied now, the necessary conditional distribution functions become simple one-dimensional unconditional distribution functions since the components of a point uniformly distributed in the above rectangle are independent. Thus the individual coordinates can separately be produced, independently of each other and furthermore uniformly on the corresponding edges.

Let the rectangle T be given by its $(n+1)$ vertices $\mathbf{0} = (0,\ldots,0)$, $\mathbf{x}_1 = (a_1, 0, \ldots, 0)$, $\mathbf{x}_2 = (0, a_2, 0, \ldots, 0), \ldots,$ $\mathbf{x}_n = (0, \ldots, 0, a_n)$, that is $T = \{\mathbf{x} \mid 0 \leqq x_i \leqq a_i, \quad i = 1, \ldots, n\}$. Then uniformly distributed points can be produced in T by the following algorithm.

VRE Uniform vector in rectangle

1. Generate the numbers $u_1, \ldots, u_n$.
2. Compute $x_i \longleftarrow u_i a_i$, $i = 1, \ldots, n$ and deliver the vector $\mathbf{x}$ as uniformly distributed in T.

The algorithm is simple, can very easily be coded in high level languages. If we want to generate points uniformly in a rectangle located in a different way from T described above then the generated points have to be translated and rotated correspondingly. Similarly to obtain points uniformly distributed in a parallelepiped the points obtained by algorithm VRE have to be subjected to a linear transformation.

5.3. Uniform distribution in simplex

For the sake of simplicity we restrict our treatment to discussing the case of the n-dimensional simplex spawned by the origin and the unit vectors $\mathbf{e}_i$, $i = 1, \ldots, n$. In simplexes with different placements one can obtain uniformly distributed points after suitable translations and scalings.

For generating points uniformly in a simplex the general algorithm V1 will be applied, and the algorithm VSI will be developed, but prior to this we need the following result. Consider the n-dimensional simplex

$$H(n,a) = \{\mathbf{x} \mid x_1 + \ldots + x_n \leqq a, \quad x_1, \ldots, x_n \geqq 0\},$$

where a is a positive scalar constant, its volume will be denoted by $V(n,a)$.

Theorem 1. The volume of the simplex $H(n,a)$ is given by

$$V(n,a) = \frac{a^n}{n!}.$$

Proof. Consider one of the vertices of the simplex $H(n,a)$, different from the origin, let it be the vertex P_{n+1} given by the vector $a\mathbf{e}_n$ (Fig. 28).

The remaining n vertices $P_1, \ldots, P_n$ define an $(n-1)$ dimensional simplex $H(n-1,a)$. Draw a line from the point P_{n+1} perpendicular to the plane of the simplex $H(n-1,a)$, (it is actually the line $\overline{P_{n+1}P_1}$ where $P_1 = (0,0,\ldots,0)$) the distance of P_{n+1} from $H(n-1,a)$ is exactly a. If now the simplex $H(n,a)$ is intersected by a hyperplane which is

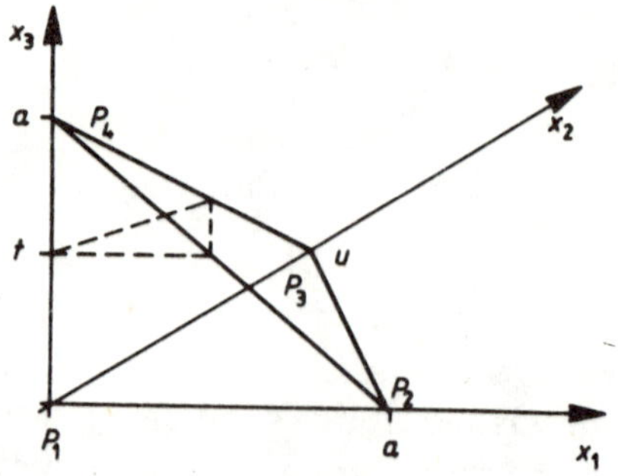

Fig. 28. Determining the volume of the simplex $H(3, a)$

parallel to the simplex $H(n-1, a)$ and the distance of the hyperplane from $H(n-1, a)$ is t ($t < a$), then the $(n-1)$ dimensional intersection is a simplex similar to $H(n-1, a)$. Denote the $(n-1)$ dimensional volume of the intersection by V', then we can write because of the similarity

$$\frac{V'}{V(n-1, a)} = \frac{(a-t)^{n-1}}{a^{n-1}}.$$

From this equation one can easily evaluate the volume of the intersection as

$$V' = \left(\frac{a-t}{a}\right)^{n-1} V(n-1, a).$$

The complete volume $V(n, a)$ of the original simplex can be obtained by integrating the volume V' of the intersected politope

$$V(n, a) = \int_0^a \left(\frac{a-t}{a}\right)^{n-1} V(n-1, a) dt = \frac{a}{n} V(n-1, a).$$

Applying this equation recursively $(n-1)$ more times to the volumes $V(i, a)$, $i = n-1, n-2, \ldots, 2, 1$, standing on the right-hand side of the equations we arrive at the equation looked for

$$V(n, a) = \frac{a}{n} V(n-1, a) = \frac{a^2}{n(n-1)} V(n-2, a) = \ldots = \frac{a^n}{n!}.$$

Let $\xi = (\xi_1, \ldots, \xi_n)$ be a random point uniformly distributed in the simplex $H(n, 1)$, its density function $p(\mathbf{x})$ is a constant, which is according to Theorem 1

$$p(x_1, \ldots, x_n) = C = \frac{1}{V(n, 1)} = n!$$

The conditional density functions required in the course of generation are evaluated next. The unconditional density function of the first component is

$$p_1(x) = \int_{\substack{x_2+\ldots+x_n \leqq 1-x_1, \\ (x_1,\ldots,x_n)\in H(n,1)}} \cdots \int C\, dx_3 \ldots dx_n = CV(n-1, 1-x_1) = n(1-x_1)^{(n-1)}$$

since the domain of integration is an $(n-1)$ dimensional simplex, given by the inequality $x_2 + \ldots + x_n \leqq 1 - x_1$.

Hence the distribution function of ξ_1 is $P_1(x_1) = 1 - (1-x_1)^n$, $0 \leqq x_1 \leqq 1$. Random numbers subject to this distribution function can be sampled by making use of the inverse of the distribution function $P_1(x_1)$ according to the direct method of random number generation, that is if u is uniformly distributed in $[0,1)$ then the density function of the value $x_1 = 1 - \sqrt[n]{1-u}$ is $p_1(x_1)$. Next the conditional density function of the second component, given the $\xi_1 = x_1$ equation can be expressed as

$$p_2(x_2) = \frac{1}{n(1-x_1)^{n-1}} \int_{\substack{x_3+\ldots+x_n \leqq 1-x_1-x_2 \\ 0 \leq x_2 \leqq 1-x_1}} \cdots \int C\, dx_3 \ldots dx_n =$$

$$= \frac{(n-1)!}{(1-x_1)^{n-1}} V(n-2,\ 1-x_1-x_2)$$

$$= \frac{n-1}{(1-x_1)^{n-1}} (1-x_1-x_2)^{n-2}.$$

Thus the corresponding conditional distribution function of the component ξ_2 is $P_2(x_2 \mid x_1) = 1 - \dfrac{(1-x_1-x_2)^{n-1}}{(1-x_1)^{n-1}}$, $0 \leqq x_2 \leqq 1 - x_1$, which is the usual power distribution $(1-y)^{n-1}$ transformed into the interval $[0, x_1)$. Random variables with distribution function $P_2(x_2 \mid x_1)$ can be generated by inverting the distribution function; expressing the variable x_2 from the equation $P_2(x_2 \mid x_1) = u$ one obtains the formula

$$x_2 = (1-x_1)(1 - \sqrt[n-1]{1-u}),$$

where x_2 is distributed with distribution function P_2 if u is a uniformly distributed random variable.

One can similarly determine the conditional density function $p_i(x_i \mid x_1, \ldots, x_{i-1})$ of the ith coordinate of the vector $(\xi_1, \ldots, \xi_n)$. It is given as

$$(5.4)\quad P_i(x_i \mid x_1, \ldots, x_{i-n}) = \frac{(n-j+1)}{(1-x_1-\ldots-x_{i-1}^{n-i+1})} (1-x_1-\ldots-x_i)^{n-1},$$

$$i = 2, 3, \ldots, n$$

n from which the distribution function can be derived as

$$(5.5) \qquad P_i(x_i \mid x_1, \ldots, x_{i-n}) = 1 - \frac{(1 - x_1 - \ldots - x_i)^{n-i+1}}{(1 - x_1 - \ldots - x_{i-1})^{n-i+1}}.$$

$$i = 2, 3, \ldots, n.$$

By inverting the distribution function P_i one obtains, that the random variable

$$(5.6) \qquad x_1 = (1 - x_1 - \ldots - x_{i-1})(1 - \sqrt[n-i+1]{1-u}), \quad i = 2, 3, \ldots, n$$

is distributed according to the distribution function $P_i(x_i \mid x_1, \ldots, x_{i-1})$, u is uniformly distributed in $[0, 1)$. Thus the following formal algorithm can be developed (where $1-u$ is substituted by the statistically equivalent uniform random number u):

VSI Uniform vector in simplex

1. [Initialize.] Set $i \longleftarrow 1$, generate u, compute $x_1 \longleftarrow 1 - u^{1/n}$, $s \longleftarrow 1 - x_1$.
2. Increase the index $i \longleftarrow i + 1$. If now $i > n$, then go to Step 4.
3. [Generate the ith component according to the density p_i.] Generate u, evaluate $x_i \longleftarrow s[1 - u^{1/(n-i+1)}]$, $s \longleftarrow s - x_i$ and return to Step 2.
4. Deliver the point $(x_1, x_2, \ldots, x_n)$ as a sample from the uniform distribution in simplex $H(n, 1)$.

Theorem 2. The point produced by Algorithm VSI is uniformly distributed in $H(n, 1)$.

Proof. The validity of the theorem is an immediate consequence of the general procedure V1 and the formulas employed, especially those in (5.5) and (5.6).

The procedure is simple enough and amenable to coding. The amount of computations for generating one vector is a linear function of the dimensionality n. As a drawback of the method one should note that powers are to be computed (see (5.6)) with non integer exponents, which is a relatively slow operation on most computers.

5.4. Uniform distribution in and on a hypersphere

Consider the n-dimensional unit hypersphere S and its surface F that can be given as

$$S = \left\{ \mathbf{x} \,\middle|\, \sum_{i=1}^{n} x_i^2 \leq 1 \right\},$$

$$F = \left\{ \mathbf{x} \,\middle|\, \sum_{i=1}^{n} x_i^2 = 1 \right\}.$$

For the sake of simplicity our algorithms will be described for the hypersphere S and its surface F. If uniformly distributed points are needed in (or on) a hypersphere with radius R which has its centre at the point $\mathbf{p}$, then the following transformations are to be carried out:

[Let $\mathbf{x}$ be uniform in S (or on F).] Deliver the point $\mathbf{x}$ with components $x_i \longleftarrow Rx_i + p_i$, $i = 1, \ldots, n$, as a uniform sample.

The random vector ξ which is uniformly distributed in the unit hypersphere S does not have a favoured direction, each direction is equally probable, thus the projection of ξ to the surface F will be uniformly distributed on F. Consequently all the methods constructed to generate uniformly distributed points inside S (interior point algorithms) can be transformed into algorithms generating points uniformly distributed on the surface F (surface algorithms). This can be done by attaching the next transformation step to the interior algorithm.

S5 Subalgorithm. Converting an interior algorithm to surface algorithm

[Let $\mathbf{x}$ be the outcome of an interior algorithm.] Compute the quantity $s \longleftarrow (x_1^2 + \ldots + x_n^2)^{-1/2}$ and deliver the vector $\mathbf{y}$ with components $y_i \longleftarrow sx_i$, $i = 1, \ldots, n$ as uniform on F.

Similarwise surface algorithms can be converted into interior algorithms if the following random radius generating step is appended to the surface algorithm (the validity of doing so will be shown later at the polar method):

S6 Subalgorithm. Converting a surface algorithm to an interior algorithm

[Let $\mathbf{y}$ be the outcome of a surface algorithm.] Generate u and compute $r \longleftarrow u^{1/n}$. Deliver the vector $\mathbf{x}$ composed of components $x_i \longleftarrow ry_i$, $i = 1, \ldots, n$ as a sample from the uniform distribution in S.

Because of these conversion possibilities the surface algorithms will be treated together with the interior algorithms in this section.

The acceptance-rejection technique is applied to the case of a hypersphere, in particular the unit hypersphere S will be enclosed in the cube $K = \{\mathbf{x} \mid \mid x_i \mid \leqq 1,\ i = 1, \dots, n\}$ and out of the points generated uniformly in K only those falling into S will be accepted.

VSA Uniform vector in sphere, acceptance-rejection

1. Generate u_i uniformly in $[0, 1)$ and transform the same into interval $[-1, +1)$: set $x_i \longleftarrow 2u_i - 1,\ i = 1, \dots, n$.
2. [Test.] Compute $s \longleftarrow \sum_{i=1}^{n} x_i^2$. If now $s > 1$, then go back to Step 1 (rejection).
3. [Acceptance.] Deliver the vector $(x_1, \dots, x_n)$ as uniform in the hypersphere S.

The algorithm is simple, easy to code, does not require special library routines. Evidently the probability of accepting a generated vector is the ratio of the volumes of the hypersphere S and the hypercube, namely

$$p_n = \frac{\dfrac{2\pi^{n/2}}{n\Gamma(n/2)}}{2^n} = \frac{\pi^{n/2}}{2^{n-1} n\Gamma(n/2)}. \tag{5.7}$$

Thus the probability of the acceptance (the efficiency of the method) in $n = 2$ dimensions is $p_2 = 0.785$, in $n = 3$ dimensions is $p_3 = 0.523$, in $n = 5$ dimensions is $p_5 = 0.164$ and in $n = 10$ dimensions it shrinks to $p_{10} = 0.0025$. Hence a serious drawback of the algorithm is that the probability of acceptance p_n rapidly decreases with growing number of dimensions n, consequently its use may be recommended in small $n \leqq 5$ dimensions only.

Consider the random variable $\boldsymbol{\xi}$ uniformly distributed in S, then denoting the volume of S by V_S its density function can be given as $f(\mathbf{x}) = \dfrac{1}{V_S}$. Let us use the vector $\boldsymbol{\psi} = (\psi_1, \dots, \psi_{n-1}, r)$ for representing the point $\boldsymbol{\xi}$ by polar coordinates and denote by T the transformation from Cartesian coordinate system to polar coordinates, that is $\boldsymbol{\psi} = T(\boldsymbol{\xi})$. The inverse transformation $\boldsymbol{\xi} = T^{-1}(\boldsymbol{\psi})$ is described by the following equations

$$
\begin{aligned}
\xi_1 &= r\cos\psi_1\cos\psi_2\ldots\cos\psi_{n-2}\cos\psi_{n-1},\\
\xi_2 &= r\cos\psi_1\cos\psi_2\ldots\cos\psi_{n-2}\sin\psi_{n-1},\\
\xi_3 &= r\cos\psi_1\cos\psi_2\ldots\sin\psi_{n-2},\\
&\;\;\vdots\\
\xi_{n-1} &= r\cos\psi_1\sin\psi_2,\\
\xi_n &= r\sin\psi_1,
\end{aligned}
\tag{5.8}
$$

where $-\pi/2 \leqq \psi_i \leqq \pi/2$, $i = 1,\ldots,n-2$, $0 \leqq \psi_{n-1} \leq 2\pi$ and $0 \leqq r \leq 1$. The Jacobian of this transformation can be obtained after a short calculation as

$$
\left|\frac{\partial T^{-1}(\psi)}{\partial\psi}\right| = r^{n-1}\cos^{n-2}\psi_1\cos^{n-3}\psi_2\ldots\cos\psi_{n-2}. \tag{5.9}
$$

Thus according to the second transformation theory in Section 3 of the first chapter the density function of the random vector ψ can be expressed as

$$
\begin{aligned}
g(\psi) &= f(T^{-1}(\psi))\left|\frac{\partial T^{-1}(\psi)}{\partial\psi}\right|\\
&= \frac{1}{V_S}\, r^{n-1}\cos^{n-2}\psi_1\cos^{n-3}\psi_2\ldots\cos\psi_{n-2}.
\end{aligned}
\tag{5.10}
$$

From the expression one can easily see that the components of the vector ψ (the angles and the radius, that is the polar coordinates) are independent, since the density function is decomposed into a product of the functions of the individual coordinates. The density function of the radius length is cr^{n-1}, where c is a suitable constant, the distribution function is $\frac{c}{n}r^n$. This implies that the random radius generation step in the second conversion step described above is functioning properly since the distribution function of $u^{1/n}$ is exactly $\frac{c}{n}\cdot r^n$. Based on the density function (5.10) the following procedure can be given for generating the polar coordinates of a uniformly distributed vector in S.

S7 Subalgorithm. Polar coordinates, preliminary

1. [Random radius length generation.] Generate u and compute $r \longleftarrow u^{1/n}$.

2. Generate u and set $\psi_{n-1} \longleftarrow 2\pi u$.
3. Generate the angles ψ_i, $i = 1, \dots, n-2$ in the interval $[-\pi/2, +\pi/2)$ according to the density function $f_i(x) = c_i \cos^{n-i-1}(x)$, where the constant c_i is given by $1/c_i = \int_{+\pi/2}^{+\pi/2} \cos^{n-i-1}(x)dx$. Compute and deliver the vector $\mathbf{x} \longleftarrow T(\psi)$ according to the formulas in (5.8).

The third step of the algorithm can be modified to obtain a more manageable version. Instead of the angle ψ_i let us consider the random variable $\varepsilon_i = D(\psi_i) = \cos^2 \psi_i$, $i = 1, \dots, n-2$ obtained in a transformation D. According to the transformation theorem its density function $h_i(z)$ can be derived as

$$
\begin{aligned}
h_i(z) &= f_i(D^{-1}(z)) = \left| \frac{\partial D^{-1}(z)}{\partial z} \right| = \\
&= c_i' \cos^{n-i-1}(\text{arc cos}\sqrt{z}) \left| \frac{\partial \text{arc cos}\, z}{\partial z} \right| = \\
&= c_i' z^{(n-i-2)/2}(1-z_i)^{-1/2}
\end{aligned}
$$

with another suitable constant c_i'. This can immediately be recognised as the density function of the beta $((n-i)/2, 1/2)$ distribution. In the final polar algorithm the values $\varepsilon_i = \cos^2 \psi_i$ will be generated according to these beta densities but since eventually what we need are the coordinates expressed in (5.8), thus we will evaluate the quantities $\cos \psi_i = \sqrt{\varepsilon_i}$ and $\sin \psi_i = \sqrt{1-\varepsilon_i}$ as well. The formal algorithm is given below.

VSP Uniform vector in sphere, polar coordinates

1. [Random radius-length generation.] Generate u and compute $r \leftarrow u^{1/n}$.
2. Generate u and set $\psi_{n-1} \leftarrow 2\pi u$, $\varepsilon_{n-1} \leftarrow \cos^2 \psi_{n-1}$.
3. Generate the sample ε_i from the beta $((n-i)/2,\ 1/2)$ distribution for $i = 1, 2, \dots, n-2$.
4. Compute the components $x_1, \dots, x_n$ according to the formulas in (5.8) using $\cos \psi_i = \sqrt{\varepsilon_i}$ and $\sin \psi_i = \sqrt{1-\varepsilon_i}$ and then deliver vector $\mathbf{x}$ as a sample from the uniform distribution in S.

This version of polar method evades the use of functions sinus and cosinus none the less the algorithm did not become simpler due to the $n(n+1)/2$ square root and the same number of multiplications required herein. By the same token the computational work is a quadratic func-

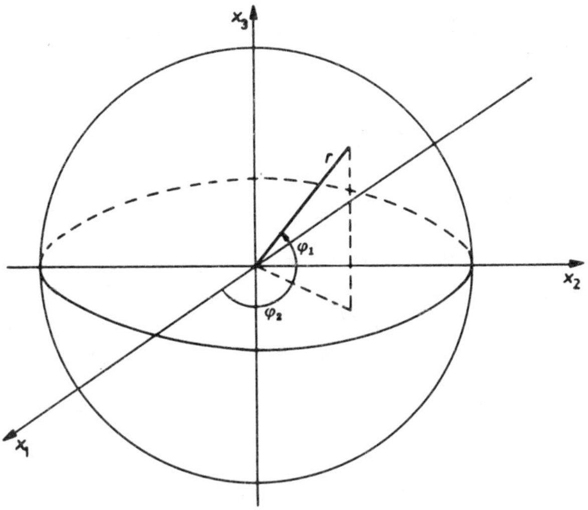

Fig. 29. A point in the sphere with polar coordinates $(r, \varphi_1, \varphi_2)$

tion of the number of dimensions, thus application of the algorithm is recommended in low dimensions, at most up to $n = 10$ dimensions.

Example. The case of $n = 3$ will separately be discussed. The generation of points uniformly distributed in the three-dimensional sphere is required (Fig. 29). Denote the polar coordinates of such a point by (ψ_1, ψ_2, r). One can see from the expression in (5.10) that the distribution function of r is now x^3, it can be inverted, thus $r = \sqrt[3]{u_3}$ has the required properties (u_1, u_2 and u_3 are uniform). The angle ψ_2 is uniform in $[0, 2\pi]$, correspondingly the expression looked for is $\psi_2 = 2\pi u_2$. Finally the density function of the angle ψ_1 is $c_1 \cos x$, where the constant c_1 can be evaluated from the equation $\int_{-\pi/2}^{+\pi/2} c_1 \cos x \, dx = 1$ as $c_1 = \frac{1}{2}$. Hence the distribution function of ψ_1 is $\frac{1}{2} \int_{-\pi/2}^{x} \cos z \, dz = \frac{1}{2} \sin x + \frac{1}{2}$ and this latter can be inverted as well, yielding the expression $\psi_1 = \arcsin(2u_1 - 1)$. Since in the final computations the values $\cos \psi_1$ and $\sin \psi_1$ are required instead of ψ_1, thus we compute $\sin \psi_1 = 2u_1 - 1$ and $\cos \psi_1 = \sqrt{1 - (2u_1 - 1)^2}$ respectively. (The same could be obtained from the beta $(1, 1/2)$ density function which is the density function of $\cos^2 \psi_1$.) Summing up all the details we can state that if u_1, u_2 and u_3

are uniformly distributed in interval $[0,1]$, then using the equations

$$\sin\psi_1 = 2u_1 - 1, \qquad \psi_2 = 2\pi u_2, \qquad r = \sqrt[3]{u_3},$$

we can compute the point $\mathbf{x}$ with (descartesian) coordinates

$$x_1 = r\cos\psi_1\cos\psi_2, \quad x_2 = r\cos\psi_1\sin\psi_2, \quad x_3 = r\sin\psi_1$$

which is uniformly distributed in the three-dimensional unit sphere around the origin.

The last algorithm in this section, which is the most useful in practice is based on the following well-known fact in probability theory. If ξ is an n-dimensional random vector whose components are normally distributed and independent of each other, then the direction determined by this vector is of uniform distribution on the surface of the unit hypersphere (this assertion will be shown to be true at the treatment of the multidimensional normal distribution in Section 5.7). The formal surface algorithm utilizing this fact is the following:

VSN Uniform vector on sphere, normal approach

1. Generate the samples $x_1, \dots, x_n$ from the standard (one-dimensional) normal distribution, independently of each other.
2. Compute the value $s \leftarrow (x_1^2 + \dots + x_n^2)^{-1/2}$ and deliver the vector $\mathbf{x}$ with components $x_i \leftarrow sx_i,\ i = 1, \dots, n$ as a sample from the uniform distribution on F.

The method is simple, it can easily be converted into computer code, the necessary computational work increases only linearly with the increase of the number of dimensions and is the fastest algorithm for $n \geqq 5$ dimensions. Because of its overall good performance it is to be preferred even in lower dimensions over the other methods unless one has specific reasons to do otherwise.

5.5. Uniform distribution in hyperellipsoid

Consider the hyperellipsoid E drawn around the origin and given by the positive definite quadratic matrix $\mathbf{R}$, that is

(5.11) $$E = \{\mathbf{x} \mid \mathbf{x}\mathbf{R}^{-1}\mathbf{x} \leqq 1\}.$$

It is known from linear algebra that the matrix **R** can be decomposed into a product of a lower triangular matrix **T** (all the element above the diagonal are equal to zero) and of its transposed, that is we can write

$$\mathbf{R} = \mathbf{T}\mathbf{T}' \tag{5.12}$$

for a matrix **T**. From the matrix **R** we can obtain matrix **T** through the following equations (Cholesky transformation)

$$\begin{aligned} t_{ij} &= 0, \qquad \text{if } i < j, \\ t_{i1} &= r_{i1}/\sqrt{r_{11}}, \qquad \text{if } 1 \leqq i \leqq n, \\ t_{ii} &= \sqrt{r_{ii} - \sum_{k=1}^{i-1} t_{ik}^2} \qquad \text{if } 1 < i \leqq n, \\ t_{ij} &= \left[r_{ij} - \sum_{k=1}^{j-1} t_{ik}t_{jk} \right] / t_{jj}\,, \qquad \text{if } 1 < j < 1 \leqq n. \end{aligned} \tag{5.13}$$

The ellipsoid E can be obtained from the unit sphere $S = \left\{ \mathbf{x} \mid \sum_{i=1}^{n} x_i^2 \leqq 1 \right\}$ by transforming the latter one with matrix **T**, since introducing the notation $\mathbf{y} = \mathbf{T}\mathbf{x}$ one can obtain

$$1 \geqq \mathbf{x}'\mathbf{x} = (\mathbf{T}^{-1}\mathbf{y})'(\mathbf{T}^{-1}\mathbf{y}) = \mathbf{y}'(\mathbf{T}\mathbf{T}')^{-1}\mathbf{y} = \mathbf{y}\mathbf{R}^{-1}\mathbf{y},$$

which means $\mathbf{y} \in E$ if $\mathbf{x} \in S$. Since the linear transformation (multiplication by **T**) does not deform the uniform distribution, the following algorithm is sound for producing vectors uniformly distributed in E.

VEN Uniform vector in ellipsoid, normal approach

1. Generate u and compute $r \longleftarrow u^{1/n}$.
2. [Sampling a uniform point in a sphere.] Generate the numbers $x_1, \ldots, x_n$ from the standard normal $N(0,1)$ distribution, independently of each other. Compute the multiplicator $s \longleftarrow (x_1^2 + \ldots + x_n^2)^{-1/2}$ and determine the vector **y** with components $y_i \longleftarrow srx_i$, $i = 1, \ldots, n$.
3. Deliver the vector $\mathbf{y} \longleftarrow \mathbf{T}\mathbf{y}$ as uniform in E.

It goes without saying that for generating a point uniformly in a sphere S in the above algorithm VEN we could use other methods as well not only the normal approach.

It is appropriate now to make a definition that will be used at discussing the generation of normally distributed random vectors.

Definition. If the random vector η is uniformly distributed on the surface F of the unit hypersphere S, then the vector $\mathbf{T}\eta$ is called uniformly distributed with respect to the volume of the ellipsoid E.

This definition is implied by the fact that the probability of the vector $\mathbf{T}\eta$ falling into a specified area on the surface of E is proportional to the volume from the origin to this area. A method of generating vectors with this distribution is an immediate consequence of the definition: the outcome of a sphere-surface algorithm has to be multiplied by $\mathbf{T}$.

Note here, that the direction given by a vector ξ uniformly distributed in the ellipsoid E (that is the projection of ξ to the surface of E) has precisely the distribution in the above formulated definition; it is uniform with respect to the volume of E. Vice versa, if the vector $\mathbf{q}$ is uniformly distributed with respect to the volume of E, then the vector $u^{1/n}\mathbf{q}$ is uniformly distributed in the ellipsoid E if u is uniform in $[0,1)$, since

$$u^{1/n}\mathbf{q} = u^{1/n}\mathbf{T}\eta = \mathbf{T}(u^{1/n}\eta),$$

where η is a vector uniformly distributed on the surface F of the sphere S.

Finally we should remark that the direction of a random vector which is distributed according to a normal distribution with correlation matrix $\mathbf{R}$ is uniformly distributed with respect to the volume of the ellipsoid E.

5.6. Chord method for generating points in hyperellipsoid

A procedure is outlined in this section which generates points with an approximate uniform distribution in hyperellipsoid.

Let us denote by S_r the n-dimensional hypersphere with radius r and with centre at the origin. Consider a random chord whose terminal points are Q_1 and Q_2 uniformly distributed on the surface of S_r. Then we have the following:

Theorem 1. Let h be the length of the random chord Q_1Q_2 then the density function of h is

$$f(h) = \frac{\Gamma(n/2)}{2^{n-3}r^{2n-4}\Gamma((n-1)/2)\pi^{1/2}} h^{n-2}(4r^2h^2)^{(n-3)/2}. \tag{5.14}$$

Making use of this theorem one can obtain the following result by straightforward integration

Theorem 2. The expected value of $\sigma_n = h^{4-n}$ is $h_0 = E\{h^{4-n}\} = = \frac{4}{n} r^{4-n}$.

The theoretical justification of the chord method is given in:

Theorem 3. Consider a random chord in S_r with length h. Generate uniformly distributed random points R_i on this chord, the number of points is a random variable $\nu = h^{4-n}/\lambda$, where λ is a constant. Let S_ρ be another n-dimensional sphere in S_r with the same centre and with radius ρ, $\rho \leqq r$. Then the expected number of points R_i lying in S_ρ is proportional to ρ^n, (the volume of S_ρ) that is

$$E\{\mu\} = \rho^n C(n, r, \lambda),$$

where $C(n, r, \lambda)$ is a constant independent of ρ and μ is the number of points in S_ρ, $\mu = \#\{R_i \mid R_i \in S_\rho\}$.

Proof. First we make some remarks. A chord in S_r with length $h < h_\rho$, where $h_\rho = 2(r^2 - \rho^2)^{1/2}$ does not intersect the hypersphere S_ρ. A chord in S_r with length $h \geqq h_\rho$ has an interval of length $(h^2 - h_\rho^2)^{1/2}$ in S_ρ. The probability that a point generated uniformly on the chord with length h is in S_ρ can be given as $\frac{(h^2 - h_\rho^2)^{1/2}}{h}$. Thus the expected number of points in S_ρ equals

$$E(\mu) = \int_{h_\rho}^{2r} f(h) \frac{h^{4-n}}{\lambda} \cdot \frac{(h^2 - h_\rho^2)^{1/2}}{h} \, \mathrm{d}h,$$

where h^{4-n}/λ is the number of the points to be generated on the chord with length h. Making use of the density function in (5.14), introducing a new variable $y = (4r^2 - h^2)^{1/2}$, then another variable α by the relation $\sin\alpha = y/2\rho$ one can obtain

$$E(\mu) = C_f \frac{1}{\lambda} \int_{h_\rho}^{2r} h(4r^2 - h^2)^{(n-3)/2} (h^2 - h_\rho^2)^{1/2} \, \mathrm{d}h =$$

$$= C_f \frac{1}{\lambda} \int_0^{2\rho} y^{n-2} (4\rho^2 - y^2)^{1/2} \, \mathrm{d}y = C_f \frac{(2\rho)^h}{\lambda} \int_0^{\pi/2} \cos^2\alpha \sin^{n-2}\alpha \, \mathrm{d}\alpha,$$

where C_f denotes the constant part in (5.14). This equation proves the assertion of the theorem.

Basically the chord method generates chords and then a number of uniformly distributed points on each chord. Thus the method is approximative because of two reasons. First, points generated on the same chord are correlated, the outcomes of the procedure are not independent samples from the uniform distribution in the hypersphere. (One can compare this situation to the one-dimensional congruential generators, that produce points in a hypercube, that lie on hyperplanes.) Second the chords as produced and used in the chord method are not random because of making multiple use of the terminal points of previous chords.

The formal description of the chord method is given for the case of the unit sphere S_1 and includes an initialization step as well.

Preparation

0. Generate $\mathbf{y}^{(1)}, \ldots, \mathbf{y}^{(m)}$ the so-called basic points (the points $\mathbf{y}^{(i)}, \mathbf{y}^{(j)}$ span the (i, j)th chord) uniformly distributed on the surface of S_1. Initialize the counters $c \leftarrow 0,\ i \leftarrow 1,\ n_c = 1,\ s \leftarrow h_0 \leftarrow 4/n$.

VSC Uniform vector in sphere, chord method

1. If $i = 0$, then go to 4.
2. [Compute the length of the (i, m)th chord.] Set $w_j \leftarrow y_j^{(i)} - y_j^{(m)}$, $j = 1, \ldots, n$ and $c \leftarrow 1$.
3. Compute $S \leftarrow (w_1^2 + \ldots + w_n^2)^{(4-n)/2}$.
4. If $S < s$ then go to 6.
5. [Generate a "sure" point on the (i, m)th chord.] Set ind $\leftarrow 1$, generate u, compute $x_j \leftarrow y_j^{(m)} + uw_j,\ j = 1, \ldots, n$ and decrease $S \leftarrow S - s$. Go to Step 13.
6. Generate u and set $v \leftarrow su$.
7. [Test for an "additional" point on the (i, m)th chord.] If $S < v$, then set ind $\leftarrow 0$ and go to Step 9.
8. [Generate the "additional" point on the (i, m)th chord.] Generate u, set ind $\leftarrow 1$ and $x_j \leftarrow y_j^{(m)} + uw_j,\ j = 1, \ldots, n$.
9. [Change the current chord to $(i+1, m)$th.] If $i \neq m - 1$, then change $i \leftarrow i + 1,\ c \leftarrow 0$ and go to Step 12.
10. [Generate a new basic point if necessary.] Generate $\mathbf{z}$ uniformly on the surface of S_1.
11. If $n_i = m - 1$, then reset $n_i \leftarrow 0$.
12. [Initialize for the changed system of basic points.] Set $n_i \leftarrow n_{i+1}$, $i \leftarrow 1$, $\mathbf{y}^{(n_c)} \leftarrow \mathbf{y}^{(m)}$, $\mathbf{y}^{(m)} \leftarrow \mathbf{z}$. If ind $= 0$ then go back to Step 2.
13. Deliver $\mathbf{x}$ as a uniformly distributed point in S_1.

The algorithm VSC generates points uniformly on the chord between the basic points $\mathbf{y}^{(1)}$ and $\mathbf{y}^{(m)}$, then on the $(2, m)$th chord, etc. After $m-1$ chords have been considered, one of the basic points is replaced by $\mathbf{y}^{(m)}$ and a new point is generated in place of $\mathbf{y}^{(m)}$. The variable n_i shows the index of the basic point to be replaced by $\mathbf{y}^{(m)}$, $c = 1$ if the length of the chord is already computed ($= 0$ otherwise), ind $= 0$ if no point is generated yet (this is necessary to be checked since it might happen that no point was generated on several consecutive chords).

In a computer implementation the following values can be proposed for the parameters: $m = 30$ basic points may be used, the choice of $s = h_0$ (in Theorem 3 the same quantity was denoted by λ) ensures that on the average one point is generated on each chord, but $s = h_0/4$ is also possible.

Implemented versions of the algorithm show significant instability in dimensions $n > 15$ the reason of which is the following. The random variable h^{4-n} (where h is a random chord length which determines the number of points to be generated has infinite variance for $n > 4$. Roughly speaking we can say that chords with small length occur rarely but they have great proportions of all the points generated. For example in $n = 15$ dimensions more than half of the points were to be generated on chords, where more than 32 points were produced and more than 10 percent of the points were to be generated on chords with more than 3×10^4 points. To improve this deficiency a random order in use of the generated points may be desirable similarly to the one-dimensional case, that is we generate M points with the VSC chord method, store the same in a matrix and a randomly chosen point will actually be used (see Section 2.7 for the description of the shuffling).

The modification of the chord method to generate uniform points in a hyperellipsoid is quite straightforward. Denote by $\mathbf{T}$ the triangular matrix for which $\mathbf{TT}' = \mathbf{R}$. The points generated by the chord method are of the form $\mathbf{y} = \alpha\mathbf{y}^{(i)} + (1-\alpha)\mathbf{y}^{(m)}$, with some constant α thus the point $\mathbf{Ty} \in E$ can be expressed as

$$\mathbf{Ty} = \alpha\mathbf{Ty}^{(i)} + (1-\alpha)\mathbf{Ty}^{(m)}.$$

Consequently it is sufficient to transform the basic points $\mathbf{y}^{(i)}$, $i = 1, \ldots, m$ instead of transforming each generated point $\mathbf{y}$. The following changes should be made in algorithm VSC in order to yield an ellipsoid algorithm:

As preparation the points $\mathbf{y}^{(i)*} = \mathbf{T}\mathbf{y}^{(i)}$, $i = 1, \ldots, m$ have to be computed and stored. Steps 2, 5, 8, 10 and 12 in Algorithm VSC should be replaced by the corresponding modified versions:

VEC Uniform vector in ellipsoid, chord method (changes in VSC)

2*. Set $w_j \leftarrow y_j^{(i)} - y_j^{(m)}$, $w_j^* \leftarrow y_j^{(i)*} - y_j^{(m)*}$, $j = 1, \ldots,$ $c \leftarrow 1$.

5*. Generate u, set $x_j \longleftarrow y_j^{(m)*} + uw_j^*$, $j = 1, \ldots, n$, $\text{ind} \longleftarrow 1$, $S \leftarrow S - s$ and go to 13.

8*. Generate u, set $x_j \leftarrow y_j^{(m)*} + uw_j^*$, $j = 1, \ldots, n$.

10*. Generate a point $\mathbf{z}$ uniformly on the surface of S_1 and compute $\mathbf{z}^* \leftarrow \mathbf{T}\mathbf{z}$.

12*. Set $n_i \leftarrow n_c + 1$, $i \leftarrow 1$, $\mathbf{y}^{(n_c)} \leftarrow \mathbf{y}^{(m)}$, $\mathbf{y}^{(n_c)*} \leftarrow \mathbf{y}^{(m)*}$, $\mathbf{y}^{(m)} \leftarrow \mathbf{z}$, $\mathbf{y}^{(m)*} \leftarrow \mathbf{z}^*$. If $\text{ind} = 0$ then go back to Step 2.

The additional programming effort (as compared to the normal approach in VEN), due to the complexity of the chord method, the greater storage requirements and the approximate character of the chord method are compensated by its speed. In computer tests the VSC algorithm is of the same speed as the VSN in case of the choice $\lambda = h_0$ (one point is generated on each chord on the average), and VEC is from two to three times faster than VEN in dimensions $n = 5 - 15$ if $\lambda = h_0/4$ was chosen.

On the other side because of the relatively sophisticated initialization procedure and set up cost, the speed properties and the instability of samples our recommendations are that VEC is to be used only when a great sample size is required (10^5 or so), VSN is always to be preferred to VSC otherwise.

5.7. Normal distribution

Consider the n-dimensional normal distribution with expected value $\mathbf{O}$ and with correlation matrix $\mathbf{R}$, its density function is

$$\varphi(\mathbf{x}) = (2\pi)^{-n/2} \, |R|^{-1/2} \exp\Big\{-\frac{1}{2}\mathbf{x}'\mathbf{R}^{-1}\mathbf{x}\Big\}, \tag{5.15}$$

where the matrix $\mathbf{R}$ is positive definite and symmetric. Let $\mathbf{T}$ be a matrix elaborated upon in the previous section, for which $\mathbf{R} = \mathbf{T}\mathbf{T}'$. The following procedure generates random vector variables with density function φ.

VNT Normal vector, matrix transformation

1. Generate the numbers x_i, $i = 1, \ldots, n$ from the one-dimensional standard normal distribution, independently of each other.
2. Compute the vector $\mathbf{y} \leftarrow \mathbf{Tx}$, and deliver the same as a sample with density function φ.

Theorem 1. The vector $\mathbf{y}$ produced by algorithm VNT has density function φ.

Proof. Let the density functions of the random variables x_i be denoted by $f_i(z_i) = (2\pi)^{-1/2} \exp\{-\frac{1}{2} z_i^2\}$, $i = 1, \ldots, n$. Then the density function of the vector $\mathbf{x} = (x_1, \ldots, x_n)$ is their product

$$g(\mathbf{z}) = \prod_{i=1}^{n} f_i(z_i) = (2\pi)^{-n/2} \exp\Big\{-\frac{1}{2}(z_1^2 + \ldots + z_n^2)\Big\}.$$

The vector $\mathbf{y}$ is obtained from vector $\mathbf{x}$ via the transformation $\mathbf{y} = \mathbf{Tx}$ (or we can write $\mathbf{x} = T^{-1}\mathbf{y}$). According to the transformation theorem on densities of vectors one can obtain the density function of the random vector $\mathbf{y} = \mathbf{Tx}$ as

$$h(\mathbf{s}) = g(T^{-1}\mathbf{s}) \left| \frac{\partial T^{-1}\mathbf{s}}{\partial \mathbf{s}} \right| = (2\pi)^{-n/2} \exp\Big\{-\frac{1}{2}(T^{-1}\mathbf{s})'(T^{-1}\mathbf{s})\Big\} \left|T^{-1}\right|,$$

$$= (2\pi)^{-n/2} |R|^{-1/2} \exp\Big\{-\frac{1}{2}\mathbf{s}' R^{-1} \mathbf{s}\Big\} = \varphi(\mathbf{s}).$$

We can produce normally distributed vectors according to the general method V1 as well, the conditional distribution functions are all normal distribution functions. Compared to the VNT transformation method this latter method is more sophisticated and a little slower, partly because of the need to determine the necessary conditional densities.

Here a result will be proven that validates the algorithm VSN producing points on the surface of a sphere.

Theorem 2. Let $\boldsymbol{\xi}$ be a random vector, whose components are independent normally distributed (**0** mean, **1** variance) random variables and denote $\sigma^2 = \xi_1^2 + \ldots + \xi_n^2$. The direction $\eta = (\xi_1/\sigma, \ldots, \xi_n/\sigma)$ defined by $\boldsymbol{\xi}$ on the surface F of the unit sphere is uniformly distributed on F.

Proof. One can obtain similarly to the proof of the previous theorem that the density function of the random vector ξ is

$$(5.16) \qquad g(z) = (2\pi)^{-n/2} \exp\left\{-\frac{1}{2}(z_1^2 + \ldots + z_n^2)\right\}.$$

Now the Cartesian coordinates are replaced by polar ones using transformation (5.8). From (5.16) making use of the transformation theorem and the Jacobian (5.9) of the polar transformation (5.8) the density function of the random vector $\boldsymbol{\psi} = (\psi_n, \ldots, \psi_{n-1}, r)$ can be obtained as

(5.17)
$$h(\psi_1, \ldots, \psi_{n-1}, r) = (2\pi)^{-n/2} \exp\left\{-\frac{1}{2}r^2\right\} r^{n-1} \cos^{n-2}\psi_1 \ldots \cos\psi_{n-2}.$$

Since in this density function h the densities of the angles $\psi_n, \ldots, \psi_{n-1}$ are identical with those in the density function (5.10), the theorem is proven.

Note that the length r of the normal vector with density function φ in (5.15) is a random variable with density function $c \cdot r^{n-1} \exp\{-\frac{1}{2}r^2\}$ from equation (5.17) for some constant c, that is r is χ distributed with n degree of freedom.

Consequently the following multidimensional polar algorithm, which generates normally distributed vectors with density function φ in (5.15) is correct.

VNP Normal vector, polar transformation

1. [Generation of direction.] Generate $\mathbf{y}$ from the uniform distribution on the surface F of the unit hypersphere S.
2. [Generation of length.] Generate a sample r from the χ^2 distribution with n degree of freedom.
3. [Transformation.] Compute $\mathbf{x} \leftarrow \sqrt{r}\mathbf{T}\mathbf{y}$ and deliver $\mathbf{x}$ as a sample with density φ.

The matrix $\mathbf{T}$ in the last step of the algorithm is defined by the equation $\mathbf{R}^{-1} = \mathbf{T}\mathbf{T}'$. This algorithm VNP is basically of the same speed as the matrixtransformation method VNT. The VNP algorithm of the polar transformation can be rendered faster only in the case we could generate vectors uniformly distributed with respect to the volume of the ellipsoid $E = \{\mathbf{x} \mid \mathbf{x}'R^{-1}\mathbf{x} \leqq 1\}$ faster than the known procedures (presently the fastest exact method to do so is the sphere-surface algorithm VSN plus the matrix multiplication by T; these two add up to the same time what is required by VNT essentially).

There is a possibility of fast generation of stationary normal vectors, though, via the multidimensional polar method. Below a method is described for generating stationary uniform vectors in a hyperellipsoid; when supplemented by a length generation step of the polar method a generator of stationary normal vectors is obtained.

The proposed method will first be described for the case of a hypersphere next it is modified for generating vectors uniformly distributed with respect to the volume of the hyperellipsoid. The main features of the algorithm are the following. As a preparation a set of vectors $\mathbf{y}^{(1)}, \ldots, \mathbf{y}^{(m)}$ is generated by the VSN sphere algorithm on the surface F of the unit sphere S, where m is arbitrary but fixed. One of these vectors plays a distinguished role in the generation, it will be called the pivot vector and denoted by $\mathbf{b}$, while the other members of the set are the so called basic vectors. The role of the pivot vector is shifting from one basic vector to the other during the course of generation. At the very beginning let $\mathbf{y}^{(1)}$ be the pivot vector. Making use of the pivot vector and the remaining $m-1$ basic vectors, $m-1$ new vectors will be produced by some rules, these vectors will be the looked for generated vectors. Next, the vectors are "refreshed": a vector from the basic vectors is abandoned and a replacement vector is generated by algorithm VNT independently of the previous vectors. Using this new vector as pivot, the procedure is repeated. The formal description of the algorithm is the following.

Preparation

0. [Preparation.] Initialize the index of the pivot vector $q \leftarrow 1$, the index of the basic vector $i \leftarrow 1$, which is presently of interest and generate the vectors $\mathbf{y}^{(1)}, \ldots, \mathbf{y}^{(m)}$ uniformly on F by VNT. Set $\mathbf{b} \leftarrow \mathbf{y}^{(q)}$.

VSS Uniform vector in sphere, stationary

1. [Test of indices.] Increase $i \leftarrow i+1$, if $i = m+1$, then reset $i \leftarrow 1$.
2. [Test: have all the basic points been used up?] If $i = q$, then go to Step 4.
3. [Generate a point $\mathbf{y}$ by making use of the basic vector $\mathbf{y}^{(i)}$ and the pivot vector $\mathbf{b}$, that is produce $\mathbf{y} = c_1\mathbf{y}^{(1)} + c_2\mathbf{b}$ for some constants c_1, c_2.] Compute $s^2 \leftarrow \sum_{j=1}^{n} (y_j^{(i)} - b_j)^2$, $c \leftarrow 1/(4-s^2)^{1/2}$, and evaluate $\mathbf{y} \leftarrow c(y^{(i)} + \mathbf{b})$. Go to Step 5.

4. [Change of basic point; generation of a new pivot vector.] Increase $q \leftarrow q+1$, if $q = m+1$, then reset $q \leftarrow 1$. Initialize $i \leftarrow q$ and generate a new uniformly distributed vector $\mathbf{z}$ in place of $\mathbf{y}^{(q)}$, as a new pivot vector, that is $\mathbf{y}^{(q)} \leftarrow \mathbf{z}$, $\mathbf{b} \leftarrow \mathbf{y}^{(q)}$. Go back to Step 1.
5. Deliver the vector $\mathbf{y}$ as uniform on F.

The vectors $\mathbf{y}$ produced by the above procedure are not independent but we state without proof the next:

Theorem 3. The outcomes of algorithm VSS are forming a stationary process.

The above procedure computes the halving point of the vectors $\mathbf{y}^{(i)}$ and $\mathbf{b}$ in Step 3, other deterministic and nondeterministic possibilities as well, exist for determining the constant, c_1 and c_2 in Step 3. Stationarity of the process holds because of the rule of changing the pivot vector. The generated point $\mathbf{y}$ is the linear combination of two basic vectors, $y = c_1\mathbf{y}^{(i)} + c_2\mathbf{y}^{(q)}$, thus $\mathbf{Ty} = c_1\mathbf{Ty}^{(i)} + c_2\mathbf{Ty}^{(q)}$. This in turn means that production of points on the surface of the hyperellipsoid E can be done in the same economical way as was the case with the Chord method. In particular we need an inventory of the *transformed* basic points $\mathbf{Ty}(i)$ and to obtain the transformed generated points we do not multiply $\mathbf{y}$ by $\mathbf{T}$ but compute the linear combination of the transformed basic points.

When independence is not strictly required, then the use of stationary uniform or normal vectors may be preferred, especially if one takes into account that the stationary generators are three to seven times faster than the VNT generator in dimensions n =20–50.

We close this section with an application of a variant of the ratio of uniform method to the multinormal distribution.

Introduce some notation; let $t(v_0, v_1, \ldots, v_n) = \left(\frac{v_1}{\sqrt{v_0}}, \ldots, \frac{v_n}{\sqrt{v_0}}\right)$ and $\gamma(v_0, v_1, \ldots, v_n) = \gamma(v_0, \mathbf{v}) = \log v_0 + \frac{1}{n+2} \cdot \frac{1}{v_0}\mathbf{v}' R^{-1}\mathbf{v}$ and define the domain $D \in R^{n+1}$ as follows $D = \{(v_0, v_1, \ldots, v_n) \mid \gamma(v_0, \mathbf{v}) \leqq 0\}$. Then we have

Theorem 4. If $(v_0, v_1, \ldots, v_n)$ is uniformly distributed in D then $\mathbf{x} = t(v_0, v_1, \ldots, v_n)$ is distributed with density function $\varphi(\mathbf{x})$.

To render a practical algorithm we need to determine the smallest possible rectangle entailing the domain D. This can be obtained using

Lagrangian multipliers and solving the optimum problem as

$$-\sqrt{\frac{n+2}{e}\sigma_{ii}} \leqq 0 \leq v_0 \leqq 1 \leqq v_i \leqq +\sqrt{\frac{n+2}{e}\sigma_{ii}}, \quad i = 1, 2, \ldots, n.$$

5.8. Decomposition method for normal distribution

The decomposition method proved to be extremely useful in constructing one-dimensional random number generators, in fact all fast generators are based on decomposition principle. Here we discuss the possible implementation of the decomposition technique to n-dimensional cases through an example where a decomposition type method is given for multidimensional normal distribution.

Similarly to the decomposition procedure established in one dimension we try to present the density function $\varphi(\mathbf{x})$ of the n-dimensional normal distribution in (5.15) as a mixture of densities $e_i(\mathbf{x})$, $i = 1, \ldots, k$, $r_i(\mathbf{x})$, $r_2(\mathbf{x})$ to be specified later. Let us consider the equation

$$\varphi(\mathbf{x}) = p_1 e_1(\mathbf{x}) + \ldots + p_k e_k(\mathbf{x}) + p_{k+1} r_1(\mathbf{x}) + p_{k+2} r_2(\mathbf{x}), \tag{5.18}$$

where $p_i \geqq 0$, $i = 1, \ldots, k+2$, $p_1 + \ldots + p_{k+2} = 1$. A very important but hardly answerable question in practice is how to choose the density functions on the right-hand side of equation (5.18) in order to ensure that (i) the generation is done according to a simple density function most of the time, (ii) generation of random samples according to the other (remaining) densities is also possible with not too great difficulty, (iii) the handling of the decomposition and the generation is computationally feasible.

The trivial extension of the one-dimensional case to n-dimensions, that is to divide the domain of definition of the density function $\varphi(\mathbf{x})$ into small subdomains and defining uniform distributions in these subdomains does not work, as one can conclude from the following reasoning. Consider the truncated variant of the normal density $\varphi(\mathbf{x})$, which is defined on the (bounded) cube $K = \{\mathbf{x} \mid |x_i| \leqq 5, \ i = 1, \ldots, n\}$ as

$$\varphi'(\mathbf{x}) = \begin{cases} c\varphi(\mathbf{x}), & \mathbf{x} \in K, \\ 0, & \text{otherwise} \end{cases}$$

where c is a constant ensuring $\int_{R^n} \varphi'(\mathbf{x})\,\mathrm{d}\mathbf{x} = 1$. If each interval of interpretation on a coordinate axis was divided into ℓ points (choose $\ell \sim 10$) then the number of n-dimensional subcubes would be ℓ^n. To store the data describing the cubes, the corresponding probabilities, generate random samples with this discrete density is a formidable task even in $n = 5$ dimensions. And this would yield only a rough approximation to the modified density function $\varphi'(\mathbf{x})$.

Our proposal is a little bit different from this straightforward generalization: instead of the above described "vertical" decomposition consider the following "horizontal" decomposition. Split up the interval of the values of the density function $\varphi(\mathbf{x})$ into k part (Fig. 30). Introduce some notations. Let

$$E_i = \{\mathbf{x} \mid \varphi(\mathbf{x}) \geqq m_i\},\ i = 1, \ldots, k$$

be a set of hyperellipsoid, where

$$(5.19) \qquad 0 < m_1 < \ldots < m_k < m_{k+1} = \varphi(\mathbf{0})$$

is a division of the interval $[0, \varphi(\mathbf{0})]$. The sets E_i, $i = 1, \ldots, k$ are similar hyperellipsoids, their common center is the origin since we have $E_i = \{\mathbf{x} \mid \varphi(\mathbf{x}) \geqq m_i\} = \left\{\mathbf{x} \mid \mathbf{x}'R^{-1}\mathbf{x} \leqq -2\ell n[m_i(2\pi)^{n/2}\,|R|^{1/2}]\right\}$, where R is positive definite.

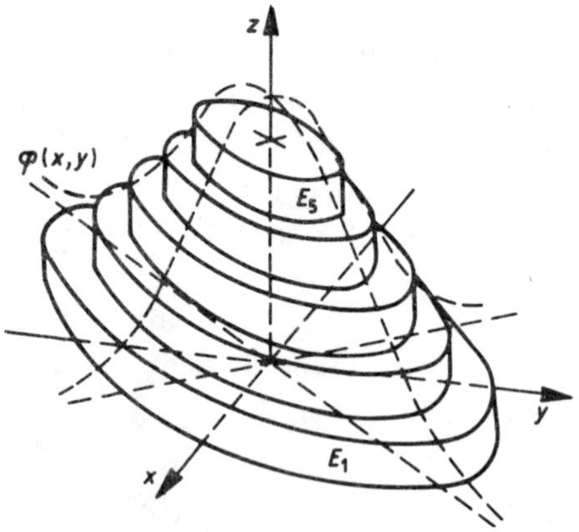

Fig. 30. Decomposition of the multidimensional normal density function

Now define the function e_i in the hyperellipsoid E_i to be the density function of the uniform distribution in E_i, that is let

$$e_i(\mathbf{x}) = \begin{cases} 1/V_i, & x \in E_i, \\ 0, & \text{otherwise}, \end{cases}$$

where V_i is the volume of the hyperellipsoid E_i, $V_i = \int_{E_i} d\mathbf{x}$. The probabilities p_i can be obtained from the equations

$$p_i = V_i(m_i - m_{i-1}), \quad i = 1, \ldots, k$$

since this choice will ensure that the equality $p_1 e_1(\mathbf{x}) + \ldots + p_i e_i(\mathbf{x}) = m_i$ holds on the surface of E_i, that is if $\mathbf{x} \in \{\mathbf{x} \mid \varphi(\mathbf{x}) = m_i\}$.

An important feature of this decomposition that the data describing the ellipsoids E_i are easy to store. Define the multiplicators μ_i by the following equations

$$\mu_i = \frac{c_i}{c_1}$$

where c_i is the constant appearing in the definition of the hyperellipsoid E_i, that is $c_i^2 = -2\ell n[m_i(2\pi)^{n-2}\,|R|^{1/2}]$. Making use of these multiplicators one can write

$$E_i = \{\mathbf{x} \mid \mathbf{x} = \mu_i \mathbf{y}, \ \mathbf{y} \in E_1\}, \quad i = 1, \ldots, k,$$

consequently the hyperellipsoids can be specified by the hyperellipsoid E_1 and the multiplicators μ_i. Thus instead of the selecting with probability p_i the density function e_i — what we need in the generation procedure based on the decomposition (5.18) — equivalently we can consider the random choice of the scalar multiplicators μ_i that is done with probability p_i.

The functions $p_{k+1}r_1(\mathbf{x})$ and $p_{k+2}r_2(\mathbf{x})$ are determined to ensure the soundness of decompositional equation (5.18): the former will be the wedge of the density function $\varphi(\mathbf{x})$, the latter will be the tail. More precisely define the functions

$$p_{k+1}r_1(x) = \begin{cases} \varphi(\mathbf{x}) - \sum\limits_{i=1}^{n} p_i e_i(\mathbf{x}), & \text{if } \mathbf{x} \in E_1 \\ 0, & \text{otherwise,} \end{cases} \tag{5.20}$$

$$p_{k+2}r_2(\mathbf{x}) = \begin{cases} 0, & \text{if } \mathbf{x} \in E_1, \\ \varphi(\mathbf{x}), & \text{if } \mathbf{x} \notin E_1. \end{cases} \tag{5.21}$$

The constant p_{k+2} can be determined from the condition that the function $r_2(\mathbf{x})$ is a density function, thus

$$p_{k+2} = \int\limits_{R^n - E_1} \varphi(\mathbf{x})\, dx = \int\limits_{c_1^2}^{+\infty} k_n(z)\, dz$$

where $k_n(z)$ is the density function of the χ^2 distribution with n degree of freedom. The last term on the right-hand side is the easily available residual probability of a χ^2 distribution. The last probability p_{k+1} can be determined from the equation of $\sum_{i=1}^{k+2} p_i = 1$ as

$$p_{k+1} = 1 - p_{k+2} - \sum_{i=1}^{n} p_i.$$

We now outline the ellipsoid algorithm based on decomposition (5.17) generating normally distributed random vectors.

VNE Normal vector, ellipsoid decomposition

1. [Selecting one of the densities e_i.] Generate u. If $u > \sum_{i=1}^{k} p_i$, then proceed to Step 4, otherwise choose one of the indices i according to the probabilities p_i, $i = 1, \ldots, k$, that is select a multiplicator μ_i.
2. [Generate a point with density e_i.] Generate a uniformly distributed vector $\mathbf{z}$ in the hyperellipsoid E_1.
3. Compute the vector $\mathbf{x} \leftarrow \mu_i \mathbf{z}$ and advance to Step 6.
4. [Generate a vector from the wedge of the distribution.] If $u > 1 - p_{k+2}$, then go to Step 5, otherwise generate a vector $\mathbf{x}$ according to the density function r_1 and go to Step 6.
5. [Generate a vector from the tail of the distribution.] Generate a point uniformly in E_1, project this point to the surface of E_1, the projection will be denoted by $\mathbf{z}$ (its direction is uniform with respect to the volume of E_1). Generate a random number q from the χ distribution with n degree of freedom, which is greater than the constant c_1 and compute $\mathbf{x} \leftarrow (q/c_1)\mathbf{z}$.
6. Deliver the vector $\mathbf{x}$ as a sample with density function $\varphi(\mathbf{x})$.

The algorithm VNE is sketchy in several respects. In order to build an algorithm utilizable in practice a fast procedure (table method or economical technique) should be inserted for sampling according to the discrete distribution p_i in the first step, — generally the number k of possible values is great. The acceptance-rejection technique applied for generating samples with density function r_1 has to be detailed because of the tangled nature of $r_1(\mathbf{x})$, similarly the tail-sampling procedure according to $r_2(\mathbf{x})$ should be elaborated upon, though it is simpler than the former. Finally, a point very important in the computer implementation and numerical efficiency is the problem of the type of decomposition we apply to split up the interval $[0, \varphi(\mathbf{0})]$. Particularly the sum $p_1 + \ldots + p_k$ depends in an essential way on the system of dividing points $m_1, \ldots, m_k$; the sum should be near to 1 and the value of k that determines the quantity of datas to be stored should be small which are partially contradictory desires. We can proceed with equal partitioning that defines $m_i = \frac{\varphi(\mathbf{0})}{k} \cdot i$ but it yields unsatisfactory outcomes even in $n = 4$ dimensions. The most suitable procedure proved to be the following adaptive technique shaped up in lengthy computer experimentations. The idea was to keep the probabilities of approximately equal magnitude. The partitioning was started with a fixed steplength from the $\varphi(\mathbf{0})$ terminal point of the interval $[0, \varphi(\mathbf{0})]$, until the corresponding probabilities did not exceed a certain limit, then the steplength was decreased to its tenth and we kept advancing. The resulting division had much denser located points near to 0 than near to $\varphi(\mathbf{0})$; the disproportion between the different ends of the interval was increasing with increasing the number of dimensions (because of the greater volume of the n-dimensional hyperellipsoids). Using this procedure ensured up to $n = 20$ dimensions, that the probability $p_1 + \ldots + p_k$ was greater than 0.95 for a moderate value of k (generally we had $500 \leqq k \leqq 1000$).

Note that application of the Chord method for producing uniformly distributed stationary points in the hyperellipsoids (that is with density e_i) would result in a fast procedure for generating stationary normal vectors.

This algorithm VNE is not faster than the matrix transformation method VNT and essentially can be considered as a discrete version of the multidimensional polar technique VNP. The reason why we present and investigate this method is the related question of multidimensional decomposition technique. Application of this "horizontal" decomposition described above may give rise to fast generation procedures in case of other, non uniformly distributed vectors. Such distributions of interest may be the elliptically symmetric distributions, whose density functions depend on the quadratic function of the variables. These densities can be decomposed similarly to the procedure applied to the normal distribution above, since the level sets of these distributions are ellipsoids.

One of the best knowns among the elliptically symmetric distributions is the multidimensional t distribution, whose density function is

$$t(\mathbf{x}) = \frac{\Gamma((\nu+n)/2)}{(\pi\nu)^{n-2}\Gamma(\nu/2)\,|R|^{1/2}}(1+\nu^{-1}\mathbf{x}'R^{-1}\mathbf{x})^{-(\nu+n)/2}.$$

This density function is that of the vector $\mathbf{y}$, which has components $y_j = x_j(\chi_\nu/\sqrt{\nu})^{-1}$, $j = 1, \ldots, n$, where $(x_1, \ldots, x_n)$ is a normally distributed random vector with density function $\varphi(\mathbf{x})$ and correlation marix $\mathbf{R}$, χ_ν is a random variable from the χ distribution with n degree of freedom and χ_ν and $\mathbf{x}$ are mutually independent. Of course details of a sampling procedure for generating numbers with density function $t(\mathbf{x})$ should carefully be worked out. The speed of the generator would ultimately depend on the speed of generating uniformly distributed points in a hyperellipsoid, similarly to the case of the polar method VNP and the ellipsoid method VNE.

5.9. Notes

5.1. For description of these general methods see the books of Sobol, Hammersley and Handscomb.

5.4. This section is based on Deák (1979). The preliminary version of the polar coordinates method was published by Hicks and Wheeling (1958) without giving full background and specifying the generation procedure in Step 3. The algorithm VSP in this form is from Deák (1979). Though the basics of the VSN algorithm are common knowledge we mention that Mueller (1958) published it first, recommended for random number generation.

5.6. The chord method was published in Deák (1980). For a proof of the first Theorem see Kendall and Moran (1963), pp. 53–54.

5.7. The generation procedure of stationary uniform and normal vectors was developed in Deák (1980), where some properties of the generated vectors were also examined. For other possibilities of determining a point in Step 3 of algorithm VSS see also Deák (1980), where the proof of the third Theorem of the section can be found as well.

The generalization of the ratio of uniform technique to n-dimensional distributions is due to Vaduva (1983). In fact he presented several variants with transformations

$$t_1(v_0, v_1, \ldots, v_n) = \left(\frac{v_1}{v_0}, \ldots, \frac{v_n}{v_0}\right),$$

$$t_2(v_0, \mathbf{v}) = \left(\frac{v_0 + v_1}{v_0}, \ldots, \frac{v_0 + v_n}{v_0}\right), t_3(v_0, \mathbf{v}) = (v_0 + v_1, \ldots, v_0 + v_n),$$

and showed how to apply these general methods for Dirichlet and t distributions. Also the proofs of Theorem 4 and more general propositions can be found there.

5.8 The method was published in Deák (1978).

CHAPTER 6

MONTE CARLO METHODS

6.1. Introduction

The theoretical background of the simulational methods is exposed in this chapter. Deep knowledge and thorough understanding of this background, the available methods and their versions may render considerable help in planning a simulational experimentation, executing the simulational computation or in better appraisal of the results yielded.

In most cases of Monte Carlo computations the task is either of evaluating a multidimensional integral or it can be considered as such. Assume that in course of computations N uniformly distributed random numbers $u_1, \dots, u_N$ were used up and the final result of the computations was the random variable

$$G(u_1, \dots, u_N) \tag{6.1}$$

for some N dimensional function G. The expected value of this random variable can be given as

$$\int_0^1 \dots \int_0^1 G(u_1, u_2, \dots, u_N)\, du_1\, du_2 \dots du_N, \tag{6.2}$$

thus we actually aim at the evaluation of the expected value in (6.2). Since $G(u_1^i, \dots, u_N^i)$ itself (for some realizations $u_1^i, u_2^i, \dots, u_N^i$) is an unbiased estimator of the expected value (6.2) thus taking the arithmetic average of such realizations we can approximate the integral.

Reducing the original problem to the form of the multidimensional integral in (6.2) is not always worth all of the effort and very frequently because of computational difficulties cannot practically be carried out, nonetheless through the evaluation of such integrals the different Monte Carlo methods can very expressively be demonstrated. We also follow this way presenting methods for evaluating the one-dimensional integral

$$I = \int_0^1 f(x)\, dx, \tag{6.3}$$

where for sake of simplicity we assume $f(x) > 0$, $x \in [0, 1]$. Certainly the conventional numerical techniques are generally more effective for computing the integral I than the Monte Carlo methods nevertheless this example is used for demonstration purposes only.

The multidimensional generalization of the various procedures to be described below is not always trivial, sometimes even causes insurmountable difficulties. Nevertheless the application of Monte Carlo methods is justified in multidimensional cases being there better than the conventional methods. In the next chapter the evaluation of the distribution function of the multidimensional normal distribution, which is of great practical importance is presented as an example of the Monte Carlo methods' successful application for multidimensional integration.

6.2. Crude Method

The crude or simple method evaluates the integral in (6.3) in the following way. Let $u_1, \ldots, u_N$ be uniformly distributed in $[0,1)$ and independent, then the random variables

$$f(u_i), \qquad i = 1, \ldots, N \tag{6.4}$$

are independent and their expected value is I, since because of the definition of the expected value

$$E\{f(u_i)\} = \int_0^1 f(u_i)\,\mathrm{d}u_i = I.$$

Consequently the arithmetic average of the random variables in (6.4) is an unbiased estimator of the integral I, that is for the expression

$$\theta_1 = \frac{1}{N} \sum_{i=1}^{N} f(u_i) \tag{6.5}$$

we have $E(\theta_1) = I$.

Equation (6.5) means that we have to take a couple of randomly chosen points in the interval $[0,1]$, compute the function value of f at these points, evaluate their arithmetic average and this average function value is multiplied by the area of the integration domain (presently it is the length of the interval $[0,1)$, that is 1).

The formal algorithm realizing the estimator θ_1 given in (6.5) is the following:

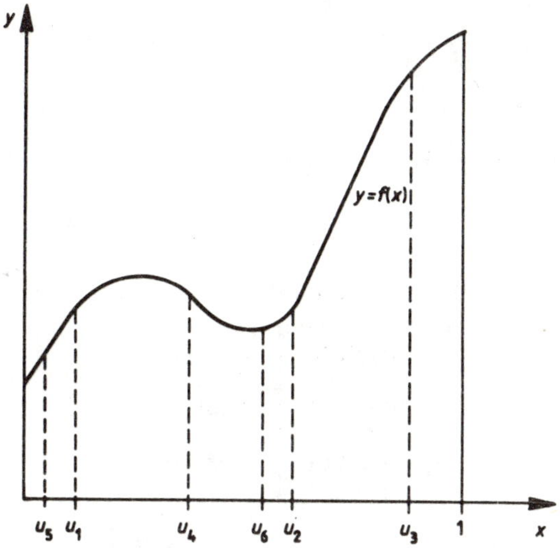

Fig. 31. The crude method

MC1 Crude Method

1. [Initialize.] Set the initial values $s \leftarrow 0$, $i \leftarrow 0$.
2. If $i = N$ then go to Step 4, otherwise increase the index $i \leftarrow i + 1$.
3. Generate the number u_i uniformly in interval $[0, 1)$ and sum up $s \leftarrow s + f(u_i)$. Go back to Step 2.
4. Deliver the number $s \leftarrow s/N$ as an approximation to the value I of the integral in (6.3).

Example. Consider the function $f(x) = e^x$, $0 \leqq x \leqq 1$. The integral of this function over the interval $[0, 1]$ can be computed as

$$\int_0^1 e^x \, dx = [e^x]_0^1 = e - 1 = 1.71828.$$

In the forthcoming sections this function and integral will be used to demonstrate the numerical application of the particular Monte Carlo methods.

Consider the numbers in Table 16, that were taken from a random number table and can be considered random. By the aid of these numbers evaluate the approximate value of the integral of the function e^x as suggested by the crude method

$$\frac{1}{16}\left[e^{0.70795} + e^{0.76575} + \ldots + e^{0.87225}\right] = \frac{29.368}{16} = 1.835.$$

The difference between the accurate and approximate value is

$$s = 0.117,$$

which is surprisingly small compared to the number of random points used.

Table 16

Random numbers from a table

0.70795	0.97796
0.58846	0.76575
0.95114	0.35801
0.57598	0.49687
0.54313	0.03569
0.70837	0.87225
0.16325	0.76151
0.65877	0.32764

6.3. The Acceptance-Rejection Method

Assume that the function $f(x)$ to be integrated is bounded in the domain of integration, that is for some constant $c > 0$ we have

$$f(x) \leqq c, \qquad x \in [0,1). \tag{6.6}$$

An estimation of the integral I of the function $f(x)$ can be obtained in the following way. Consider the rectangle given by the vertices $(0,0)$, $(1,0)$, $(1,c)$, $(0,c)$. Generate uniformly distributed points $(u_1, v_1), \ldots, (u_N, v_N)$ in the rectangle (u_i is uniformly distributed in $[0,1)$, v_i is uniformly distributed in $[0,c)$ and u_i and v_i are independent, see Section 5.2). Multiply the area of the rectangle by ratio of the points falling in, that is the quotient of the number of points below the graph of the function $f(x)$ and the number N of all the points, the number obtained is an approximate value of the integral I. Since the ratio of generated points falling in the area of integral is the same as the ratio of the area below of $f(x)$ and the area of the whole rectangle, thus the above procedure graphically yields the approximate area below $f(x)$. The method is very widely used because of its illustrativeness and easily understandable nature though this method has the worst efficiency among the Monte Carlo methods as it will be shown in the sequel.

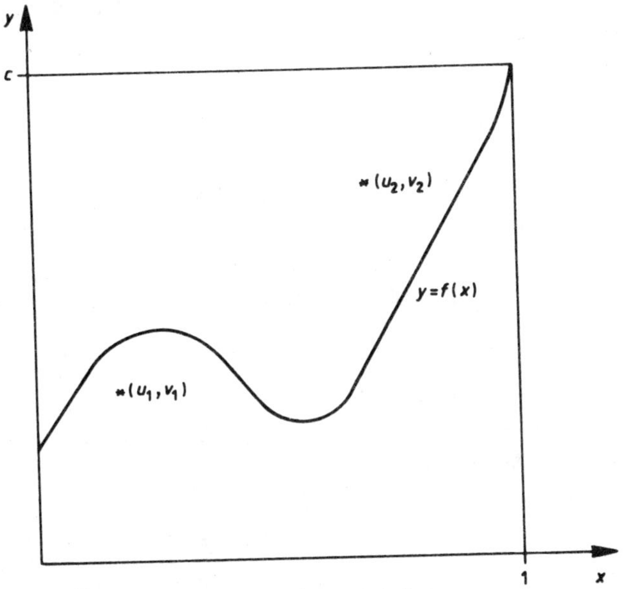

Fig. 32. An accepted point (u_1, v_1) and a rejected one (u_2, v_2) in the acceptance-rejection method

Formally, in a mathematically precise way the method can be presented as follows. Let us define the function g as

$$g(x,y) = \begin{cases} 0 & \text{if } f(x) < y, \\ 1 & \text{if } f(x) \geqq y. \end{cases}$$

The acceptance-rejection technique produces an estimate for the expected value of the random variable $cg(u,v)$, where c is the area of the rectangle, u is uniformly distributed in $[0,1)$, v is uniformly distributed in $[0,c)$. Using these notations the expected value of the random variable $cg(x,v)$ (here v is random only, x is a parameter) is $f(x)$ since

$$f(x) = \frac{1}{c}\int_0^c cg(x,v)\,\mathrm{d}v.$$

Thus the original integral I can be written as

$$I = \int_0^1 \int_0^c g(x,y)\,\mathrm{d}y\,\mathrm{d}x, \tag{6.8}$$

according to which applying the crude method for evaluating the integral I, an unbiased estimator of I is

$$\theta_2 = c\frac{1}{N}\sum_{i=1}^{N} g(u_i, v_i) = c\frac{N_1}{N}, \tag{6.9}$$

where u_i, v_i, $i = 1, \ldots, N$ are independent realizations of the random variables u, v, N_1 is the number of the points (u_i, v_i) below the graph of the function $f(x)$ and c is the area of the rectangle containing the function $f(x)$. The estimator (6.9) is produced by the following algorithm:

MC2 Acceptance-rejection method

1. Initialize the counters $s \leftarrow 0$, $i \leftarrow 0$.
2. If $i = N$ then got to Step 4, otherwise increase the index $i \leftarrow i + 1$.
3. Generate u_i uniformly in $[0, 1)$, v_i uniformly in $[0, c)$. If now $f(u_i) \geqq v_i$ then set $s \leftarrow s + 1$. Go back to Step 2.
4. Deliver the value $s \leftarrow cs/N$ as an approximation to the value I.

Example. Apply the acceptance-rejection technique for integrating the function $f(x) = \mathrm{e}^x$, $0 \leqq x \leqq 1$. The uniformly distributed points are generated in the rectangle spanned by the vertices $(0,0)$, $(0,1)$, $(1,\mathrm{e})$, $(0,\mathrm{e})$ in such a manner, that the first coordinates will be the numbers given in Table 16, the second coordinates will be the numbers depicted in Table 17, multiplied by e.

Table 17

Random numbers in $[0, 1)$

0.51435	0.17381
0.61870	0.31766
0.14783	0.07876
0.05312	0.66515
0.56774	0.41675
0.96929	0.22700
0.58481	0.16809
0.91730	0.03675

Since 13 out of the generated 16 points arc falling below the graph of the function $f(x)$, thus the approximate value of the integral according to this acceptance-rejection implementation is

$$\frac{13}{16} \cdot 2.71 = 2.209.$$

This value carries an error 0.491 which is significantly greater than the difference in case of the crude method (it is not accidental, see Section 6.5).

6.4. Error estimation

In case of the traditional deterministic integration formulas, giving approximate values for integrals of functions an upper bound can usually be given on the error, that is a number δ is secured which is greater than the difference between the exact value and the evaluated approximate value. This value δ depends on the properties of the function f and the number N of the points employed in the particular integration formula and in most cases it is much greater than the actual error. This kind of error estimation (deterministic upper bound on the error) cannot be obtained for the case of the Monte Carlo methods, because of their probabilistic nature, since the outcome of the computations done by Monte Carlo methods is a realization of a random variable. For instance the slightly probable case may happen when employing the crude Monte Carlo method, that all N generated uniformly distributed numbers $u_1, \ldots, u_N$ are in the interval $[0, \varepsilon)$ for an $\varepsilon > 0$ near to zero, or the equally improbable event may occur of having $u_i \in [1-\varepsilon, 1)$, $i = 1, \ldots, N$. In both cases the result of the computations may highly diverge from the correct value and it may happen regardless of the sample size N. Of course by increasing N the probability of such an event will decrease nevertheless it is still greater than zero. Consequently an error estimation for Monte Carlo methods must be formulated in probabilistic terms.

For characterizing a method the variance is used generally. The goodness of the crude method is thus measured by the variance

$$D^2(\theta_1).$$

If u is a uniform random variable, and the dispersion of the random variable $f(u)$ is denoted by σ_1^2 for which one can derive

$$\sigma_1^2 = D^2\left(f(u)\right) = \int_0^1 f^2(u)\,du - \left[\int_0^1 f(u)\,du\right]^2 = \int_0^1 f^2(u)\,du - I^2, \tag{6.10}$$

then an expression for $D^2(\theta_1)$ can be arrived at easily

$$D^2(\theta_1) = \frac{1}{N^2} D^2\left(f(u_1) + \ldots + f(u_N)\right) = \frac{\sigma_1^2}{N}. \tag{6.11}$$

Thus the standard deviation of the estimator is $D(\theta_1) = \sigma_1/\sqrt{N}$. From this we can perceive one of the most substantial features of the Monte Carlo methods. If the number of the sample points N is increased, the error that is basically the standard deviation decreases proportionately to $1/\sqrt{N}$ only. In other words we say that in order to halve the error one has to use four times more points and in order to obtain a result with one more accurate digit one has to complete one hundred times more computational work. Based on this fact we can state, that in simulational computations the accuracy is hardly improved by increasing the sample size. Generally it is much more expedient to increase the accuracy by decreasing the standard deviation σ, that is better we construct such estimators that contain random variables with less variance. The Monte Carlo methods described here follow exactly this path that is accuracy is increased via improved estimators.

In actual computations one cannot obtain the value $D(\theta_1)$ instead the sample variance s^2 is used for obtaining an estimation of the variance. It is

$$
\begin{aligned}
(6.12)\qquad s^2 &= \frac{1}{N-1}\sum_{i=1}^{N}(f_i - \bar{f})^2 = \\
&= \frac{1}{N-1}\sum_{i=1}^{N} f_i^2 + \frac{2\bar{f}}{N-1}\sum_{i=1}^{N} f_i + \frac{N}{N-1}\bar{f}^2 = \\
&= \frac{1}{N-1}\sum_{i=1}^{N} f_i^2 - \frac{N}{N-1}\bar{f}^2
\end{aligned}
$$

which is computed from the samples $f_i = f(u_i)$, $i = 1, \ldots, N$ and the average $\bar{f} = \frac{1}{N}\sum_{i=1}^{N} f_i$. The evaluation of the sum in (6.12) requires one additional multiplication only as compared to the computation of integral-estimation $\bar{f}$, thus it is expedient to evaluate s^2 in all cases. This property is one of the advantages of the simulational method: with little additional effort almost accurate error estimation can be produced simultaneously with the other calculations.

Example. In the case of the function $f(x) = \mathrm{e}^x$ the variance can be determined as

$$
\begin{aligned}
\sigma^2 = D^2\left(f(u)\right) &= \int_0^1 \mathrm{e}^{2x}\,\mathrm{d}x - \left[\int_0^1 \mathrm{e}^x\,\mathrm{d}x\right]^2 = \frac{1}{2}\mathrm{e}^2 - \frac{1}{2} - (\mathrm{e}-1)^2 = \\
&= 2\mathrm{e} - \frac{3}{2} - \frac{1}{2}\mathrm{e}^2 \sim 0.2420.
\end{aligned}
$$

It means that applying the crude method to compute the integral of the function e^x the standard deviation of the calculated value will be $D(\theta_1) = 0.4919/\sqrt{N}$. In case of $N = 16$ the standard deviation is 0.123 which is near to the error 0.117 of the value computed in Section 6.2. Using equation (6.12) and the random numbers from Table 16 the sample variance is ($N = 16$):

$$s^2 = \frac{1}{N-1}\sum_{i=1}^{N} f_i^2 - \frac{N}{N-1}\bar{f}^2 = \frac{57.287}{15} - \frac{16}{16}\cdot 3.369 =$$
$$= 3.819 - 3.594 = 0.225 = (0.475)^2,$$

which is also a rather good approximation of the exact theoretical variance. Note also the error according to s^2 would be 0.119 which is almost exactly the actual error 0.117.

A little bit different, but also utilizing the variance of the estimator, way of assessing the error of the estimator is the following. We can assume, that the distribution of the random variable θ_1 is normal, then it follows that the random variable $\eta = \dfrac{\theta_1 - E(\theta_1)}{D(\theta_1)} = \dfrac{\theta_1 - I}{(\sigma/\sqrt{N})}$ is standardized normal variable with distribution $N(0,1)$. This gives rise the possibility to characterize the error of the computations by giving a probabilistic bound on the absolute value of the difference $\theta - I$. Denoting the distribution function of the standard normal distribution by Φ we can write

$$P\{|\theta_1 - I| \leqq \lambda D(\theta_1)\} = P\{I - \lambda D(\theta_1) \leq \theta_1 \leq I + \lambda D(\theta_1)\} =$$

$$= \frac{1}{\sqrt{2\pi}D(\theta_1)}\int_{I\lambda D(\theta_1)}^{I+\lambda D(\theta_1)} e^{-\frac{(x-I)^2}{2D^2(\theta_1)}}\,dx = \tag{6.13}$$

$$= \frac{1}{\sqrt{2\pi}}\int_{-\lambda}^{+\lambda} e^{-\frac{z^2}{2}}\,dz = 2\Phi(\lambda) - 1,$$

where λ is any positive number. According to this equation the difference between our estimator θ_1 and the true value I will be less than twice the dispersion the standard deviation $D(\theta_1)$ with probability $2\Phi(2)-1 \sim 0.95$. In other words if we evaluate the estimator θ_1 for a fixed sample size N twenty times, independently of each other, than the error in nineteen cases out of twenty will not be greater than $2D(\theta_1)$ (with great probability). The difference $|I-\theta_1|$ can only very rarely be greater than three times the standard deviation (once in three hundred cases), since $2\Phi(3)-1 \sim 0.997$.

In practice generally we are satisfied with 95% error bound. It means that if we want to permit an error δ only, then the sample size N should be chosen in such a way that the equation

$$\delta = 2D(\theta_1) = 2\frac{\sigma}{\sqrt{N}}$$

is satisfied, that is for N we have

$$N = 4\frac{\sigma^2}{\delta^2}.$$

Greater reliability is rarely required. In practice the value N is computed through a preliminary stage: a small value is assigned to N and s^2 is determined, in turn σ^2 is substituted by s^2 in the formula of N that yield the required value of $N = 4s^2/\delta^2$.

6.5. Comparison of Methods

When comparing two methods the variances of the methods play an important role but the computational effort required to accomplish the calculations are also to be taken into account. the computational effort is generally equated with the running time of the computer program. If the two method evaluate one sample in roughly the same time then the comparison of the variances suffices.

Consider the method of acceptance-rejection. Since the evaluation of the function g figuring in the estimator θ_2 of the acceptance-rejection means basically the evaluation of the function f thus we can safely assume that the necessary computational work is equal for the computation of the estimators θ_1 and θ_2. In order to compare the crude method and the acceptance-rejection technique we need the variances $D^2(\theta_1)$ and $D^2(\theta_2)$ only. To determine the value of $D(\theta_2)$ note that the random variable $g(u_i, v_i)$ takes on the value 1 with probability $(1 - I/c)$ that is we can write

$$D^2\left(g(u_i, v_i)\right) = \frac{I}{c}\left(1 - \frac{I}{c}\right) = \frac{I}{c} - \frac{I^2}{c^2}.$$

After some calculations the following expression can be derived for the variance of the acceptance-rejection method

$$D^2(\theta_2) = D^2\left(\frac{1}{N}\sum_{i=1}^{N} cg(u_i, v_i)\right) = \tag{6.14}$$

$$= \frac{c^2}{N^2}\cdot N\left(\frac{I}{c} - \frac{I^2}{c^2}\right) = \frac{1}{N}(cI - I^2).$$

The difference between the variances of the two estimators can be worked out of the expressions in (6.11) and (6.14) respectively

$$D^2(\theta_2) - D^2(\theta_1) = \frac{1}{N}[cI - I^2] - \frac{1}{N}\left[\int_0^1 f^2(u)\,du - I^2\right] =$$

$$= \frac{1}{N}\left[cI - \int_0^1 f^2(u)\,du\right] = \frac{1}{N}\int_0^1 f(u)\,(c - f(u))\,du.$$

Since for the difference in the last integral one has $c - f(u) \geq 0$ (because of the definition of the upper bound c) it implies that the difference between the variances is nonnegative as well, that is

$$D(\theta_2) \geqq D(\theta_1), \tag{6.15}$$

and we have equality only in the case of $c \equiv f(u)$, $u \in [0,1)$. It means that the acceptance rejection technique for evaluating integrals always gives worse results (with greater variance) than the application of the crude method, furthermore the more significant is the deviation of $f(x)$ from the constant c the greater the difference between the efficiencies of the methods becomes.

According to another possible interpretation the following is meant by the inequality in (6.15). If the role of the randomness in the simulational computations is decreased (the number of the free variables in the integrand is reduced), that is the value $cg(x, v)$ is replaced by $f(x)$ which is its expected value and instead of the double integral (6.8) the simple integral (6.3) is evaluated by simulational method (θ_1 is applied instead of estimator θ_2), then we obtain better results. Or to put it in still other words if we know something, we should incorporate this knowledge into our computations. In this sense we have to strive to decrease the role of the randomness in simulational methods, reduce the number of parameters to be simulated, since less number of parameters in simulations means that the integral to be approximated is of lesser dimensions.

Example. Let $f(x) = \mathrm{e}^x$, $c = \mathrm{e}$, where $f(x) \leqq c$ if $x \in [0,1)$ clearly and denote

$$g(x,y) = \begin{cases} 0, & \text{if } f(x) < y, \\ 1, & \text{if } f(x) \geqq y. \end{cases}$$

Apply now the acceptance-rejection technique for computing the integral

$$I = \int_0^1 f(x)\,\mathrm{d}x = \int_0^1 \int_0^c g(x,y)\,\mathrm{d}y\,\mathrm{d}x.$$

Making use of the result $I = \mathrm{e} - 1$ the variance of the acceptance-rejection estimator can be obtained as

$$D^2(\theta_2) = \frac{1}{N}(\mathrm{e}I - I^2) = \frac{1}{N}(e-1) \sim \frac{1.7182}{N}.$$

Comparing the error of the acceptance-rejection method to that of the crude method one obtains

$$D(\theta_1) = \frac{0.4919}{\sqrt{N}} < \frac{1.3108}{\sqrt{N}} = D(\theta_2)$$

which is of the same magnitude as the actual error and the difference in deviations respectively obtained during the actual numerical computations.

To conclude this section we outline the procedure, by the aid of which two methods of different work requirements can be compared. Assume, that the first method in t_1 time produces an estimation with standard deviation σ_1, whereas the second method produces in t_2 time an estimation with standard deviation σ_2. Decrease in the standard deviation is possible at the expense of the square of the computational work, as we have already seen. In order to produce an estimator with variance σ_1^2 by the second method we have to achieve the $\frac{\sigma_1}{\sigma_2}$-th part of the original standard deviation σ_2 of the estimator by the second method, that is the original execution time t_2 increases by a factor $\frac{\sigma_2^2}{\sigma_1^2}$. Consequently while the first method produces an estimator with standard deviation σ_1 in t_1 time, a result with the same standard deviation by the second method is obtained in $t_2 \frac{\sigma_2^2}{\sigma_1^2}$ time. The two method can be compared if the times required to obtain the results of the same goodness are set against each other, that is t_1 and $t_2 \frac{\sigma_2^2}{\sigma_1^2}$ respectively. The quotient

$$h = \frac{t_1 \sigma_1^2}{t_2 \sigma_2^2} \tag{6.16}$$

is called the efficiency of the second method compared to the first method and it indicates how many times is the second method faster than the first method.

Example. Applying the crude method and the acceptance-rejection method for evaluating the integral of the function $f(x) = \mathrm{e}^x$ the efficiency of the second method compared to the crude method can be obtained as

$$h = \frac{D^2(\theta_1)}{D^2(\theta_2)} = \frac{0.242}{1.718} \sim 0.14.$$

It means that the acceptance-rejection technique is considerably less efficient than the crude method, actually the crude method is seven times faster.

6.6. Importance sampling

Here and in the subsequent sections methods are described, whose estimators have smaller variances as compared to the variance of the crude method's estimator. Of course this is possible only in those cases if we possess some additional information on the function to be integrated beyond computability and integrability and based on this information a better estimator can be constructed.

The essence of the importance sampling is that the function values are not chosen with equal probability but the more important values, that represent the greater part in the final total, will be selected with greater probability. By rare sampling of the small function values computational effort can be saved. Of course the function values thus obtained are not simply to be summed up in the arithmetic average instead they should be totalled according to some weights.

Consider the task of evaluating the integral

$$I = \int_0^1 f(x)\,\mathrm{d}x$$

and let $h(x)$ be a density function in the interval $[0,1]$ that can be 0 only where $f(x)$ is 0 as well, that is we assume that for any $x \in [0,1)$ for which

$f(x) > 0$ we have $h(x) > 0$. The integral to be evaluated can be rewritten in the form

(6.17) $$I = \int_0^1 \frac{f(x)}{h(x)} h(x)\, \mathrm{d}x = \int_0^1 \frac{f(x)}{h(x)}\, \mathrm{d}H(x),$$

where $H(x) = \int_0^x h(y)\mathrm{d}y$ is the distribution function corresponding to $h(y)$. The right-hand side of this equation shows that if the sample points y_i in interval $[0, 1)$ are generated with density function $h(x)$ then an estimator of the integral is the arithmetic average of the ratios $\frac{f(y_i)}{h(y_i)}$. More precisely denoting the samples of the random variable with density function $h(x)$ by y_i, $i = 1, \ldots, N$, then the importance sampling estimator of I is

(6.18) $$\theta_3 = \frac{1}{N} \sum_{i=1}^{N} \frac{f(y_i)}{h(y_i)}.$$

MC3 Importance Sampling

1. Initialize $s \leftarrow 0$, $i \leftarrow 0$.
2. If $i = N$ then go to Step 4, otherwise increase index $i \leftarrow i = 1$.
3. Generate the random number y_i according to the density function h, let $s \leftarrow s + f(y_i)/h(y_i)$ and go back to Step 2.
4. Deliver the value $s \leftarrow s/N$ as an approximation to the integral I.

Obviously the expected value of the estimator θ_3 is the integral I to be evaluated, that is θ_3 is unbiased, its standard deviation can be determined in the following way:

(6.19) $$\begin{aligned} D^2(\theta_3) &= \frac{1}{N} \left\{ \int_0^1 \left(\frac{f(x)}{h(x)} \right)^2 \mathrm{d}H(x) - \left[\int_0^1 \frac{f(x)}{h(x)}\, \mathrm{d}H(x) \right]^2 \right\} = \\ &= \frac{1}{N} \left\{ \int_0^1 f^2(x)/h(x)\, \mathrm{d}x - I^2 \right\}. \end{aligned}$$

This standard deviation becomes minimal if the density function $h(x)$ is proportional to the function $f(x)$, that is if $h(x) = cf(x)$, where $1/c = \int_0^1 f(x)\, \mathrm{d}x = I$. In this case we have

$$D^2(\theta_3) = \frac{1}{N}\left\{\frac{1}{c}\int f(x)\,dx - I^2\right\} = 0,$$

a most desirable result. Unfortunately in practice it is hardly realizable since to derive the density function h we need the value $c = 1/I$ and we already had possessed the value of c (and by the same token I) we would not have to apply a simulational method for computing I. This whole argument nonetheless is not worthless. The reasoning and the expression (6.19) exhibit when the estimator θ_3 is good, when the standard deviation $D(\theta_3)$ is small: the better the density $h(x)$ approximates (imitates) the integrand $f(x)$ (the smaller is the fluctuation of the ratio $f(x)/h(x)$ in the interval of integration) the smaller the variance $D^2(\theta_3)$ becomes. Since the ratio $f(x)/h(x)$ cannot be constant because of the reasons given above we may only attempt to produce a nearly constant $f(x)/h(x)$. When selecting a density function $h(x)$ from among the possible approximating functions $h'(x)$ we have to take into account another restriction as well: not only the computability of the integral $\int_0^1 h'(x)dx$ is to be assumed (in order to derive the corresponding density function) but we should be able to generate random variables with density function $h(x)$, possibly easy.

Example. Consider the function $f(x) = e^x$ and try to select an appropriate density function. The choice of $h(x) = 1$ gives only the crude method. As a next try we take $h(x) = 2x$, but this does not satisfy the positivity assumption, since $h(0) = 0$ though $f(0) \neq 0$. (The application of this density $h(x)$ for importance sampling technique yields an estimator, whose variance $\sigma^2_{f/h}$ is infinite: $\sigma^2_{f/h} = \frac{1}{2}\int_0^1 e^{2x}/x\,dx - (e-1)^2$.)

Knowing that $f(x)$ is monotone and the fluctuation of $f(x)/h(x)$ should be small we may try a linear function which is halving both the values of $f(0)$ and $f(1)$, that is let $f(0)/h(0) = 2$ and $f(1)/h(1) = 2$ where h is linear. This gives us $h'(x) = 0.5 + (e-1)x/2$. The density function $h(x)$ proportional to $h'(x)$ can be computed with the value

$$\int_0^1 [0.5 + 0.86x]\,dx = 0.93,$$

(where the coefficient $\frac{e-1}{2}$ of x in $h'(x)$ was rounded to 0.86), so that

$$h(x) = 0.5373 + 0.9247x.$$

Construct now an importance sampling estimator with this density function $h(x)$. Its variance can be expressed as

$$\sigma^2_{f/h} = \frac{1}{3}\frac{D^2(\theta_1)}{D^2(\theta_3)} = \frac{1}{3}\frac{0.2420}{0.00526} \sim 15.$$

6.7. Stratified Sampling

The method is similar to the importance sampling. In the earlier described method the positions of the points to be generated were determined by a density function, now the interval of the function's definition is split up into smaller regions and the number of points to be generated according to the uniform distribution is fixed in each region. So this can be regarded as a discrete version of the importance sampling.

Let a division of the interval $[0, 1)$ be given by the splitting points $0 = a_0 < a_1 < \ldots < a_k = 1$. Apply now in the intervals $[a_{i-1}, a_i)$, $i = 1, \ldots, k$ the crude method separately, generating n_i points in the ith interval. The estimator of the integral can be given in the following formula

$$\theta_4 \sum_{i=1}^{k}\sum_{j=1}^{n_i}(a_i - a_{i-1})\frac{1}{n_i}f\left(a_{i-1} + (a_i - a_{i-1})u_{ij}\right) \tag{6.20}$$

where u_{ij} is a sample from the uniform distribution in the ith interval $[a_{i-1}, a_i)$, $i = 1, \ldots, k$, $j = 1, \ldots, n_i$. The formal algorithm comprising the estimator θ_4 is given below.

MC4 Stratified Sampling

1. Let $s \leftarrow 0$, $i \leftarrow 0$.
2. If $i = k$ then go to Step 4, otherwise set $i \leftarrow i+1$, $s_i \leftarrow 0$, $t_i \leftarrow a_i - a_{i-1}$.
3. Generate n_i numbers u_{ij}, $j = 1, \ldots, n_i$ uniformly in $[0, 1)$ and compute the sum $s_i \leftarrow f(a_i + t_i u_1) + \ldots + f(a_i + t_i u_{n_i})$, then compute $s_i \leftarrow t_i s_i / n_i$ and sum up $s \leftarrow s + s_i$. Go back to Step 2.
4. Deliver the number s as an approximate value of the integral I.

On the individual subintervals $[a_{i-1}, a_i)$ the method approximates evidently the integral $\int_{a_{i-1}}^{a_i} f(x)dx$, which means that θ_4 is an unbiased estimator of the integral I. The error of the method is computed in the following formula:

$$(6.21)\quad D^2(\theta_4) = \sum_{i=1}^{k} \frac{a_i - a_{i-1}}{n_i} \int_{a_{i-1}}^{a_i} f^2(x)\,\mathrm{d}x - \sum_{i=1}^{k} \frac{1}{n_i} \left[\int_{a_{i-1}}^{a_i} f(x)\,\mathrm{d}x \right]^2 .$$

Several possibilities are open with regard to the choice of the partitioning points a_i and the number n_i of the points in the particular subintervals. If the points a_i are already fixed then the n_i^2 can be chosen proportionally to the standard deviation of the crude method in that subinterval, that is to the quantity

$$\left\{ (a_i - a_{i-1}) \int_{a_{i-1}}^{a_i} f^2(x)\,\mathrm{d}x - \left[\int_{a_{i-1}}^{a_i} f(x)\,\mathrm{d}x \right]^2 \right\}^{1/2} .$$

This means that where the computed result is going to have greater uncertainty there the error is decreased by increasing the number of sample points.

Taking into consideration the fact, that in most practical problems the variances in the subintervals neither are known, nor can be computed by direct integration thus the following two-stage procedure is advised to be followed. First make an estimation of the variances in the subintervals employing a small number ($n_i' \sim 10$) of sample points and then choose the sample numbers n_i proportionally to the obtained standard deviation. In the case we have the liberty of choosing the splitting points then the best choice would be to obtain a division where the variance of the function f is the same in all subintervals. Since it is hard to execute in practice, a more simple way may be applied: the interval of integration is divided into equal parts.

Example. Let $f(x) = \mathrm{e}^x$ and consider the partitioning of interval $[0,1)$ into four equal subintervals with points $a_1 = \frac{1}{4}$, $a_2 = \frac{2}{4}$, $a_3 = \frac{3}{4}$. Apply the stratified sampling generating 2, 4, 6 and 8 points respectively in the subsequent subintervals. Then making use of the equation

$$(a_i - a_{i-1}) \int_{a_{i-1}}^{a_i} e^{2x}\,\mathrm{d}x - \left[\int_{a_{i-1}}^{a_i} e^x\,\mathrm{d}x \right]^2 =$$

$$=\frac{1}{4}\left[\frac{1}{2}\,e^{2x}\right]_{a_{i-1}}^{a_i}-\left\{[\,e^x]_{a_{i-1}}^{a_i}\right\}^2=$$
$$=\frac{1}{8}\,e^{2a_i}-\frac{1}{8}\,e^{2a_{i-1}}-\,e^{2a_i}-\,e^{2a_{i-1}}+2\,e^{a_i+a_{i-1}}=$$
$$=2e^{a_i+a_{i-1}}-\frac{9}{8}\,e^{2a_i-1}-\frac{7}{8}\,e^{2a_i}$$

we can express the variance $D^2(\theta_4)$ of the stratified sampling's estimator as

$$ND^2(\theta_4)=\frac{1}{2}(0.00042)+\frac{1}{4}(0.00069)+\frac{1}{6}(0.00114)+\frac{1}{8}(0.00188)$$
$$=0.000808=(0.0284)^2$$

that shows a decrease in the error similar to the decrease for the importance sampling (one must note that here we have basically 20 points, not just one).

6.8. Control Variates Method

We want to define a function g imitating possibly well the integrand function f, furthermore we suppose that we can compute exactly the integral

$$J=\int_0^1 g(x)\,\mathrm{d}x$$

by mathematical theory. In case of finding such a function g, if u is uniformly distributed random variable in interval $[0,1)$ then the random variable $f(u)$ is well approximated by the random variable $g(u)$ so one can say that $g(u)$ controls the behaviour of $f(u)$, this is the reason for the name of the method.

In order to obtain a formal description of the method the original integral is decomposed into two parts

$$I=\int_0^1 f(x)\,dx=\int_0^1 g(x)\,\mathrm{d}x+\int_0^1[f(x)-g(x)]\,\mathrm{d}x. \tag{6.22}$$

The first of the two integrals standing on the right-hand side can be evaluated accurately according to the assumption made, its value is J. For the second integral the crude method of Monte Carlo integration is applied. The estimator is

$$\theta_s=J+\frac{1}{N}\sum_{i=1}^{N}[f(u_i)-g(u_i)], \tag{6.23}$$

where u_i, $i = 1, \ldots, N$ are realizations of u. The algorithm corresponding to the estimator θ_5 can be the following:

MC5 Control Variates

1. Initialize $s \leftarrow 0$, $i \leftarrow 0$.
2. If $i = N$ then go to Step 4, otherwise increase the index $i \leftarrow i + 1$.
3. Generate u_i and set $s \leftarrow s + f(u_i)$. Go back to Step 2.
4. Deliver the value $s \leftarrow s/N + J$, as an approximate value of the integral I.

The estimator θ_5 is evidently unbiased, that is $E(\theta_5) = I$, for its variance one can obtain

(6.24)
$$D^2(\theta_5) = \frac{1}{N} \cdot \sigma_5^2 = \frac{1}{N} \left\{ \int_0^1 [f(x) - g(x)]^2 \, dx - \left[\int_0^1 f(x) - g(x) \, dx \right]^2 \right\};$$

from which after a little calculation one can obtain

$$\begin{aligned} \sigma_5^2 =& \int_0^1 f^2(x) \, dx - \left[\int_0^1 f(x) \, dx \right]^2 - 2 \int_0^1 f(x) g(x) \, dx + \\ &+ 2 \int_0^1 f(x) \, dx \int_0^1 g(x) \, dx + \int_0^1 g^2(x) \, dx - \left[\int_0^1 g(x) \, dx \right]^2 = \\ =& D^2(f(i)) - 2E(f(u)g(u)) + D^2(g(u)). \end{aligned}$$

Since the first term $D^2(f(u))$ is the variance of the crude method, thus the method of control variates is an improvement (has lesser variance) over the crude method if the inequality

(6.25)
$$D^2(g(u)) < 2E(f(u)g(u))$$

holds. According to this the construction of an estimator based on control variates technique can be formulated in another way. Having a random variable $f(u)$ we are looking for another random variable $g(u)$ the expected value of which is known (using the earlier introduced notation $E(g(u)) = J$), furthermore there is a strong positive correlation between $f(u)$ and $g(u)$ (they change into the same direction). If the inequality (6.25) is not satisfied then the given function $g(u)$ is unsuitable for reducing variance in the estimator. In other words not every function is good for constructing control variates estimator, actually, without respecting (6.25) we may end up with more computational effort *and* greater variance.

We give still another interpretation of the method. Consider equality (6.22) that can be regarded as if it represents the following evaluation procedure: from the area below the graph of the integrand an easily integrable part is taken out, it is calculated theoretically and only the remaining area is estimated by Monte Carlo method (Fig. 33). In this context the method of control variates can be called the method of integrating the main part.

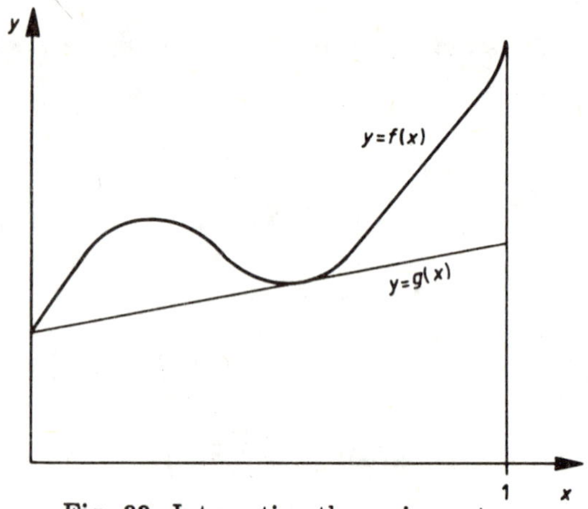

Fig. 33. Integrating the main part

We would like to draw the reader's attention to understand that the control variates method does nothing else but realizes the principle discussed in Section 6.5 where it was stated when comparing different methods: the role of randomness in Monte Carlo computations should be decreased if possible. If a part of the integral can be computed analytically (because of an available approximating function, or other knowledge), then having done so the precision of the remaining calculations do increase.

Example. Let $f(x) = e^x$ and consider an approximating function

$$g(x) = \begin{cases} 1, & \text{if } 0 \leqq x \leqq 0.5, \\ 1.5, & \text{if } 0 < x \leqq 1. \end{cases}$$

Using this function in the control variates method the variance of the estimator can be computed (suppose we have only one sample point $N = 1$):

$$\sigma_5^2 = \int_0^{1/2} (e^x - 1)^2 \, dx \int_{1/2}^{1} \left(e^x - \frac{3}{2}\right)^2 dx - \left(e^x - \frac{9}{4}\right)^2 =$$

$$= \frac{1}{2}e^2 - 3e + \sqrt{e} + \sqrt{e} + \frac{25}{8} - \left(e - \frac{9}{4}\right)^2 =$$

$$= -\frac{1}{2}e^2 + \frac{3}{2}e + \sqrt{e} - \frac{31}{16} \sim 0.0941 = (0.306)^2 .$$

The function $f(x)$ is probably better approximated by a linear function, consider for example the function $g'(x) = 2x$. In this case we have

$$\sigma_5'^2 = \int_0^1 (e^x - 2x)^2 \, dx - (e-2)^2 =$$

$$= \left[\frac{1}{2}e^{2x} + \frac{4}{3}x^3\right]_0^1 - 4\int_0^1 xe^x \, dx - (e-2)^2 = \frac{1}{2}e^2 - \frac{19}{6} - (e-2)^2 \sim$$

$$\sim 0.001193 = (0.108)^2$$

which indicates a substantial decrease of variance in relation to both the crude method and the approximating function $g(x)$. Employing $g'(x)$ in the control variates method may mean a three times increase in computational work as compared to that of the crude method, thus the efficiency of the estimator θ_5 can be given as

$$h = \frac{1}{3}\frac{\sigma_1^2}{\sigma_5^2} \sim 6.76 .$$

6.9. Antithetic Variates Method

The variance of the estimator can also be decreased by somehow compensating the big jumps in the function value, that is by using random variables changing into the opposite direction. Consider the random variables $f(u)$ and $g(u)$, assume that we know the integral of the function $g(x)$ and $\int_0^1 g(x)\, dx = I$. The expected value of the arithmetic average of these two random variables is equal to I as well, that is an unbiased estimator for the value of I can be given as:

$$\theta_6 = \frac{1}{N}\sum_{i=1}^{N}\frac{1}{2}(f(u_i)+g(u_i)). \tag{6.26}$$

The formal algorithm of producing the estimator θ_6 is the following one.

MC6 Antithetic Variates

1. Set $s \leftarrow 0$, $i \leftarrow 0$.
2. If $i = N$, then go to Step 4.
3. Generate u_i, set $s \leftarrow s + f(u_i) + g(u_i)$ and go back to Step 2.
4. Deliver the value $s \leftarrow s/(2N)$ as an approximation to the integral value I.

The variance of the random variable θ_6 can be evaluated by the next equation

$$D^2(\theta_6) = \frac{1}{N}D^2\left(\frac{1}{2}(f(u)+g(u))\right) = \frac{1}{N}\cdot\int[f(x)+g(x)]^2\,\mathrm{d}x - \frac{1}{N}\cdot I^2.$$

This can be rewritten for the variance σ_6^2 of one sample as

$$\begin{aligned}\sigma_6^2 &= D^2\left(\frac{1}{2}(f(u)+g(u))\right) = \\ &= \frac{1}{4}D^2(f(u)) + \frac{1}{2}\left[E\left(f(u)g(u)\right) - \right. \\ &\left. - E\left(f(u)\right)E\left(g(u)\right)\right] + \frac{1}{4}D^2(g(u)),\end{aligned} \tag{6.27}$$

where the middle term is the covariance between the random variables $f(u)$ and $g(u)$. The first term on the right-hand side is one-fourth of the variance of the crude method thus (taking into account the fact that in evaluating θ_6 two function values have to be computed for one sample point meaning by this that for θ_6 the computational work is double that of the estimator θ_1) the sum of the second and third term should be less than 0 in order to ensure that θ_6 is better than θ_1. This requirement is expressed in the following inequality

$$D^2(g(u)) \leqq -2\left[E\left(f(u)g(u)\right) - E\left(f(u)\right)E\left(g(u)\right)\right]. \tag{6.28}$$

A significant difference can be encountered between $D^2(\theta_6)$ and $D^2(\theta_1)$ if there is a strong negative correlation between the random variables $f(u)$ and $g(u)$ since the right-hand side of (6.28) contains the $\mathrm{cov}(f(u), g(u))$.

If the function $f(x)$ is monotone, then the function $g(x) = f(1-x)$ satisfies our requirement of $\int g(x)\,dx = I$ and the evaluation of the estimator θ_6 consists of the determination of the sum

$$\frac{1}{N}\sum_{i=1}^{N}\frac{1}{2}[f(u_i) + f(1-u_i)]. \tag{6.29}$$

In this particular case we speak about the symmetrization of the function (or the estimator) and this version is the most frequently used among the antithetic techniques.

The procedure is well illustrated in Fig. 34: the averaged function is much more smooth (of less variance), than the original function. A little bit more general variant of the previous symmetrized estimator (6.24) is the following

$$\frac{1}{N}\sum_{i=1}^{N}\frac{1}{m}\sum_{j=0}^{m-1} f\left(\frac{u_i + j}{m}\right). \tag{6.30}$$

This can be considered as a randomised version of an equidistant deterministic numerical integration formula.

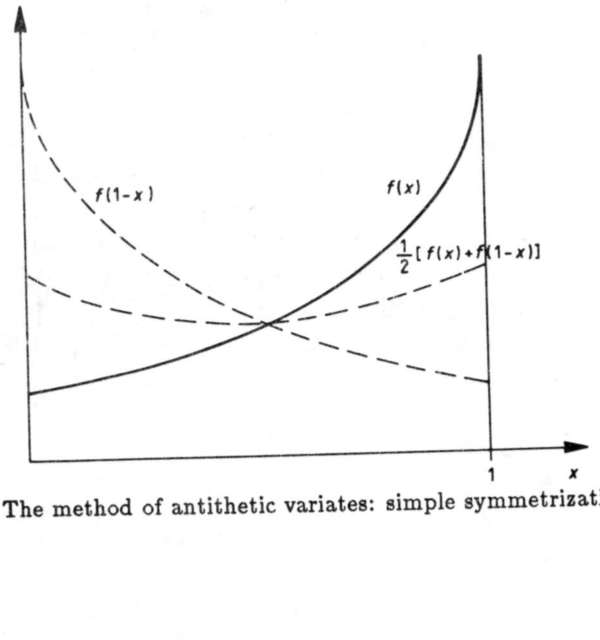

Fig. 34. The method of antithetic variates: simple symmetrization

Example. Let $f(x) = e^x$ since it is a monotone function, the antithetic variates technique can be applied with the function $g(x) = f(1-x)$. The estimator becomes of the form

$$\frac{1}{N}\sum_{i=1}^{N}\frac{1}{2}[f(u_i) + f(1-u_i)], \tag{6.31}$$

the variance of which can be computed according to (6.27) as

$$\begin{aligned} D^2(\theta_6) &= \frac{1}{N}\left[\frac{1}{4}D^2(f(u)) + \frac{1}{2}E(f(u)f(1-u)) - \right. \\ &\quad \left. - \frac{1}{2}E(f(u))E(f(1-u)) + \frac{1}{4}D^2(f(1-u))\right] = \\ &= \frac{1}{2N}D^2(f(u)) + \frac{1}{2N}\int_0^1 e^x\, e^{1-x}\, dx - \frac{(e-1)^2}{2N} = \\ &= \frac{1}{2N}D^2(f(u)) + \frac{1}{2N}e - \frac{1}{2N}(e-1)^2 \sim \frac{1}{2N}0.007789. \end{aligned}$$

6.10. Multidimensional Monte Carlo Methods

In this section the multidimensional generalizations of the one-dimensional techniques described in the foregoing sections of this chapter are demonstrated. This demonstration is sketchy on two counts. First the multidimensional methods are much less elaborated than the one-dimensional procedures. Second in more than one dimension there are much more possibilities in adapting the general methods to the individual problem than we could list here all of them. The methods we are going to present here are the applications of the ideas described for one-dimensional cases and this application (or adaptation) has to be done for each concrete problem anew. In the next chapter such an application is described in case of evaluating an n-dimensional integral, the integral of the multinormal density function.

The n-dimensional application of the crude method for evaluating the integral

$$I = \int_D f(\mathbf{x})\, d\mathbf{x} \tag{6.32}$$

is the following. Assume that the domain D of integration is bounded, its volume is of unit measure, that is $\int_D \mathrm{d}\mathbf{x} = 1$, $\mathbf{u}$ is uniformly distributed in D, then the estimator of the integral I can be given as

$$T_1 = \frac{1}{N} \sum_{i=1}^{N} f(\mathbf{u}_i), \tag{6.33}$$

where $\mathbf{u}_i$, $i = 1, \ldots, N$ are independent realizations of the random vector $\mathbf{u}$. Practical applicability of the method depends on the assumptions made (is D bounded or not) and how easily can uniformly distributed points in D be generated.

The n-dimensional variant of the acceptance-rejection technique for the case of the bounded function $f(\mathbf{x})$ can be given as follows. Let $f(\mathbf{x}) \leqq c$, $\mathbf{x} \in D$ for a constant $c > 0$, and define the $(n + 1)$ dimensional function

$$g(\mathbf{x}, y) = \begin{cases} 0, & \text{if } f(\mathbf{x}) < y, \\ 1, & \text{if } f(\mathbf{x}) \geqq y. \end{cases}$$

Using this notation the corresponding estimator is

$$T_2 = \frac{1}{N} \sum_{i=1}^{N} g(\mathbf{x}_i, y_i),$$

where the point $(\mathbf{x}_i, y_i)$ is the realization of the $(n+1)$ dimensional random variable which is uniformly distributed in the domain $D' = \{\mathbf{x}, y, | \mathbf{x} \in D, 0 \leqq y \leqq c\}$.

To describe the importance sampling technique consider an n-dimensional density function $h(\mathbf{x})$. The modification of the integral is the following

$$I = \int_D f(\mathbf{x})\, \mathrm{d}\mathbf{x} = \int_D \frac{f(\mathbf{x})}{h(\mathbf{x})}\, \mathrm{d}H(\mathbf{x}),$$

where $H(\mathbf{x})$ is the distribution function corresponding to the density function $h(\mathbf{x})$. The estimator based on this equation is

$$T_3 = \frac{1}{N} \sum_{i=1}^{N} \frac{f(\mathbf{x}_i)}{h(\mathbf{x}_i)}$$

where the points $\mathbf{x}_i$ where generated according to the density function $h(\mathbf{x})$. The snag in applying this method in n-dimensions is to find or to construct an n-dimensional density function according to which one can relatively easily generate random points and still gives a reasonable approximation to the function $f(\mathbf{x})$.

The stratified sampling technique can generally be recommended for application if the domain D of integration can easily be decomposed. This is the case if D is an n-dimensional rectangle or the unit cube $K = \{\mathbf{x}|\mathbf{0} \leqq \mathbf{x} \leqq \mathbf{1}\}$. Clearly it suffices to deal with the latter. The ith edge of the cube K is split up by the points $a_{i1}, a_{i2}, \ldots, a_{ik_i}$, the "product" of these partitions, the $k_1k_2 \ldots k_n$ rectangles will form the strata. The estimator is

$$T_4 = \sum_{i=1}^{\prod_{j=1}^{n} k_j} \sum_{j=1}^{n_j} \frac{1}{n_j} t_i f(\mathbf{u}_{ij}),$$

where t_i is the volume of the ith rectangle, u_{ij} is a sample from the uniform distribution in the ith rectangle.

The control variate technique can easily be applied if we have a readily integrable part. Formally the integral is rewritten as

$$I = \int_D f(\mathbf{x})\, d\mathbf{x} = \int_D g(\mathbf{x})\, d\mathbf{x} + \int_D [f(\mathbf{x}) - g(\mathbf{x})]\, d\mathbf{x},$$

where it is assumed that the first integral on the right-hand side, can be evaluated analytically, the second integral is computed by crude Monte Carlo method. The estimator is

$$T_5 = \int_D g(\mathbf{x})\, d\mathbf{x} + \frac{1}{N} \sum_{i=1}^{N} [f(\mathbf{x}_i) - g(\mathbf{x}_i)],$$

where $\mathbf{x}_i$ is a realization of the random variable, uniformly distributed in D.

Out of the great variety of possible versions of the antithetic variates method generally only the symmetrization technique is used. Let the integration domain be the cube K previously used and consider its partitioning as given at the stratified sampling. Each small rectangle is symmetric to its center, denote by $\mathbf{u}_{ij}$ the jth sample from the uniform distribution in the ith rectangle and by $\mathbf{u}'_{ij}$ its reflection to the centre of the ith rectangle. Then the corresponding estimator is:

$$T_6 = \sum_{i=1}^{\prod_{j=1}^{n} k_j} \frac{1}{n_i} \sum_{j=1}^{n_i} t_i \frac{1}{2} [f(\mathbf{u}_{ij}) + f(\mathbf{u}'_{ij})].$$

6.11. Notes

In this chapter — except the concluding section 6.10 — we essentially followed the treatment in the book of Hammersley and Handscomb which is now being regarded as classical.

As a simple example of simulation's application one should consider Buffon's needle problem, described in the first section of the Appendix.

CHAPTER 7

COMPUTING THE DISTRIBUTION FUNCTION OF THE MULTIDIMENSIONAL NORMAL DISTRIBUTION

7.1. Formulation of the problem

In mathematical statistics, stochastic programming and several branches of applied mathematics the following problem is frequently encountered. Let us denote by $\varphi(\mathbf{x})$ the density function of the n-dimensional normal distribution with expected value 0 and correlation matrix $\mathbf{R}$, and by $\Phi(\mathbf{x})$ its distribution function, that is let

(7.1) $$\varphi(\mathbf{x}) = \frac{1}{(2\pi)^{n/2}|\mathbf{R}|^{1/2}} e^{-\frac{1}{2}x'\mathbf{R}^{-1}x},$$

(7.2) $$\Phi(\mathbf{x}) = \int_{-\infty}^{x} \varphi(\mathbf{y})\, d\mathbf{y}.$$

Our task is to evaluate the values of the distribution function Φ, that is to develop suitable numerical methods for computing the integral (Fig. 35)

(7.3) $$p = I = \Phi(\mathbf{h}) = \int_{-\infty}^{h_n} \dots \int_{-\infty}^{h_1} \varphi(\mathbf{x})\, d\mathbf{x}.$$

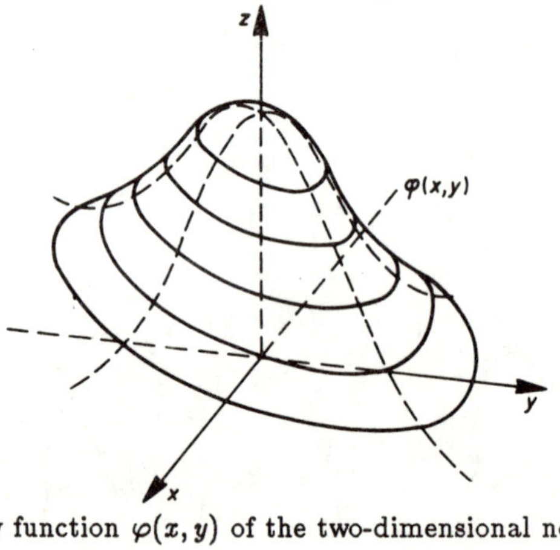

Fig. 35. The density function $\varphi(x, y)$ of the two-dimensional normal distribution

A related problem of interest is the computation of the integral

(7.4) $$q = \int \dots \int_{Q} \varphi(\mathbf{x})\, dx$$

where Q is an n-dimensional rectangle $Q = \{\mathbf{x}|\mathbf{a} \leqq \mathbf{x} \leqq \mathbf{b}\}$ for any vectors $\mathbf{a}$ and $\mathbf{b}$, that is the computation of the probability content of the rectangle. For the sake of simplicity we assume that the normal distribution is non-degenerated, that is $|\mathbf{R}| \neq 0$. Mainly we will discuss the integral (7.3) but most of the results will be applicable to the integral (7.4) as well.

Applying the traditional numerical methods in greater ($n > 4$) dimensions for evaluating the integral (7.3) involves quite a few problems. The computational work is rapidly swelling with increasing n: if the integration is carried out over a unit hypercube for instance, and on one edge of the cube k splitting points are located then altogether k^n function values are to be evaluated. The conventional deterministic methods for computing n-dimensional integrals are as a rule characterized by this feature called dimensional effect. A further handicap of these methods is their sensitiveness for great correlations; if $|\mathbf{R}|$ is near to zero the convergence is usually very slow. The reason of this is that $\varphi(\mathbf{x})$ in the origin becomes higher as $|\mathbf{R}|$ tends to zero: $\varphi(\mathbf{0}) = \frac{1}{|\mathbf{R}|^{1/2}} \frac{1}{(2\pi)^{n/2}}$, and in most other places it becomes smaller (one must not forget that the integral of $\varphi(\mathbf{x})$ over the whole space R^n is always 1), producing thus a more and more distorted shape. Having understood the inherent difficulties of the deterministic integration methods and demonstrating the application of Monte Carlo methods to evaluating the integral I our aim was not simple a demonstration of the methods but to illustrate and prove the point that there are problems in practice that best can and should be solved by Monte Carlo methods. This argument is not refuted by the fact that in the past simulational methods were frequently applied for several problem classes where the application was not efficient and thus should not take place.

In case of a concrete problem the question emerges when to use Monte Carlo methods when conventional deterministic technique. In our view, when there is a conventional method for solving the problem and the speed, accurary and efficiency of the method is satisfactory, than there is no need of simulation. If we do not possess such a method then the simulational approach is considered to be the *first but not last* solution technique for the problem. By the aid of Monte Carlo technique we always get some kind of result — this is why the simulation is considered the *first* solution technique —, but this result is generally burdened with quite a big error. Introducing variance reducing techniques or applying at least partially deterministic methods the error can be decreased, but the improvement in accuracy never will match that of the conventional numerical methods, — this is why we do not, can *not* consider it the *last* method. The optimal way of applying Monte Carlo method is attained in a sequence of methods, where the simulation plays less and less a role. In case of large sophisticated systems or great dimensions the reduction of the randomness' role is difficult thus in these problems the simulational method is almost entirely the only possible solution technique. The evaluation of the n-dimensional integrals (7.3) and (7.4) discussed in this chapter is of this sort, though the original simulational method was improved in several ways, the best method for evaluating the integral (7.3) still remains a simulational method.

7.2. Monte Carlo Integration

The crude Monte Carlo integration method is applied for evaluating the integral I. Denote the integration domain of the integral (7.3) by D, that is let

$$D = \{\mathbf{x} | \mathbf{x} \leqq \mathbf{h}\}.$$

For the sake of simplicity we assume $\mathbf{h} \geqq 0$. When the rewriting of the given formulas for $\mathbf{h} \not\geqq 0$ are not self-evident then the necessary reformulations are provided. Assume that $\mathbf{b}$ is a constant vector whose components are equal to a positive constant c, $b_i = c$, $i = 1, \dots, n$, and also consider the following n-dimensional cube

$$K = K(c) = \{\mathbf{x} | -\mathbf{b} \leqq \mathbf{x} \leqq \mathbf{b}\}.$$

We also assume for the sake of simplicity, that $h_i < c$, $i = 1, \dots, n$. By the aid of the vector $\mathbf{b}$ the unbounded integration domain D is truncated as follows.

Define $h_i^* = \min(h_i, c)$, $i = 1, \dots, n$ and the truncated domain can be given now as

$$D^* = \{\mathbf{x} | -\mathbf{b} \leqq \mathbf{x} \leqq \mathbf{h}^*\}.$$

In what follows we will deal with the evaluation of the integral over the truncated area

$$I^* = \int_{D^*} \dots \int \varphi(\mathbf{x})\, d\mathbf{x} \tag{7.5}$$

instead of the original integral I. This truncation changing the integration domain into a bounded one is usual in this type of computations and by a suitable chosen constant c the error due to the truncation will be negligible small. The greater is c the smaller the difference between I

and I^* but on the other hand the uniformly distributed points have to be generated in a greater domain, which in turn decreases the accuracy of the computations (the same number of points over a greater area). Consequently when selecting the value c both of these partially contradictory requirements are to be taken into consideration.

A possible choice is the value $k_{n,1-\varepsilon}$, which is defined by the equation

$$P\{k_{n,1-\varepsilon}^2 \leqq \chi_n^2\} = \varepsilon,$$

where ε is a suitable small positive number and χ_n^2 is a χ^2 distributed random variable with n degree of freedom. (Some of the numerical values $k_{5,0.99} = 3.88$, $k_{5,0.999} = 4.53$, $k_{10,0.99} = 4.81$, $k_{10,0.999} = 5.44$). The reasoning behind choosing $k_{n,1-\varepsilon}$ for the value c is the following. Using the notation

$$S(r) = \left\{ \mathbf{x} \,\middle|\, \sum_{i=1}^{n} x_i^2 \leqq r^2 \right\}$$

for the hypersphere around the origin with radius r and assuming for sake of simplicity that we deal with the density function $\tilde{\varphi}(\mathbf{x})$ of the normal distribution with independent components (that is we have $R = I$) the inequality

$$1 - \varepsilon = \underset{S(k_{n,1-\varepsilon})}{\int \cdots \int} \tilde{\varphi}(\mathbf{x})\, d\mathbf{x} \leqq \underset{K(k_{n,1-\varepsilon})}{\int \cdots \int} \tilde{\varphi}(\mathbf{x})\, d\mathbf{x} \leqq \underset{S(\sqrt{n}k_{n,1-\varepsilon})}{\int \cdots \int} \tilde{\varphi}(\mathbf{x})\, d\mathbf{x}$$

holds. In other words it means that the integral of the density function $\tilde{\varphi}(x)$ outside the cube K is less than ε, that is the relation $|I - I^*| \leqq \varepsilon$ holds as well.

Apply the crude Monte Carlo method for evaluating the integral (7.5). Consider the random variables $\varphi(\mathbf{u}^{(i)})$ where $\mathbf{u}^{(i)}$, $i = 1, \ldots, N$ are uniformly distributed random vectors in the domain D^*. Denote the area of the domain D^* by T^* then the expected value of the random variable $T^*\varphi(\mathbf{u}^{(i)})$ is I^*. Consequently the sum

$$\theta_1 = \frac{1}{N} T^* \sum_{i=1}^{N} \varphi(\mathbf{u}^{(i)}) \tag{7.6}$$

is an unbiased estimator of the integral I^*. The variance of this estimator can be given as

$$D^2(\theta_1) = \frac{1}{N} \underset{D^*}{\int \cdots \int} \varphi^2(\mathbf{x})\, dx - \frac{{I^*}^2}{N}.$$

The formal algorithm of the Monte Carlo integration method based on the formula in (7.6) can be stated as follows:

NP1 Normal probability by direct integration

1. Generate the points $\mathbf{u}^{(i)}$, $i = 1, \ldots, N$ uniformly in the domain D^*.
2. Compute the sum $s \leftarrow \sum_{i=1}^{N} \varphi(\mathbf{u}^{(i)})$ and deliver the value $s \leftarrow sT^*/N$ as an approximation to the integral I^*.

The variance of the random variable $\varphi(\mathbf{u}^{(i)})$ occurring in the estimator (7.6) can be decreased generally in case of $\mathbf{h} \geqq 0$, if the integral I^* is not evaluated directly through estimator θ_1 but in a way indirectly. If the value of the integral

$$I^{**} = \int_{K-D^*} \ldots \int \varphi(\mathbf{x})\, d\mathbf{x} \tag{7.7}$$

is already available, then the value I^* can be determined from the equation

$$I^* = \int_K \ldots \int \varphi(\mathbf{x})\, d\mathbf{x} - \int_{K-D^*} \ldots \int \varphi(\mathbf{x})\, d\mathbf{x} \sim 1 - I^{**},$$

where $\sim$ indicates an approximative equality; the error of approximation results from the truncation and for practical reasons it can be neglected (generally the variance of the estimator is several times greater than the difference $|I^* - (1 - I^{**})|$).

For computing the integral I^{**} the random variable $T^{**}\varphi(\mathbf{u}^{(i)})$ is employed, where T^{**} denotes the area of the domain $K - D^* = \{\mathbf{x} | \mathbf{x} \in K, \mathbf{x} \notin D^*\}$ furthermore the random variable $\mathbf{u}^{(i)}$ is uniformly distributed in the domain $K - D^*$. The estimator for computing I^{**} is the following

$$\theta_2 = \frac{1}{N} T^{**} \sum_{i=1}^{N} \varphi(\mathbf{u}^{(i)}),$$

whose standard deviation that is the standard error of the estimator can be given as

$$D(\theta_2) = T^{**} D(\varphi(\mathbf{u}^{(i)}))/\sqrt{N}.$$

The formal algorithm is very similar to that of the estimator θ_1.

NP2 Normal probability, by indirect integration

1. Generate the vectors $\mathbf{u}_N^{(i)}$, $i = 1, \ldots, N$ uniformly in the domain $K - D^*$.

2. Compute the sum $s_2 \leftarrow \sum_{i=1}^{N} \varphi(\mathbf{u}_i)$ and let $s_2 \leftarrow s_2 T^{**}/N$.
3. Deliver the value $s_1 \leftarrow 1 - s_2$ as an estimation of the integral I^*.

If the value of the integral I^* is sufficiently large (greater than 0.6 as it is the case in most problems of stochastic programming), than the variance of the estimator θ_2 is generally less than the variance of the estimator θ_1. A heuristic explanation for this fact may be that the peak of the density function located in the origin, which sometimes (in case of small $|R|$) is very high, is inside the domain D^* and not in the integration domain $K - D^*$ of the integral I^{**}, so switching from the evaluation of I^* to I^{**} the violent changes in the integrand are avoided. Another indication of the smaller variance is given in the next section.

7.3 Comparison of integration techniques

The relative efficiency of the estimators θ_1 and θ_2 is illustrated in this section making use of a different problem involving integrals similar to the integrals of (7.5) and (7.7). Consider the task of evaluating the integral

$$J_1 = J_1(r_1) = \int_{S(r_1)} \tilde{\varphi}(\mathbf{x})\,d\mathbf{x} \tag{7.8}$$

which is the integral of the normal density function $\tilde{\varphi}$ of the normal distribution with independent components and $S(r_1)$ is the hypersphere with centre in the origin and radius r_1. To this problem will be applied the direct integration technique related to the estimator θ_1 and the indirect integration technique described in the estimator θ_2. The variances s_1^2 and s_2^2 of the random variables in the resulting estimators will be determined and compared. Assume that $\mathbf{u}$ is a uniformly distributed point in the hypersphere $S(r_1)$, then we have

$$s_1^2 = s_1^2(r_1) = D^2(V(n_1, r_1)\tilde{\varphi}(\mathbf{u})) = E(V^2(n_1, r_1)\tilde{\varphi}^2(\mathbf{u})) - J_1^2,$$

where $V(n, a)$ denotes the volume of the hypersphere $S(a)$, in particular $V(n, a) = \dfrac{2\pi^{n/2}a^n}{n\Gamma(n/2)}$. Consider now the integral

$$J_2 = J_2(r_1, r_2) = \int_{S(r_2)-S(r_1)} \tilde{\varphi}(\mathbf{x})\,d\mathbf{x},$$

which is the counterpart of the integral I^{**} in (7.7) (computed by the estimator θ_2), where r_2 is a sufficiently large radius (we can choose for instance $r_2 = k_{n,1-\varepsilon}$ for a suitable small ε), furthermore $r_2 \geqq r_1$ and $S(r_2) - S(r_1) = \{\mathbf{x}|\mathbf{x} \in S(r_2), \mathbf{x} \notin S(r_1)\}$. The hypersphere $S(r_2)$ plays now the role of the truncating hypercube K appearing in connection with the integrals I^* and I^{**}. Making use of the value of the integral J_2 one can determine J_1 through

$$J_1(r_1) = J_1(r_2) - J_2(r_1, r_2) \sim 1 - J_2(r_1, r_2)$$

just in the same way we did in connection with the integrals I^* and I^{**}. The estimators of the integrals J_1 and J_2 are

$$t_1 = \frac{1}{N} V(n, r_1) \sum_{i=1}^{N} \tilde{\varphi}(\mathbf{v}_i),$$

$$t_2 = \frac{1}{N} [V(n, r_2) - V(n, r_1)] \sum_{i=1}^{N} \tilde{\varphi}(\mathbf{u}_i),$$

respectively, where $\mathbf{v}_i$, $i = 1, \ldots, N$ are independent samples of the uniform distribution in $S(r_1)$, $\mathbf{u}_i$, $i = 1, \ldots, N$ are independent samples of the uniform distribution in $S(r_2) - S(r_1)$. Let $\mathbf{u}$ be uniformly distributed in the domain $S(r_2) - S(r_1)$ then the variance of the random variable occuring in the estimator of the indirect integration is

$$\begin{aligned} s_2^2 = s_2^2(r_1, r_2) &= D^2\left([V(n, r_2) - V(n, r_1)]\, \tilde{\varphi}(\mathbf{u})\right) \\ &= E\left([V(n, r_2) - V(n, r_1)]^2\, \tilde{\varphi}^2(y)\right) - J_2^2(r_1, r_2). \end{aligned}$$

It is easily seen that $s_2(0, r_1) = s_1(r_1)$ thus what we need for comparing the values of the dispersions s_1 and s_2 respectively is an expression for evaluating s_2.

Assume that $\mathbf{u}$ is uniform in $S(r_2) - S(r_1)$ and $r_1 \leqq r \leqq r_2$. Then we know that

$$P\left\{\tilde{\varphi}(\mathbf{u}) \geqq (2\pi)^{-n/2} e^{-r^2/2}\right\} = \frac{V(n, r) - V(n, r_1)}{V(n, r_2) - V(n, r_1)},$$

since the inequality in the paranthesis on the left hand side holds if $\mathbf{u}$ is in the hypersphere $S(r)$. From this one can derive

$$E\left([V(n, r_2) - V(n, r_1)]^2\, \tilde{\varphi}(u)\right) =$$

$$= \int_{r_1}^{r_2} [V(n, r_2) - V(n, r_1)]\, (2\pi)^{-n} e^{-r^2} \frac{2\pi^{n/2} r^{n-1}}{\Gamma(n-2)}\, dr.$$

Now apply the substitution $r^2 = y/2$ to obtain after some formal manipulations

$$E\left(\left[V(n,r_2) - V(n,r_1)\right]^2 \tilde{\varphi}(\mathbf{u})\right)$$

$$= \frac{V(n_1 r_2) - V(n_1 r_1)}{2^n \pi^{n/2}} \int_{2r_1^2}^{2r_2^2} \frac{e^{-y/2} y^{n/2-1}}{\Gamma(n/2) 2^{n/2}} \, dy.$$

Denote the distribution function of the χ^2 distribution with n degree of freedom by G_n, then the last integral can be given as

$$G_n(2r_2^2) - G_n(2r_1^2).$$

Consequently the two variances looked for can be obtained as

$$s_2^2(r_1, r_2) = \frac{V(n,r_2) - V(n,r_1)}{2^n \pi^{n/2}} \left[G_n(2r_2^2) - G_n(2r_1^2)\right] -$$

$$- \left[G_n(r_2^2) - G_n(r_1^2)\right],$$

$$s_1^2(r_1) = \frac{V(n,r_1)}{2^n \pi^{n/2}} G_n(2r_1^2) - G_n(r_1^2),$$

where we made use of the equalities $J_1(r_1) = G_n(r_1^2) = J_2(r_1, r_2) = = G_n(r_2^2) - G_n(r_1^2)$ expressing the values of integrals in terms of χ^2 distribution functions.

Some numerical values of the standard deviations s_1 and s_2 are given in Tables 18 and 19. These are in conformity with the computer experiences showing that the estimators θ_1 and θ_2 can be used up to $n = 6$ dimensions generally if we do not require too great accuracy (neither are the correlations between the individual components too great). If we have a correlated case, $R \neq E$, the above reasoning can be followed exactly the same way employing hyperellipsoids instead of hyperspheres, but the variances will be greater because of the factor $|R|^{-1/2}$ figuring in the density function $|R|^{-1/2}$.

From the comparison of the numerical values s_1 and s_2 it can be seen that in some cases great decreases (as great as tenfold) can be obtained in the standard deviations if the estimator θ_2 is employed instead of θ_1.

7.4. Importance Sampling

It is an obvious thought to generate the points normally and not uniformly, in particular we generate points according to the density function

Table 18

Numerical values of the standard deviation $s_1(r_1)$

n	ϵ	0.01	0.001
2	$k_{2,1-\epsilon}$	3.03	3.71
	$s_1(k_{2,1-\epsilon})$	1.13	1.56
6	$k_{6,1-\epsilon}$	4.10	4.73
	$s_1(k_{6,1-\epsilon})$	3.37	5.33
10	$k_{10,1-\epsilon}$	4.81	5.43
	$s_1(k_{10,1-\epsilon})$	7.27	13.43

Table 19

Numerical values of the standard deviations $s_1(r_2)$ and $s_2(r_1, r_2)$

		$J_1(r_2) = 0.7$	$J_1(r_2) = 0.9$
n=2	r_1	0.03	3.03
	r_2	1.55	2.14
	$s_1(r_2)$	0.24	0.57
	$s_2(r_1, r_2)$	0.26	0.058
n=6	r_1	4.10	4.10
	r_2	2.68	4.10
	$s_1(r_2)$	0.67	1.52
	$s_2(r_1, r_2)$	0.53	0.15
n=10	r_1	4.81	4.81
	r_2	3.43	4.00
	$s_1(r_2)$	1.15	2.77
	$s_2(r_1, r_2)$	0.67	0.20

φ to be integrated. This way the integral of concern becomes

$$I = \int_D \dots \int \varphi(\mathbf{x})/\varphi(\mathbf{x})\, d\Phi(\mathbf{x}) = \int_D \dots \int d\Phi(\mathbf{x}). \tag{7.9}$$

Denoting the characteristic function of the integration domain D by $f(\mathbf{y})$, that is defining

$$f(\mathbf{y}) = \begin{cases} 1, & \text{if } \mathbf{y} \leqq h, \\ 0, & \text{otherwise} \end{cases}$$

we can write the integral (7.9) in another form

$$I = p = \int_{R^n} \dots \int f(\mathbf{x})\, d\Phi(\mathbf{x}), \tag{7.10}$$

where the notation p was introduced for the value of the integral simultaneously. Now assume, that the random vector variables $\mathbf{x}_1, \dots, \mathbf{x}_N$ are normally distributed with density function φ, then the integral (7.10) can be estimated by the random variable $f(\mathbf{x}_i)$ or more precisely by the arithmetic average of such variables, the random variable

$$\theta_3 = \frac{1}{N}\sum_{i=1}^{N} f(\mathbf{x}_i). \tag{7.11}$$

The realization of this estimator is embodied in the following formal algorithm.

NP3 Normal probability by importance sampling ("acceptance-rejection")

1. Generate the vectors $\mathbf{x}_i$, $i = 1, \dots, N$ according to the density function φ and initialize $i \leftarrow 0$, $s \leftarrow 0$.
2. Increase the index $i \leftarrow i + 1$. If $i > N$ then go to Step 3. If $\mathbf{x}_i \leqq \mathbf{h}$ then add $s \leftarrow s + 1$. Repeat this step from the beginning.
3. Deliver the value $s \leftarrow s/N$ as an approximation to the value p.

This algorithm is called acceptance rejection algorithm, too, since the evaluation of the function value $f(\mathbf{x}_i)$ is equivalent to examining the relation $\mathbf{x}_i \in D$ (or in a little different form $\mathbf{x}_i \leqq \mathbf{h}$), which expresses that the generated vector falls into domain D. But the resemblance is only superficial. Algorithm NP3 is similar to the acceptance-rejection algorithm described in the previous chapter (in Section 6.5) only formally, in essence the evaluation of the integral (7.3) by the estimator (7.11) is an importance sampling procedure.

The random variable $f(\mathbf{x}_i)$ is binomially distributed with parameter p, since $P\{f(\mathbf{x}_i) = 1\} = p$, consequently the standard error of the estimator θ_3 is

$$D(\theta_3) = \frac{1}{\sqrt{N}}\sqrt{p(1-p)}.$$

Comparing the estimator θ_3 to the previous estimators θ_1 and θ_2 improvement can be noticed in two respects. First the standard deviation $\sigma_3 = \sqrt{p(1-p)}$ of the random variable $f(\mathbf{x}_i)$ is always less than 0.5 which is not the case for the estimators θ_1 and θ_2. What is more as it has been remarked θ_1 and θ_2 are very sensitive with regard to the value of $|R|$; if $|R|$ is near to zero, then the variances of θ_1 and θ_2 may become quite large, whilst the variance $D^2(\theta_3)$ is not dependent on $|R|$. Second the variance of the estimator θ_3 is independent of the number of dimensions though the computational effort required to generate a vector $\mathbf{x}_i$ is dependent on the value of n. Because of these two good features, lacking other reasons (speed, accuracy), for computing normal probabilities algorithm NP3 is used generally and is recommended to use. This algorithm is simple, easy to implement (using algorithm VNT for generating normally distributed vectors) and of reasonable speed in most cases.

We remark here that for probabilities p near to 1 or 0 the standard error is substantially smaller than the value 0.5 for example for $p = 0.99$ we have $\sigma_3 = 0.095$.

An expression is needed for determining the sample size N of the estimator in case of a prescribed error ε. It can be developed from the Chebyshev inequality that has the form

$$P\{|p - \theta_3| \geqq \varepsilon\} \leqq \frac{p(1-p)}{N\varepsilon^2}$$

in our case and gives an upper bound for the probability of the event, that the error will be greater than an ε in terms of p, N and ε. Given a reliability level δ (the right-hand side of the inequality) the required sample size is

$$N = \frac{p(1-p)}{\delta\varepsilon^2}. \tag{7.12}$$

Generally the value of p is not known in advence (this is the value to be computed) thus a practical approach would require to evaluate an approximate value p^* by using a small sample size $N = 25$, say, and then in turn this p^* can be substituted into (7.12) to obtain a value for N. Assuming the same level δ of reliability and the same error ε we can obtain a smaller sample size if we use the a version of the Chebyshev inequality obtained by Bernstein, which is of the form

$$P\{|p - \theta_3| \geqq \varepsilon\} \leqq 2\exp\left[-\frac{N\varepsilon^2}{2p(1-p)\left[1 + \frac{\varepsilon}{2p(1-p)}\right]^2}\right]. \tag{7.13}$$

The right-hand side of the inequality is made equal to the reliability δ, generally one is satisfied with $\delta = 0.05$. From (7.13) given ε and δ — the sample size can be obtained as

$$N = -\frac{1}{\varepsilon^2}2p(1-p)\left[1 + \frac{\varepsilon}{2p(1-p)}\right]^2 \ln(0.5\delta). \tag{7.14}$$

Some numerical values are given here with reliability level $\delta = 0.05$ (roughly it means that the error of the result will be less than the prescribed ε in 19 cases out of 20): if $\varepsilon = 0.05$ and $p = 0.5$ then $N = 890$, for $\varepsilon = 0.01$ and $p = 0.5$, $N = 19{,}143$.

Algorithm NP3 generally yields satisfactory results and its speed can also be termed adequate. By the aid of NP3 on a medium size and speed computer a two digit accurate result can be obtained in a few seconds up to $n = 20$ dimensions. Less accurate results can be evaluated up to $n = 50$ dimensions in longer but still acceptable length of time.

In special cases, if the probability p is near to 1 or 0, algorithm NP3 can be restructured and modified in order to yield a faster version. These two versions will be dealt with in the remainder of this section and in the next section.

In case of small probabilities p the amount of computational work of computing the function value $f(\mathbf{x}_i)$ or in other words testing the inequality $\mathbf{x}_i \leqq \mathbf{h}$ can be decreased by a fair part. The basic idea is that not all components of the vector $\mathbf{x}_i$ to be generated will be computed in advance and probably they will not be needed even later on in all cases. For a given vector $\mathbf{h}$ the generated vector $\mathbf{x}$ is not in the domain $D = \{\mathbf{y} | \mathbf{y} \leqq \mathbf{h}\}$ if for at least one index k the inequality $x_k > h_k$ holds. That is if one component of the vector $\mathbf{x}$ is not in D, then the whole vector $\mathbf{x}$ is outside of D which means $f(\mathbf{x}) = 0$. Thus the determination of the other components is superfluous since all we need is the function value $f(\mathbf{x})$. Consequently it is expedient to compute the components of $\mathbf{x}$ one by one and having determined x_k the test of the inequality $x_k > h_k$ should be carried out prior to computing the next component. As far as the order of the components to be computed is concerned one can obviously see that first should come the components that have greater probability of being outside of domain D. It means that an order of indices is looked for in case of which the probabilities

$$P\{x_k > h_k\}, \qquad k = 1, \ldots, n$$

are presenting a monotone decreasing series; let us denote by $k_1, \ldots, k_n$ that permutation of the indices for which

$$P\{x_{k_i} > h_{k_i}\} \geqq P\{x_{k_{i+1}} > h_{k_{i+1}}\}.$$

Using this order $k_1, \ldots, k_n$ for evaluating components of $\mathbf{x}$ it is ensured, that the probability of the event, that the first computed component will be outside of D is greater or equal the probability of the second component's falling out. The distribution of the one-dimensional random variable x_i is the marginal distribution of the random vector $\mathbf{x}$, which has a multi-dimensional normal density function φ, thus according to a well-known theorem it is a one-dimensional normal distribution with expectation 0 and variance 1 for all indices k. This indicates that the looked for order of the indices can be obtained with the increase of the components h_k, $k = 1, \dots, n$, that is $k_1, \dots, k_n$ is determined by

$$h_{k_1} \leqq h_{k_2} \leqq \dots \leqq h_{k_n}.$$

This ordering has to be done only once for each vector $\mathbf{h}$ as a preparation and then the components of each generated vector are evaluated according to this order. The modification of algorithm NP3 done according to the above remarks is given as

NP4 Normal probability, "acceptance-rejection" for small probability

1. Initialize $i \leftarrow 0$, $N_1 \leftarrow 0$.
2. Increase the index $i \leftarrow i + 1$. If $i > N$ then go to Step 5. Generate y_j, $j = 1, \dots, n$ from the standard one-dimensional normal distribution and set $j \leftarrow 0$.
3. [Compute and test the components of $\mathbf{x}_i$ one by one.] Increase $j \leftarrow j+1$ if $j > n$, then go to Step 4. Compute the sum $s \leftarrow \sum_{q=1}^{k_j} y_q t_{k_j q}$ (here t_{ij} is the (i, j) element of the matrix $\mathbf{T}$, for which $\mathbf{TT}' = \mathbf{R}$). If now $s > h_{k_j}$ then go back to Step 2 (the generated vector is not in the domain D), otherwise repeat this step from the beginning.
4. [Acceptance, the vector $\mathbf{x}^{(i)}$ is in domain D.] Increase $N_1 \leftarrow N_1 + 1$ and go back to Step 2.
5. Deliver the value $s \leftarrow N_1/N$ as an approximation to the integral I.

The adventage of algorithm NP4 over algorithm NP3 is that, roughly speaking, the whole vector $\mathbf{x}_i$ is to be computed only in pN cases not in all N cases. In the remaining $N - pN$ cases only several but at any rate not all the components are to be determined. Since when generating normally distributed vectors the multiplication by matrix $\mathbf{T}$ is the most time consuming affair, the execution time is substantially decreased for small values of p. The method is recommended for $p \leqq 0.1$.

7.5. Control variates method

For great probabilities p (near to 1) it seems worthwhile to exactly integrate a part of the density function and apply the acceptance-rejection described in the previous section only to the remaining part of the integrand. The basic idea of the algorithm makes use of the fact that the integral of the normal density function over a hyperellipsoid can be expressed exactly. We elaborate on this idea. Consider the hyperellipsoids given by $E = \{\mathbf{x}|\mathbf{x}'\mathbf{R}^{-1}\mathbf{x} \leqq k^2\}$ and denote the greatest one inside domain D by E_c. (The method to be described below works for $\mathbf{h} > \mathbf{0}$ only.) The constant related to this greatest hyperellipsoid will be denoted by k_c, that is let

$$E_c = \{\mathbf{x}|\mathbf{x}'\mathbf{R}^{-1}\mathbf{x} \leqq k_c^2\}, \qquad E_c \in D \tag{7.14}$$

and $E \notin D$ if $k > k_c$. Denote the distribution function of the χ distribution with n degree of freedom by F_n, then we can write

$$\int_{E_c} \varphi(\mathbf{x})\,\mathrm{d}\mathbf{x} = F_n(k_c)$$

and this is the part of the integral that can be expressed explicitly in terms of F_n; F_n is a one-dimensional function easily computable numerically. Making use of this equation one can obtain

$$\begin{aligned} I = \int_D \varphi(\mathbf{x})\,\mathrm{d}(\mathbf{x}) &= \int_{E_c} \varphi(\mathbf{x})\,\mathrm{d}(\mathbf{x}) + \int_{D-E_c} \varphi(\mathbf{x})\,\mathrm{d}(\mathbf{x}) = \\ &= F_n(k_c) + \int_{R^n} f_{D-E_c}(y)\,\mathrm{d}\Phi(y), \end{aligned}$$

where $f_{D-E_c}(\mathbf{y})$ is the characteristic function of the domain $D - E_c$, that is

$$f_{D-E_c}(y) = \begin{cases} 1, & \mathbf{y} \in D - E_c, \\ 0 & \text{otherwise.} \end{cases}$$

The estimator of I corresponding to this decomposition is

$$\theta_5 = F_n(k) + \frac{1}{N}\sum_{i=1}^{N} f_{D-E_c}(\mathbf{y}_i), \tag{7.15}$$

where $\mathbf{y}_i$, $i = 1, \ldots, N$ are samples from the normal distribution with density φ subject to the condition $\mathbf{y}_i \in D - E_c$. Such vectors can easily be produced by the multidimensional polar method only a modification should be inserted into the vector generating algorithm VNP restricting the "length" of the vector be greater than k_c. The formal algorithm can now be described in full details.

NP5 Normal probability, "acceptance-rejection" for great probability

1. Set $N_1 \leftarrow 0$ and $i \leftarrow 0$.
2. Increase $i \leftarrow i + 1$, if $i > N$ then go to Step 4. Generate a sample k from the χ distribution with n degree of freedom, for which $k \geqq k_c$. (It could be done by generating u uniformly between $[F_n^{-1}(k_c), 1]$ and taking $k \leftarrow F_n^{-1}(u)$.)
3. Generate a vector $\mathbf{x}$ which is uniformly distributed with respect to the volume of the ellipsoid E and compute $\mathbf{y} \leftarrow k\mathbf{x}$. If $\mathbf{y} \in D$ then increase the counter $N_1 \leftarrow N_1 + 1$. Go back to Step 2.
4. Deliver the value $s \leftarrow F_n(k_c) + N_1/N$ as an approximation to the original probability p.

Generally, if p is near to 1, say $p = 0.95$ than one can expect less computational work using NP5 as compared to NP3. One must furthermore be aware, that the numerical determination of the hyperellipsoid E_c or its constant k_c is a separate task, to which seems to be no immediate solution. Nevertheless for the normal distribution with independent components one can easily see that $k_c = \min_i h_i$.

7.6. Decreasing the number of variables

When describing the one-dimensional versions of the Monte Carlo methods we formulated the principle that the role of randomness in simulational experiences should be decreased since by the same the accuracy of the estimator is increased. This principle is applied now in evaluating the integral I; the integration is carried out according to one of the variables — we integrate along a line.

Let $\mathbf{x}$ be a normally distributed random vector with density function φ, it can be given in the form

$$\mathbf{x} = k\mathbf{T}\mathbf{z},$$

too, where k is a χ distributed random variable with n degree of freedom and with distribution function F_n, $\mathbf{z}$ is a random vector uniformly distributed on the surface

$$S_s = \left\{ \mathbf{x} \,\middle|\, \sum_{i=1}^{n} x_i^2 = 1 \right\}$$

of the unit sphere, k and $\mathbf{z}$ are independent of each other and finally $\mathbf{T}$ is the transformation matrix for which $\mathbf{TT}' = \mathbf{R}$. Integral (7.10) estimated by θ_3 can be decomposed into the double integral

$$I = \int_{R^n} f(\mathbf{x})\, d\Phi(\mathbf{x}) = \int_{E_s} \int_0^{\infty} f(k\mathbf{y})\, dF_n(k)\, dV(\mathbf{y}), \tag{7.16}$$

where $V(\mathbf{y})$ is the distribution function of the distribution which is uniform with respect to the ellipsoid $E = \{\mathbf{q}|\mathbf{q}'R^{-1}\mathbf{q} \leq 1\}$, E_s denotes the surface of the ellipsoid E, $E_s = \{\mathbf{q}|\mathbf{q}'R^{-1}\mathbf{q} = 1\}$ (or in other words E_s is the $\mathbf{T}$ transform of the hypersphere's surface S_s, i.e. $E_s = \{\mathbf{q}|\mathbf{q} = T\mathbf{x}, \mathbf{x} \in S_s\}$). The inner integral in the double integral (7.16) — assuming a given vector $\mathbf{y} \in E_s$ — can be expressed by the help of F_n. In particular the inner integral gives the probability of the event that a vector $k\mathbf{y}$ (with $\mathbf{y}$ fixed, k is χ distributed with distribution function F_n) falls into the domain D.

Define the function $e_1(\mathbf{y})$ for vectors $\mathbf{y}$ on the surface E_s of the hyperellipsoid E as follows

$$e_1(\mathbf{y}) = \int_0^{\infty} f(k\mathbf{y})\, dF_n(k), \qquad \mathbf{y} \in E_s. \tag{7.17}$$

Using this function, integral (7.16) can be rewritten as

$$I = \int_{R^n} f(\mathbf{x})\, d\Phi(\mathbf{x}) = \int_{E_s} e_1(\mathbf{y})\, dV(\mathbf{y}). \tag{7.18}$$

Note that the integral on the right-hand side consists of integration over an $(n-1)$-dimensional surface which is situated in an n-dimensional space, thus as compared to the original integral (7.16) the number of the free variables is decreased by one; we have integrated along the line $\mathbf{y}$. For evaluating the right-hand side of (7.18) the crude Monte Carlo method is employed, that is an unbiased estimator of I can be given as

$$\theta_6 = \frac{1}{N} \sum_{i=1}^{N} e_1(\mathbf{y}_i), \tag{7.19}$$

where $\mathbf{y}_i$, $i = 1, \ldots, N$ are independent samples with distribution function $V(\mathbf{y})$. The formal algorithm corresponding to the estimator θ_6 is stated below.

NP6 Normal probability, integration along a line

1. Initialize $i \leftarrow 0$, $s \leftarrow 0$.
2. Increase the index $i \leftarrow i + 1$. If $i > N$, then go to Step 4.

3. Generate the vector $\mathbf{y}_i$ according to the distribution function V (i.e. uniformly with respect to the volume of the hyperellipsoid E) and add $s \leftarrow s + e_1(\mathbf{y}_i)$. Go back to Step 2.
4. Deliver the value $s \leftarrow s/N$ as an approximation to the value of I.

A bit of explanation here seems to be appropriate. The original integral

$$p = \int_{R^n} f(\mathbf{x})\, d\Phi(\mathbf{x})$$

gives the probability of the event, that a normally distributed vector $\mathbf{x}$ falls into the domain D, that is

$$p = P\{\mathbf{x} \leqq \mathbf{h}\}.$$

Substituting $\mathbf{x} = k\mathbf{y}$ we obtain

$$p = P\{k\mathbf{y} \leqq \mathbf{h}\} = \int_{E_s} P\{k\mathbf{y} \leqq \mathbf{h} | \mathbf{y} = \mathbf{x}\}\, dV(\mathbf{x}) = \int_{E_s} e_1(\mathbf{y})\, dV(\mathbf{y}).$$

The crucial point of the algorithm is the easy computability of the function e_1. The general case will be described in a separate section.

7.7. Antithetic variates technique

We want to construct an estimator in which the arithmetic average of negatively correlated random variables is involved. For this purpose the previous estimator θ_6 is modified in the following way: the vector $-\mathbf{y}$, too, will be used in the estimator besides the vector $\mathbf{y}$, this corresponds to the simple symmetrization in one-dimensional problems. Introduce the notation

$$e_2(\mathbf{y}) = e_1(\mathbf{y}) + e_1(-\mathbf{y}) \tag{7.20}$$

by the aid of which the new estimator can be given as

$$\theta_7 = \frac{1}{N} \sum_{i=1}^{N} \frac{1}{2} e_2(\mathbf{y}). \tag{7.21}$$

The random variables $e_1(\mathbf{y})$ and $e_1(-\mathbf{y})$ are negatively correlated in fact if we assume $\mathbf{h} = \mathbf{0}$. In this special case even the variance of the estimator can be evaluated in explicit terms. Since $P\left\{\frac{1}{2}e_2(\mathbf{y})=0.5\right\}=2p$ and $P\left\{\frac{1}{2}e_2(\mathbf{y})=0\right\}=1-2p$, thus we have

$$D^2(\theta_7)=\frac{1}{N}\cdot\left(2p\frac{1}{4}-p^2\right)=\frac{p(0.5-p)}{N}<D^2(\theta_3)=\frac{p(1-p)}{N}.$$

7.8. Orthonormalized estimators

We will make use once more of the notion of antithetic variates technique by providing points "uniformly" on the hyperellipsoid surface $E_s=\{\mathbf{x}|\mathbf{x}'\mathbf{R}^{-1}\mathbf{x}=1\}$. The procedure to be described can be viewed as a multidimensional generalization of the formula (6.31) of the previous chapter yielding a multiple symmetrization.

Consider the random matrix $\mathbf{B}=(\mathbf{b}_r,\ldots,\mathbf{b}_n)$ whose components $\mathbf{b}_1,\ldots,\mathbf{b}_n$ are orthonormalized n-dimensional vectors (that is $\mathbf{b}_i'\mathbf{b}_j=\delta_{ij}$, where $\delta_{ij}=0$, if $i\neq j$ and $\delta_{ij}=1$ if $i=j$) and $\mathbf{B}$ is uniformly distributed on the group of all possible $n(n-1)/2$ dimensional orthonormalized matrices: $\mathbf{B}$ is a randomly oriented orthonormalized system of vectors. Consider its transform $\mathbf{U}=\mathbf{TB}$, the system of vectors $U=(\mathbf{u}_1,\ldots,\mathbf{u}_n)$, $\mathbf{u}_i=\mathbf{Tb}_i$ thus obtained is R^{-1} orthonormalized, that is $\mathbf{u}_iR^{-1}\mathbf{u}_j=\delta_{ij}$. Apply now the estimator θ_7 for each vector $\mathbf{u}_i$ that is let r be the following random variable

$$r=\frac{1}{n}\sum_{i=1}^{n}\frac{1}{2}e_2(\mathbf{u}_i).$$

Generating N realizations $r^{(j)}$, $j=1,\ldots,N$ of this random variable we can give the estimator θ_8 as their arithmetic average

$$\theta_8=\frac{1}{N}\sum_{j=1}^{N}r^{(j)}. \tag{7.22}$$

An orthonormalized system B can be exploited better than that, with a little additional work the number n of points on E_s can substantially be increased. Consider the vectors $\mathbf{b}_i$, $i=1,\ldots,n$, then we can choose k vectors out of them in $\binom{n}{k}$ different ways; denote the result of one of such choices by $\mathbf{b}_{i_1},\ldots,\mathbf{b}_{i_k}$. Now we construct the vectors

$$\mathbf{b}(\mathbf{s},\mathbf{i})=\frac{1}{\sqrt{k}}\left[s_1\mathbf{b}_{i_1}+s_2\mathbf{b}_{i_2}+\ldots+s_k\mathbf{b}_{i_k}\right],$$

where $\mathbf{s}$ is a sign vector, $s_j = +1$ or $s_j = -1$, $j = 1, \ldots, k$, $\mathbf{i}$ is an index vector $\mathbf{i} = (i_1, \ldots, i_k)$. The vectors $\mathbf{b}(\mathbf{s}, \mathbf{i})$ are of unit length. For a given fixed set $\mathbf{i} = (i_1, \ldots, i_k)$ there are 2^{k-1} different choices of the sign vector $\mathbf{s}$ to yield different lines determined by the vectors $\mathbf{b}(\mathbf{s}, \mathbf{i})$. This means that by choosing any k among the vectors of the system B and assigning to them all possible different signs we can produce $2^{k-1}\binom{n}{k}$ different directions.

The estimator that arises from applying the estimator θ_7 for these $2^{k-1}\binom{n}{k}$ vectors $\mathbf{b}(\mathbf{s}, \mathbf{i})$ (or more precisely for their transforms $\mathbf{Tb}(\mathbf{s}, \mathbf{i})$) will be denoted by O_k. Formally the estimator O_k can be expressed for N different realizations of the system $\mathbf{b}_1^{(\ell)}, \ldots, \mathbf{b}_n^{(\ell)}$, $\ell = 1, \ldots, N$ as follows

$$O_k = \frac{1}{N} \sum_{\ell=1}^{N} \frac{1}{2^k \binom{n}{k}} \sum_{\mathbf{s}, \mathbf{i}} e_2(\mathbf{Tb}(\mathbf{s}, \mathbf{i})).$$

The estimator θ_8 is of course the estimator O_1. Note here that the matrix multiplication $\mathbf{Tb}(\mathbf{s}, \mathbf{i})$ should not be computed for all $2^{k-1}\binom{n}{k}$ vectors but for the n basic vectors $\mathbf{b}_i$. For having computed and stored the vectors $\mathbf{u}_i = \mathbf{Tb}_i$ the vectors $\mathbf{Tb}(\mathbf{s}, \mathbf{i})$ appearing in the estimator O_k can be obtained by simple addition and subtraction instead of matrix multiplication:

$$\mathbf{Tb}(\mathbf{s}, \mathbf{i}) = \mathbf{T}\left\{\frac{1}{\sqrt{k}} \sum_{j=1}^{k} s_j \mathbf{b}_{i_j}\right\} = \frac{1}{\sqrt{k}} \sum_{j=1}^{k} s_j \mathbf{u}_{i_j}.$$

This way considerable savings can be made in the computational work of evaluating the estimator O_k.

A preliminary version is given below for evaluating the estimator O_2, it was chosen as a reasonable compromise between speed and sophistication.

NP7 Normal probability, orthonormalized-2 estimator for distribution function (preliminary version)

1. Set $q \leftarrow 0$, $p \leftarrow 0$.
2. Increase $q \leftarrow q + 1$. If $q > N$, then go to Step 5. Generate an orthonormalized system of vectors $\mathbf{b}_1, \ldots, \mathbf{b}_n$, compute $\mathbf{u}_i \leftarrow \mathbf{Tb}_i$, $i = 1, \ldots, n$ and set $r \leftarrow 0$.
3. Compute for all possible pairs (i, j), $i, j = 1, \ldots, n$, $i \neq j$ and sign combinations $\mathbf{s} = \pm\mathbf{1}$ the vectors $\mathbf{y}(i, j, \mathbf{s}) \leftarrow \dfrac{1}{\sqrt{2}}(s_1\mathbf{u}_i + s_2\mathbf{u}_j)$ and add $r \leftarrow r + e_1(\mathbf{y}(i, j, s))$.

4. Set $p \leftarrow p + r / \left[2^n \binom{n}{2}\right]$ and go back to Step 2.
5. Deliver the value $p \leftarrow p/N$ as the probability I, given by the distribution function $\Phi(\mathbf{h})$.

More details on the algorithm, including the generations of an orthonormalized system of vectors and the evaluation of function e_1 will be given in subsequent sections.

7.9. Probability of a rectangle

The last algorithm NP7 is for computing the probability of an infinite domain D given as

$$I = \Phi(\mathbf{h}) = \int_{-\infty}^{h_n} \dots \int_{-\infty}^{h_1} \varphi(\mathbf{x})\, d\mathbf{x}. \tag{7.23}$$

The same technique can be applied for evaluating the probability content q of a rectangle $Q = \{\mathbf{y} | \mathbf{a} \leqq \mathbf{y} \leqq \mathbf{b}\}$ given by the integral

$$J = q = \int_{a_n}^{b_n} \dots \int_{a_1}^{b_1} \varphi(\mathbf{x})\, d\mathbf{x}, \tag{7.24}$$

where $\mathbf{a}$ and $\mathbf{b}$ are given and $\mathbf{a} \leqq \mathbf{b}$; the only modification concerns the probability content of a line inside Q.

More precisely define $g(\mathbf{x})$ to be the indicator function of Q, that is

$$g(\mathbf{x}) = \begin{cases} 1, & \text{if } \mathbf{x} \in Q, \\ 0, & \text{otherwise,} \end{cases}$$

then we can state the following unbiased estimator of q as

$$\theta_9 = \frac{1}{N} \sum_{i=1}^{N} g(\mathbf{x}_i), \tag{7.25}$$

Where $\mathbf{x}_i$, $i = 1, \dots, N$ are independent normal samples with density function φ. The formal algorithm is given here

NP8 Normal probability, importance sampling (acceptance-rejection) for a rectangle

1. Generate the vectors $\mathbf{x}_i$, $i = 1, \dots, N$ according to the density function φ and initialize $i \leftarrow 0$, $s \leftarrow 0$.
2. Increase the index $i \leftarrow i + 1$. If $i > N$, then go to Step 3. If $\mathbf{x}_i \ngeqq \mathbf{a}$, then repeat this step from the beginning. If $\mathbf{x}_i \leqq \mathbf{b}$ then add $s \leftarrow s + 1$. Repeat this step from the beginning.

3. Deliver the value $s \leftarrow s/N$ as an approximation to the integral J.

Similarly to the decomposition of I we can write

$$q = \int_Q \varphi(\mathbf{x})\,d\mathbf{x} = \int_{R^n} g(\mathbf{x})\,d\Phi(\mathbf{x}) = \int_{E_s} \int_0^\infty g(k\mathbf{y})\,dF_n(k)\,dV(\mathbf{y}) =$$

$$= \int_{E_s} \frac{1}{2} e_3(\mathbf{y})\,dV(\mathbf{y}),$$

where the function e_3 was introduced to give the probability content of a line with respect to the domain Q, that is

$$e_3(\mathbf{y}) = \int_{r\mathbf{y}\in Q} g(k\mathbf{y})\,dF_n(k), \qquad y \in E_s. \tag{7.26}$$

In other words $\frac{1}{2}e_3(\mathbf{y})$ gives the probability of the event that given a line $r\mathbf{y}$, and generating a χ distributed random variable k with n degree of freedom, the vector $k\mathbf{y}$ is in Q. (The factor $\frac{1}{2}$ is needed to obtain probabilities; if the integration in (7.26) was restricted to $r\mathbf{y} \in Q$, $r > 0$, the resulting function $e_3'(\mathbf{y})$ would be a probability.) Thus the corresponding estimator of q is given by

$$\theta_{10} = \frac{1}{N} \sum_{i=1}^{N} \frac{1}{2} e_3(\mathbf{y}_i),$$

where $\mathbf{y}_i$, $i = 1, \dots, N$ are independent realizations of the random vector $\mathbf{y}$ uniformly distributed with respect to the volume of the hyperellipsoid E. So this estimator θ_{10} is the counterpart of the estimator θ_7.

We can develop orthonormalized estimators T_k of probability q similarly to the estimators O_k of the probability p. Below we give the formal algorithm in a preliminary version for the estimator T_2, which is the estimator arising from application of θ_{10} to all the lines obtained via adding up in all possible ways two vectors of an orthonormalized system of vectors.

NP9 Normal probability, orthonormalized-2 estimator for the probability of a rectangle (preliminary version)

1. Initialize $\ell \leftarrow 0$, $q \leftarrow 0$.
2. Increase $\ell \leftarrow \ell + 1$. If $\ell > N$ then go to Step 5. Generate an orthonormalized system of vectors $\mathbf{b}_1, \dots, \mathbf{b}_n$, compute $\mathbf{u}_i \leftarrow \mathbf{T}\mathbf{b}_i$, $i = 1 \dots, n$, and set $r \leftarrow 0$.

3. Compute for all possible index pairs (i, j), $i, j = 1, \ldots, n$, $i \neq j$ and the sign combinations ($s_1 = +1$, $s_2 = +1$ or $s_i = +1$, $s_2 = -1$) the vectors $\mathbf{y}(i, j, s_1, s_2) \leftarrow \frac{1}{\sqrt{2}}(s_1 \mathbf{u}_i, +s_2 \mathbf{u}_j)$ and add $r \leftarrow r + e_3(\mathbf{y}(i, j, s_1, s_2))$.
4. Set $q \leftarrow q + \Big/ \left[2^n \binom{n}{2}\right]$ and go back to Step 2.
5. Deliver $q \leftarrow q/N$ as an estimator of the integral J.

We would like to note here that similar techniques can obviously be devised for any n-dimensional domain to obtain its probability content. The practical viability is however restricted by the condition whether one can develop a computationally suitable procedure for computing the probability content of a line inside that domain or not (in our case functions e_2 and e_3).

7.10. Probability content of a line

Here algorithms are developed for evaluating the probability content of a ray or a line, that is how can one obtain function values $e_1(\mathbf{y})$, $e_2(\mathbf{y})$ or $e_3(\mathbf{y})$, used in previous estimators.

First consider a simpler case, the probability content of a ray $r\mathbf{y}$, $r \geqq 0$, if we assume in addition, that $\mathbf{h} \geqq 0$. This particular case ($\mathbf{h} \geqq 0$) is of special importance in practice and for this reason we develop the necessary algorithms in detail. Define the function

$$e_{1+}(\mathbf{y}) = \int_0^\infty f(k\mathbf{y})\, dF_n(k), \qquad \mathbf{y} \in E_s, \qquad \mathbf{h} \geqq 0, \tag{7.27}$$

which gives the probability of the event that $k\mathbf{y} \in D = \{\mathbf{x} | x \leqq \mathbf{h}\}$ where k is χ distributed random variable with n degree of freedom and $\mathbf{y}$ is given. (The + in the notation e_{1+} emphasizes the positiveness of $\mathbf{h}$.) It can be rewritten as

$$e_{1+}(\mathbf{y}) = F_n(\max_{\substack{r\mathbf{y} \leqq \mathbf{h} \\ r \geqq 0}} r).$$

Consider the intersections of the ray $r\mathbf{y}$ with the hyperplanes $\{\mathbf{x} | x_i = h_i\}$, $i = 1, \ldots, n$; obviously the first intersection (smallest r) is of interest for us. Thus the argument of the right-hand side function can be obtained as (Fig. 36)

$$\max_{\substack{r\mathbf{y} \leqq \mathbf{h} \\ r \geqq 0}} r = \begin{cases} \min\limits_{i: y_i > 0} \dfrac{h_i}{y_i}, & \text{if } \mathbf{y} \nleqq \mathbf{0} \\ +\infty & \text{if } \mathbf{y} \leqq \mathbf{0}. \end{cases}$$

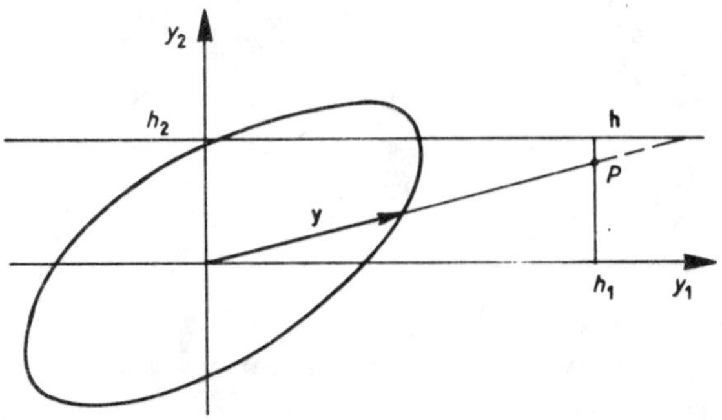

Fig. 36. Intersection of the ray $r\mathbf{y}$ with $0 = \{\mathbf{y} \leqq \mathbf{h}\}$

It means that the evaluation of $e_{1+}(\mathbf{y})$ consists of taking the minimum of at most n quotients and evaluating a one-dimensional distribution function F_n at that point.

The next step is to develop an algorithm for evaluating the function

$$e_{2+}(\mathbf{y}) = e_{1+}(\mathbf{y}) + e_{1+}(-\mathbf{y}), \tag{7.28}$$

which is the probability content of the line $r\mathbf{y}$, $-\infty < r < +\infty$ (the assumption $\mathbf{h} \geqq 0$ is still needed). Of course this value will be in the interval $[0, 2]$. Evaluation of the value $e_{2+}(\mathbf{y})$ means determining those two constant c_1 and c_2, for which $c_1 = \min\limits_{r\mathbf{y}\in D} r$, $c_2 = \max\limits_{r\mathbf{y}\in D} r$. Then $e_{2+}(\mathbf{y}) =$ $= F_n(-c_1) + F_n(c_2)$ since $-c_1 \geqq 0$, $+c_2 \geqq 0$ because of $\mathbf{h} \geqq \mathbf{0}$. The algorithm proceeds from component to component and is looking for the minimum of $h_i/y_i - s$ or their maximum dependent on whether h_i/y_i is positive or not. The formal algorithm is

S8 Subalgorithm, Computation of the function $e_{2+}(\mathbf{y})$

1. Set $c_1 \leftarrow K$, $c_2 \leftarrow K$ and $i \leftarrow 0$ (where k is a large positive constant, $K = 10^{10}$ say).
2. Increase the index $i \leftarrow i + 1$. If $i > n$, then go to Step 5. If $y_i = 0$, then change $y_i \leftarrow 1/K$.
3. [Testing the ith component, here the argument $\max r$ is computed for $e_{1+}(\mathbf{y})$.] Set $s \leftarrow h_i/y_i$, if $s < 0$ then go to Step 4. If $s < c_1$, then change $c_1 \leftarrow s$. Go back to Step 2.
4. [Preparing the evaluation of $e_{1+}(-\mathbf{y})$.] If $-s < c_2$, then set $c_2 \leftarrow -s$. Go back to Step 2.
5. Deliver the value $s \leftarrow F_n(c_1) + F_n(c_2)$.

In the case of the orthonormalized estimators there is a great number of vectors for which the function e_{2+} is to be evaluated so it seems worthwhile to improve algorithm S8. It will be shown how the division in Step 3 can be avoided at the cost of some preparations. For the sake of simplicity this modified function evaluation technique will be described in case of vectors $(\mathbf{u}_i+\mathbf{u}_j)/\sqrt{2}$ required in course of evaluating the estimator O_2.

Simultaneously with computing the vectors $\mathbf{u}_i = \mathbf{T}\mathbf{b}_i$ evaluate and store the quantities

$$v_{ij} = \frac{u_{ij}}{h_j}, \qquad i,j = 1,\dots,n,$$

where u_{ij} is the jth component of the vector $\mathbf{u}_i$. Here we assume that $h_i \neq 0$, otherwise the value of h_i is changed to a small positive value. When evaluating e_{2+} at the point $\frac{1}{\sqrt{2}}(\mathbf{u}_i + \mathbf{u}_j)$ the following minimum is required

$$\begin{aligned} r &= \min_k \left\{ \frac{h_k}{(u_{ik}+u_{jk})/\sqrt{2}} \,\middle|\, \frac{h_k}{u_{ik}+u_{jk}} > 0 \right\} \\ &= \min_k \left\{ \frac{\sqrt{2}}{v_{ik}+v_{jk}} \,\middle|\, v_{ik}+v_{jk} > 0 \right\} \qquad (7.29) \\ &= \frac{\sqrt{2}}{\max_k \{v_{ik}+v_{jk} \mid v_{ik}+v_{jk} > 0\}}. \end{aligned}$$

Of course the minimum can similarly be evaluated for indices k given by the condition $\frac{h_k}{(u_{ik}+u_{jk})} < 0$. By this modification for computing r we need only one division instead of n. Straightforward application of the algorithm S8 would need $n^2(n-1)$ division in case of estimator O_2, the above variant would require $n^2 + 2n(n-1) = 3n^2 - 2n$ divisions (n^2 divisions are needed to obtain the values v_{ij}). We do not give here a formal description of the version of algorithm S8 incorporating the modification suggested by (7.29) but we build it into the final algorithm NP10 of the orthonormalized-2 estimator instead.

Now we turn to the general case of obtaining the probability content of a line inside D, that is we are interested in evaluating the function

$$e_2(\mathbf{y}) = e_1(\mathbf{y}) + e_1(-\mathbf{y}) = \int_0^{\infty} f(k\mathbf{y})\,\mathrm{d}F_n(k) + \int_0^{+\infty} f(-k\mathbf{y})\,\mathrm{d}F_n(k). \qquad (7.30)$$

(Now the assumption on positivity of h_i is dropped.) This means determining the "entry" and "exit" constants of the line $r\mathbf{y}$ respectively, that is those values c_1 and c_2, for which $r\mathbf{y} \in D$, if $c_1 \leqq r \leqq c_2$. The probability content $e_2(\mathbf{y})$ then can be obtained from the following expression

$$(7.31) \qquad e_2(\mathbf{y}) = \begin{cases} F_n(c_2) - F_n(c_1), & \text{if } c_1, c_2 \geqq 0, \\ F_n(-c_1) + F_n(c_2), & \text{if } c_1 < 0, c_2 > 0, \\ F_n(-c_2) - F_n(-c_1), & \text{if } c_1 < 0, c_2 < 0, \\ 0, & \text{if } c_1 > c_2. \end{cases}$$

Now we set out to build up an algorithm SE2 to determine c_1 and c_2.

Consider the halfspace $D_i = \{\mathbf{x}|x_i \leq h_i\}$, given by the hyperplane $H_i = \{\mathbf{x}|x_i = h_i\}$, $i = 1, \dots, n$ and denote the entry and exit constants with regard to $\mathbf{y}$ and D_i by c_{1i} and c_{2i} respectively, that is c_{1i} and c_{2i} are defined as

$$r\mathbf{y} \in D_i, \qquad \text{if } c_{1i} \leqq r \leqq c_{2i}, \qquad i = 1, \dots, n.$$

In order to determine c_{1i} and c_{2i} for a fixed i we have to investigate four different cases. Assuming $y_i \neq 0$ and $\{D_i \cap r\mathbf{y}\}$ is not empty and denoting $s_i = h_i/y_i$ we have

(i) if $h_i \leqq 0$, $y_i > 0$, then $c_{1i} = -\infty$, $c_{2i} = s_i$,
(ii) if $h_i \leqq 0$, $y_i < 0$, then $c_{1i} = s_i$, $c_{2i} = +\infty$,
(iii) if $h_i > 0$, $y_i > 0$, then $c_{1i} = -\infty$, $c_{2i} = s_i$,
(iv) if $h_i > 0$, $y_i < 0$, then $c_{1i} = s_i$, $c_{2i} = +\infty$.

Since $D = \bigcap_{i=1}^{n} D_i$, the line $r\mathbf{y}$ is inside D if it is inside all the halfspaces D_i, $i = 1, \dots, n$, that is we have to determine the common part of the intervals (c_{1i}, c_{2i}) setting

$$c_1 = \max_i c_{1i}, \qquad c_2 = \min_i c_{2i}.$$

The formal algorithm for determining the constants c_1 and c_2 may be the following

SE2 Subalgorithm, Computation of c_1, c_2 ($r\mathbf{y} \in D$ if $c_1 \leqq r \leqq c_2$)

1. [Preparation.] Set $c_1 \leftarrow -K$, $c_2 \leftarrow K$, $i \leftarrow 0$ ($K = 10^{10}$).
2. Increase $i \leftarrow i + 1$. If $i > n$, then go to Step 9.
3. Compute the ratio $s \leftarrow h_i/y_i$. If $h_i > 0$, then go to Step 6.
4. [Case (i).] If $y_i < 0$ then go to Step 5, otherwise if $c_2 > s$, then change $c_2 \leftarrow s$. Go to Step 8.

6. [Case (iii).] If $y_i < 0$, then go to Step 7. If $c_2 > s$, then set $c_2 \leftarrow s$. Go to Step 8.
7. [Case (iv).] If $c_1 < s$ then replace $c_1 \leftarrow s$.
8. [Testing for empty intersection.] If $c_1 < c_2$ then go back to Step 2, otherwise set $c_1 \leftarrow 0$, $c_2 \leftarrow 0$.
9. Deliver c_1 and c_2, as entry and exit constants.

The constants c_1 and c_2 computed by this algorithm will be denoted as $(c_1, c_2) = E2(\mathbf{y}, \mathbf{h})$. This notation will be used to obtain an expression for the probability content of a line in a rectangle.

Consider the function e_3 defined as

$$e_3(\mathbf{y}) = \int g(k\mathbf{y})\, dF_n(k), \qquad \mathbf{y} \in E_s, \tag{7.32}$$

where $g(\mathbf{x}) = 1$, if $\mathbf{x} \in Q = \{\mathbf{z} | \mathbf{a} \leqq \mathbf{z} \leqq \mathbf{b}\}$ for two fixed vectors $\mathbf{a}$ and $\mathbf{b}$. Define two related domains $Q_1 = \{\mathbf{z} | \mathbf{z} \leqq \mathbf{b}\}$ and $Q_2 = \{\mathbf{x} | \mathbf{a} \leqq \mathbf{x}\}$, then we have

$$Q = Q_1 \cap Q_2.$$

The intersection of Q_1 and the line $r\mathbf{y}$ is characterized by the constants d_1 and d_2 for which $r\mathbf{y} \in Q_1$ if $d_1 \leqq r \leqq d_2$ holds. Using the notation defined after algorithm SE2 we can write

$$(d_1, d_2) = E2(\mathbf{y}, \mathbf{b}).$$

Denote by d_3 and d_4 the constants, for which $r\mathbf{y} \in Q_2$, if $d_3 \leqq r \leqq d_4$ holds. The intersection of $r\mathbf{y}$ and the domain Q_2 is of the same probability as the intersection of the line $-r\mathbf{y}$ and the domain $Q_2^- = \{\mathbf{z} | \mathbf{z} \leqq -\mathbf{a}\}$, because of the symmetricity of the two domains and the two lines to the origin, and furthermore we have $\varphi(\mathbf{x}) = \varphi(-\mathbf{x})$. Then we have

$$(d_3, d_4) = E2(-\mathbf{y}, -\mathbf{a}).$$

The remaining part is similar to the development of e_2. In order to obtain c_1 and c_2 for which $r\mathbf{y} \in Q$ if $c_1 \leqq r \leqq c_2$, we have to determine the common part of the intervals (d_1, d_2) and (d_3, d_4), that is

$$\begin{array}{lll} c_1 = \max(d_1, d_3), & c_2 = \min(d_2, d_4), & \text{if } c_1 \leqq c_2, \\ c_1 = 0, & c_2 = 0, & \text{otherwise.} \end{array}$$

Finally the probability content of the line $r\mathbf{y}$ in the rectangle Q can be given as

$$e_3(\mathbf{y}) = \begin{cases} F_n(c_2) - F_n(c_1), & \text{if } c_1 \geqq 0,\ c_2 \geqq 0, \\ F_n(-c_1) + F_n(c_2), & \text{if } c_1 < 0,\ c_2 \geqq 0, \\ F_n(-c_1) + F_n(-c_2), & \text{if } c_1 < 0,\ c_2 < 0. \end{cases} \tag{7.33}$$

The formal algorithm for evaluating the function $e_3(\mathbf{y})$ is the following one:

S9 Subalgorithm, Probability content of a line in a rectangle (function $e_3(\mathbf{y})$)

1. Use algorithm SE2 for obtaining $(d_1, d_2) \leftarrow E2(\mathbf{y}, \mathbf{b})$, $(d_3, d_4) \leftarrow E2(-\mathbf{y}, -\mathbf{a})$.
2. If $d_3 > d_1$ then change $d_1 \leftarrow d_3$. If $d_4 \leftarrow d_2$, then change $d_2 \leftarrow d_4$.
3. If $d_1 \geqq d_2$, then deliver $s \leftarrow 0$ as the value of $e_3(\mathbf{y})$.
4. If $d_1 < 0$, then go to Step 5, otherwise deliver $s \leftarrow F_n(d_2) - F_n(d_1)$.
5. If $d_2 \geqq 0$ then deliver $s \leftarrow F_n(-d_1) + F_n(d_2)$, otherwise deliver $s \leftarrow F_n(-d_1) - F_n(-d_2)$.

7.11. A detailed orthonormalized algorithm

Summarizing the details given in Sections 7.8 and 7.10 (especially the computation of $e_{2+}(\mathbf{y})$ and the subsequent modification) we describe a detailed algorithm suitable for evaluating the O_2 estimator in case of $\mathbf{h} \geqq 0$. More precisely let

$$O_2 = \frac{1}{N} \sum_{i=1}^{N} r^{(i)},$$

where $r^{(1)}, \ldots, r^{(N)}$ are independent realization of the random variable

$$r = \frac{1}{n(n-1)} \sum_{i=1}^{n-1} \sum_{j=i+1}^{n} e_{2+}\left(\frac{\mathbf{u}_i + \mathbf{u}_j}{\sqrt{2}}\right) + e_{2+}\left(\frac{\mathbf{u}_i - \mathbf{u}_j}{\sqrt{2}}\right),$$

($\mathbf{u}_i$ are defined by $\mathbf{u}_i = T\mathbf{b}_i$, $B = (\mathbf{b}_1, \ldots, \mathbf{b}_n)$ is a uniformly distributed orthonormalized system). We choose algorithm O_2 because on one hand it is possible to incorporate all the improvements and on the other hand it proved to be successfull in computer experiences (in dimensions n =4–20 it was of appropriate speed and in dimensions n =20–50 it was the fastest one).

As a preparation compute and store the square matrix $\mathbf{W}$ whose elements are given by $w_{ij} = t_{ij}/h_i$, $i, j = 1, \ldots, n$, where t_{ij} is the (i,j)th element of the transformator matrix $\mathbf{T}$. As a matter of course in case of $h_i = 0$ we replace its value by $\varepsilon \sim 10^{-10}$, say

NP10 Normal probability, orthonormalized-2 estimator (h $\geqq$ 0 only)

1. Initialize $p \leftarrow 0$, $q \leftarrow 0$.
2. [Produce the qth orthonormalized random system B.] Let $q \leftarrow q+1$, if $q > N$, then go to Step 14. Generate the vectors $\mathbf{b}_i$, $i = 1, \ldots, n$, whose components are independent and of standard normal distribution. Compute $s \leftarrow \sum_{i=1}^{n} b_{1i}^2$, $\mathbf{b}_1 \leftarrow \mathbf{b}_1/\sqrt{s}$ and initialize the index $i \leftarrow 1$.
3. [Orthonormalize the system B]. Increase $i \leftarrow i+1$ and evaluate the scalar products $x_j \leftarrow \sum_{k=1}^{n} b_{ik}b_{jk}$, $j = 1, \ldots, n$. Modify the vector $\mathbf{b}_i$ (to make it orthogonal to $\mathbf{b}_1, \ldots, \mathbf{b}_{i-1}$) setting $\mathbf{b}_i \leftarrow \mathbf{b}_i - \sum_{j=1}^{i-1} x_j \mathbf{b}_j$ and finally normalize it computing $s \leftarrow \sum_{j=1}^{n} b_{ij}^2$, $\mathbf{b}_i \leftarrow \mathbf{b}_i/\sqrt{s}$. If now $i \neq n$, then repeat this step from the beginning.
4. Compute the numbers $v_{ij} \leftarrow \sum_{k=1}^{n} w_{jk}b_{ik}$ for all possible index pairs (i, j), $i, j = 1, \ldots, n$. Initialize $i \leftarrow 0$, $s \leftarrow 0$.
5. [Setting the indices for producing the vectors $\mathbf{b}(\mathbf{s}, \mathbf{i})$.] Increase $i \leftarrow i+1$, if $i > n-1$, then go to Step 13. Set $j \leftarrow i$.
6. Increase $j \leftarrow j+1$, if $j > n$, then go back to Step 5. Set the initial values $k \leftarrow 0$, $c_1 \leftarrow \varepsilon$, $c_2 \leftarrow \varepsilon$, $c_3 \leftarrow \varepsilon$, $c_4 \leftarrow \varepsilon$, where ε is a small positive value, $\varepsilon \sim 10^{-10}$ say.
7. [Evaluation of the number given in (7.29).] Increase the index of the component of current interest $k \leftarrow k+1$, if $k > n$, then go to Step 12. Compute the values $x \leftarrow v_{ik} + v_{jk}$ and $y \leftarrow v_{ik} - v_{jk}$.
8. [Computing the minimum of the sums provided their positivity.] If $x < 0$, then go to Step 9. If $c_1 < x$, then change $c_1 \leftarrow x$. Go to Step 10.
9. [Computing the minimum of the negative sums.] If $c_2 < -x$, then set $c_2 \leftarrow -x$.
10. [Computing the minimum of positive differences.] If $y < 0$, then go to Step 11. If $c_3 < y$, then set $c_3 \leftarrow y$. Go back to Step 7.
11. [Computing the minimum of negative differences.] If $c_4 < -y$, then change $c_4 \leftarrow -y$. Go back to Step 7.
12. [The values of the function e_{2+} are added to the values already evaluated.] Let $s \leftarrow s + F_n(\sqrt{2}/c_1) + F_n(\sqrt{2}/c_2) + F_n(\sqrt{c_4})$ and go back to

Step 6 to compute the sum and the difference of two new vectors $\mathbf{b}_i$ and $\mathbf{b}_j$.

13. Let $p \leftarrow p + s/(2n(n-1))$ and go back to Step 2 to generate a new system B.
14. Deliver the value $p \leftarrow p/N$ as an estimator of the distribution function value $\Phi(\mathbf{h})$.

From this algorithm one can easily derive the algorithms corresponding the other estimators O_k. The algorithm of the estimator O_3 for instance can be obtained the following way. In Step 7 of algorithm NP10 compute the values $x_1 \leftarrow v_{ik} + v_{jk} + v_{rk}$, $x_2 \leftarrow v_{ik} + v_{jk} - v_{rk}$, $x_3 \leftarrow v_{ik} - v_{jk} - v_{rk}$, $x_4 \leftarrow v_{ik} - v_{jk} - v_{rk}$ (instead of computing x and y) Steps 8 through 11 should be doubled in order to evaluate eight minimums $c_1, \ldots, c_8$ (instead of the four ones $c_1, \ldots, c_4$), in Steps 5 and 6 the change of the indices i, j and r should be handled to cover all the terms in the summation $\sum_{i=1}^{n-2} \sum_{j=i+1}^{n-1} \sum_{r=j+1}^{n}$, etc. Of course to write the program of a special estimator O_k, different from the above NP10 is worthwhile only in the case, where performance of the algorithm (see next section) and repeated use in a given dimension justifies doing so.

7.12. Computer experiences

One can express an assessment on the efficiency of the Monte Carlo methods, even knowing the dispersion of the relevant estimator, only after extensive computer experiences; exceptions to this rule are the most simple problems, where the required computational effort can be estimated in advance. The variances of the methods described in this chapter for evaluating the multidimensional normal distribution function or the probability of a rectangle in case of multinormal density cannot generally be determined thus in making a comparison of the available methods we have to rely on computer experiences all the more. Taking into account that the behaviour of an algorithm is changing for each numerical example thus we had to compile a variety of examples possible very different from each other. To illustrate the performance of the algorithms some of the numerical results are presented below. For one example we give the number of the example, the number n of dimensions and the relevant exact probability p (details of the numerical examples are given in the Appendix).

First experiences with algorithms for evaluating distribution function values are presented. The estimators considered were the following:

1. Importance sampling estimator θ_3 ("acceptance-rejection" technique), with normally distributed vectors.
2. Integrating along a line, estimator θ_7, using independent directions, given by normal vectors.
3. Estimators O_k of random orthonormalized systems; the algorithms used have incorporated all the improvements, that is O_2 was exactly the algorithm NP10, and all the others were of similar type. For each example only the datas concerning the run of the best performing algorithm is given.

Computer runs on this set of examples were done on a Honeywell Bull 66/60 computer of the Central Statistical Office, Budapest. The computer is of medium speed, cycle time 1 μsec, the time necessary to evaluate a square root is 80 μsec.

In Table 20 we gave three characteristics of the above described estimators for each example (i) σ — the sample standard deviation computed from the actual sample, that is the standard error, (ii) t — the execution time in secs, (iii) eff — the efficiency of the method compared to the estimator θ_3 (see section 6.5): while computing the efficiency the variance of the estimator θ_3 was determined via the theoretically sound formula $\sqrt{p(1-p)/N}$; this is why for the first estimator the efficiency is different from 1.

Results similar to those presented in Table 20 were obtained in the course of computer experimentation with a large number of numerical examples. The following conclusions can be drawn:

1. If for a particular example the probability to be evaluated is near to 1, then not only the estimator θ_3 provides a result with smaller variance than the result for examples of medium probability but also the increase in efficiency will be greater than for examples of small probability.
2. The best performances can be observed for algorithms of the orthonormalized estimators. In $n = 5$ dimensions the estimator O_2 or O_3, in $n = 10$ dimensions O_4, in n =15–20 dimensions O_2 or O_3 whilst in n =20–50 dimensions O_2 is the fastest.
3. Two digit accurate results can be obtained up to $n = 10$ dimensions in 0.1 sec, up to $n = 20$ dimensions in 1 sec, up to n =50 dimensions in 10 sec (to obtain three digit accurate results one needs 5 secs, 60 secs and 900 secs respectively).

Table 20

Comparison of estimators in n =4–50 dimensions (sample size $N = 100$)

Examples		Estimators: acceptance rejection	integration along a line	orthonormalized	
1.	σ	0.04	0.018	0.0017	
$n = 4$	t	0.076	0.16	1.0	
$p = 0.6406$	eff	0.9	3.0	55.0	O_2
2.	σ	0.0043	0.022	0.0015	
$n = 5$	t	0.096	0.24	2.5	
$p = 0.2324$	eff	0.9	1.4	30.0	O_3
3.	σ	0.050	0.024	0.00021	
$n = 10$	t	0.25	0.40	90.2	
$p = 0.5830$	eff	0.9	2.7	140.0	O_4
4.	σ	0.035	0.015	0.000084	
$n = 10$	t	0.25	0.35	83.8	
$p = 0.8527$	eff	0.1	3.8	518.0	O_4
5.	σ	0.027	0.017	0.00046	
$n = 15$	t	0.45	0.69	140.0	
$p = 0.1105$	eff	1.3	2.0	15.0	O_3
6.	σ	0.025	0.010	0.00061	
$n = 20$	t	0.75	0.95	68.3	
$p = 0.9229$	eff	1.0	5.0	20.8	O_2
7.	σ	0.040	0.021	0.0010	
$n = 20$	t	0.76	0.97	69.3	
$p = 0.704$	eff	1.2	3.6	21.1	O_2
8.	σ	0.049	0.025	0.00091	
$n = 30$	t	1.5	1.7	214.0	
$p = 0.566$	eff	1.0	3.2	21.0	O_2
9.	σ	0.050	0.024	0.00084	
$n = 40$	t	2.5	2.8	486.0	
$p = 0.443$	eff	1.0	3.6	18.0	O_2
10.	σ	0.048	0.024	0.00094	
$n = 50$	t	3.7	4.2	914.0	
$p = 0.365$	eff	1.0	3.5	10.2	O_2

4. For small probabilities ($p < 0.1$ or even less) the algorithms dealt with in Table 20 are not very efficient generally. In these cases the variant of the acceptance-rejection technique elaborated for small probabilities (algorithm NP4) can be employed with good results (with the simultaneous use of the vectors $\mathbf{x}_i$ and $-\mathbf{x}_i$).

5. Algorithms NP3–NP7 are not sensitive for great correlations. The Monte Carlo methods presented here proved to be considerably superior to all conventional numerical techniques: already in n=4–5 dimensions they gave a result of comparable precision, in higher dimensions the Monte Carlo methods are much better. We would like to emphasize and underline the significance of the evaluation of an $n = 50$ dimensional integral, which in itself cannot be dismissed as an easy task.

Finally we made computer runs for six examples, in case of which the value of the integral, the probability p cannot be evaluated theoretically. The results of these computer runs are summarized in Table 21, offering a challenge to anybody interested (details of the examples are in the Appendix).

We would like to mention here, that the computer programs calculating distribution function values were written with main concern on the speed. The uniform generator (two shifts and an addition) and the normal generator (the FL_5 procedure of Ahrens & Dieter) were machine-coded.

Contrary to this approach when computing the probability of a rectangle we used only FORTRAN coded subroutines and standard, well-known techniques (the uniform generator was a double precision congruential generator, the method used for producing normally distributed numbers was the polar method). Here our main aim was to show that even with no special programming knowledge a satisfactory speed can be achieved in computing normal probabilities.

Table 21

Computer results for examples with unknown probabilities

Example and dimension	Probability	Error (standard dev)	Estimator	Time (sec)	Efficiency
11. n=5	0.719411	±0.000196	O_3	40.8	128
12. n=10	0.662662	±0.000043	O_4	1444.0	206
13. n=20	0.492468	±0.000301	O_2	1106.0	18
14. n=5	0.995469	±0.000016	O_3	39.4	455
15. n=10	0.976035	±0.000008	O_4	1389.0	659
16. n=20	0.966064	±0.000124	O_2	1084.0	14

The subroutines evaluating the probability q of a rectangle were written and run on an IBM 3031 of the Hungarian Academy of Sciences. The speed of the computer is similar to that of the Honeywell computer (to illustrate this point we mention that the execution time of a square root is 53 μsecs, the uniform generator required 70 μsecs, and the polar generator 193 μsecs). Computer experiences are presented in Table 22 with regard to the following two estimators:

1. The importance sampling method for computing the probability of a rectangle, that is algorithm of the estimator θ_9.
2. The orthonormalized-2 estimator T_2 as given by algorithm NP9 where the function e_3 was evaluated by algorithm S7.

Table 22

Probabilities of rectangles for estimators θ_9 and T_2

	Probability	Error (std. dev.)	Execution time (sec)	Time for two digits (sec)	Efficiency
Example 17					
$n=2$	0.0800	0.0272	0.05	0.396	28
$P_{ex}=0.02665$	0.0212	0.0026	0.21	0.014	
Example 18					
$n=4$	0.5000	0.0500	0.10	2.69	130
$P_{ex}=0.46044$	0.4593	0.0011	1.34	0.018	
Example 19					
$n=6$	0.1200	0.0326	0.16	1.77	23
$P_{ex}=0.15327$	0.1531	0.0014	3.62	0.078	
Example 20	0.9600	0.01969	0.31	1.202	40
$n=10$	0.9782	0.00045	14.6	0.029	
Example 21	0.2700	0.0446	0.30	6.04	13
$n=10$	0.2503	0.0017	14.53	0.44	
Example 22	0.18000	0.03861	0.73	10.98	6
$n=20$	0.13412	0.00135	101.0	1.85	

The row entries of Table 22 are the examples (p denotes the exact probability, if available) for each example we have two rows, the first gives the results obtained by using estimator θ_9, the second row contains the results from computations using the estimator T_2. The column entries are consecutively the computed probability, the standard deviation of the computed probability (i.e. the error of the result) and the CPU time of the computation in seconds. The last but one column contains the times necessary to obtain a two-digit accurate result (its standard deviation is 0.01) and the last column shows the efficiency of the orthonormalized estimator as compared to the θ_9 estimator. The sample size in these computations was $N = 100$.

In addition the computation of probabilities in Examples 4 and 6 was repeated with sample size $N = 4000$ and $N = 400$ respectively yielding the results $p_4 = 0.97865$ (deviation 0.0017) and $p_6 = 0.1333$ (deviation 0.00058) respectively.

According to these computer experiences probabilities of rectangles with two accurate digits can be evaluated in less than 0.5 sec up to $n = 10$ dimensions and in less than 5 sec up to $n = 20$ dimensions with this convenient and easy to implement set of algorithms.

7.13. Notes

The algorithms presented in this chapter were developed by the author (see Deák (1978), (1980), (1986)). Another Monte Carlo method, substantially different from this approach was presented in Szántai (1975). On the comparison of the methods see Deák (1979), Szántai (1986) and Gassmann (1986) who applied these techniques for evaluating conditional probabilities.

Descriptions of the multidimensional normal distribution can be found in the book of Anderson and Johnson and Kotz. A fast method for evaluating the two dimensional normal distribution function was published by Donelly. The theorem referred to in Section 7.4 (the marginal distribution of the components of a normally distributed vector is standardized normal) can be found in Anderson's book.

CHAPTER 8

SIMULATION: EXAMPLES OF APPLICATION

8.1. Introduction

The possible and actual areas of application of simulational computations are so broad that they defy description in one single book, let alone a chapter. In this chapter we only outline some of the basic problems where Monte Carlo computations were and are used in order to obtain a numerical solution, however inaccurate, but a solution.

Roughly two broad areas of simulational computations can be distinguished. One area is that of straightforward numerical calculations, where one looks for one specific result for a predetermined value, and as an answer one gets an approximate result. Generally, for this kind of problem an accurate deterministic solution procedure is available though the required computational time or some other factors render it impractical. The next nine sections present simulational examples of this kind. In problems of the other type we do not look for one specific answer but we are more interested in the overall behavior of the system, how it responds to shifts in parameters, modifications in the structure of the system, what are the crucial factors in the system, etc. These types of simulational computations are investigated in the latter part of the chapter.

First, classical applications are described and analyzed, such as solving a large system of linear equations, a differential equation, interpolating a function of very many variables, and optimizing a function. One technique merits explicit mention. A heuristic approach is elaborated which is based on ideas taken from simulating thermodynamical systems; it is especially well suited to conducting simulations where a parameter of the goodness of the result is readily available. A great many simulation models can be classified as queuing models. A section is devoted to such a model, with a sequel, where the very basics of the SIMSCRIPT simulational language are given and a program for a queuing model is outlined. Other practical system simulational models conclude the chapter, where ecological and warfare systems, economic and demographic problems are considered.

8.2. System of linear equations

Consider the following system of linear equations:

$$
\begin{aligned}
x_1 &= a_{11}x_1 + a_{12}x_2 + \ldots + a_{1n}x_n + b1, \\
x_2 &= a_{21}x_1 + a_{22}x_2 + \ldots + a_{2n}x_n + b_2 \ldots \\
x_n &= a_{n1}x_1 + a_{n2}x_2 + \ldots + a_{nn}x_n + b_n,
\end{aligned} \tag{8.1}
$$

where given the values $a_{11}, \ldots, a_{nn}$, $b_1, \ldots, b_n$ the variables $x_1, \ldots, x_n$ are to be computed. Making use of the vector notations, it can be comprised in the shorter form

$$\mathbf{x} = \mathbf{A}\mathbf{x} + \mathbf{b}, \tag{8.2}$$

where $\mathbf{A}$ is an $n \times n$ matrix with elements $a_{11}, \ldots, a_{nn}$, $\mathbf{x} = (x_1, \ldots, x_n)$ and $\mathbf{b} = (b_1, \ldots, b_n)$ are vectors of n components, $\mathbf{A}$ and $\mathbf{b}$ are given and $\mathbf{x}$ is the vector of unknown quantities. Assume that either the inequality $\sum_{i,j=1}^{n} a_{ij}^2 < 1$ or $\max_{1 \leqq i \leqq n} \sum_{j=1}^{n} |a_{ij}| < 1$ is satisfied. According to a well-known result in numerical analysis, this assumption ensures the existence of a solution obtainable by the following iterative procedure: let $\mathbf{x}^{(0)}$ be an initial solution, then the vectors $\mathbf{x}^{(k)}$ given by the equations

$$\mathbf{x}^{(k)} = \mathbf{A}\mathbf{x}^{(k-1)} + \mathbf{b} \qquad k = 1, 2, \ldots \tag{8.3}$$

converge to a solution $\mathbf{x}^*, \mathbf{x}^* = \mathbf{A}\mathbf{x}^* + \mathbf{b}$ of system (8.2), that is, $\lim_{k \to \infty} \mathbf{x}^{(k)} = \mathbf{x}^*$. Using the $\mathbf{x}^0 = \mathbf{b}$ initial solution, the approximate solution $\mathbf{x}^{(k)}$ obtained in the kth iteration can be expressed as

$$\mathbf{x}^{(k)} = (\mathbf{E} + \mathbf{A} + \mathbf{A}^2 + \ldots + \mathbf{A}^k)\mathbf{b}, \tag{8.4}$$

thus the solution of (8.2) is $\mathbf{x}^* = \lim_{k \to \infty} \mathbf{x}^{(k)} = (\mathbf{E} - \mathbf{A})^{-1}\mathbf{b}$. The ith component of the approximate solution $\mathbf{x}^{(k)}$ in (8.4) is

$$
\begin{aligned}
x_i^{(k)} = b_i &+ \sum_{j_1=1}^{n} a_{ij_1} b_{j_1} + \sum_{j_1=1}^{n} \sum_{j_2=1}^{n} a_{ij_1} a_{j_1 j_2} b_{j_2} + \ldots + \\
&+ \sum_{j_1=1}^{n} \cdots \sum_{j_k=1}^{n} a_{ij_1} a_{j_1 j_2} \ldots a_{j_{k-1} j_k} b_{j_k}.
\end{aligned} \tag{8.5}
$$

Since the evaluation of this sum even for moderate values of k and n may cause considerable difficulties in computations, thus the application of simulation may become handy. A random variable will be constructed, the expected value of which is precisely the expression in (8.5).

First, we outline the essence of the idea to be applied upon which the construction is based. Assume that we are interested in the evaluation of the sum $S = \sum_{i=1}^{n} c_i$, where $c_i > 0, i = 1, \ldots, n$. Choose a sequence of numbers $p_1, \ldots, p_n$, for which $p_i > 0, i = 1, \ldots, n$ and $\sum_{i=1}^{n} p_i = 1$ hold; that is, let the numbers p_i be the probabilities of a discrete distribution with n different values. Define these values by the equations $d_i = c_i/p_i$, $i = 1, \ldots, n$ and subsequently consider the random variable ξ that takes on the value d_i with probability p_i, i.e.

$$P\{\xi = d_i\} = p_i, \qquad i = 1, 2, \ldots, n.$$

Then we can state the following

Theorem 1. The expected value of the random variable ξ is S.

Proof. It can obviously be seen from the definition of the expected value since

$$E(\xi) = \sum_{i=1}^{n} p_i d_i = \sum_{i=1}^{n} c_i = S.$$

Thus a Monte Carlo method can be given to evaluate the expected value of ξ, that is, our original sum S. In order to do that we produce N realizations $\xi^{(i)}$ of the random variable ξ and taking the arithmetical average of the obtained value, the sum

$$S_N = \frac{1}{N} \sum_{i=1}^{N} \xi^{(i)}$$

thus yielded is an unbiased estimator of the sum S. Hence the sum S is not computed by summing up all the n terms in the sum but some terms are generated with the corresponding probability and the whole sum S is approximated with the weighted sum of these generated terms. Of course it goes without saying that the whole procedure will be computationally favourable only in cases where n is a good way greater than $N (n >> N)$, say $n/N > 10$. The variance of the random variable ξ can be evaluated as

$$D^2(\xi) = \sum_{i=1}^{n} p_i d_i^2 - E^2(\xi) = \sum_{i=1}^{n} c_i^2/p_i - E^2(\xi).$$

This suggests the choice of the distribution p_i that minimizes c_i^2/p_i, $i = 1, \ldots, n$, in order to decrease the error in the estimator S_N. Since most probably we do not know c_i^2 or $\sum_{i=1}^{n} c_i^2$, some knowledge of the behaviour of the c_i's is the only way to improve the choice of the probabilities $p_i, i = 1, \ldots, n$, in a way that makes the variance $D^2(\xi)$ as small as possible. For example, if the sequence $c_1, c_2, \ldots$ is monotone decreasing, then the probabilities $p_1, p_2, \ldots$ should behave in the same way.

Let us apply the above described idea of evaluating S to the evaluation of the sum (8.5). Consider an $n \times n$ matrix $\mathbf{P}$, the elements P_{ij} of which satisfy the conditions

$$\sum_{j=1}^{n} p_{ij} = 1, \qquad i = 1, \ldots, n,$$

(8.6)
$$p_{ij} > 0 \quad \text{if} \quad a_{ij} \neq 0,$$
$$p_{ij} = 0 \quad \text{if} \quad a_{ij} = 0.$$

Here we need n random variables and their distributions, not just one as in the illustrative example. The numbers d_{ij} are defined as

(8.7)
$$d_{ij} = \begin{cases} \frac{a_{ij}}{p_{ij}} & \text{if } a_{ij} \neq 0, \\ 0 & \text{otherwise.} \end{cases}$$

Similarly to the previous example, in order to evaluate the sum (8.5) the individual terms are to be generated with a specific probability; the value $d_{ij_1} d_{j_1 j_2} \ldots d_{j_{k-1} j_k} b_{j_k}$ is to be sampled with probability $p = p_{ij_1} p_{j_1 j_2} \cdots p_{j_{k-1} j_k}$. Obviously the generation of the sequence of indices $I = (i, j_1, \ldots, j_k)$ with probability p suffices since these indices completely determine the value $d_{ij_1} \ldots d_{j_{k-1} j_k} b_{j_k}$ to be produced. Thus we can restrict our attention to the question: what procedure is computationally suitable to generate the set of indices I with probability p?

If index j_1 is to be produced with probability p_{ij_1}, index j_2 is to be sampled with probability $p_{j_1 j_2}, \ldots$, and index j_k is to be sampled with probability $p_{j_{k-1} j_k}$, then the set of indices I is produced with probability $p = p_{ij_1} p_{j_1 j_2} \cdots p_{j_{k-1} j_k}$. To demonstrate the generation of the sequence of indices we will make use of the terminology related to the random walk (Markov chains) in a plane. The discrete random variables $\xi_1, \ldots, \xi_k$ take on the values $1, \ldots, n$, if $\xi_i = j$ then we say that the random point in the ith moment (step) is in the jth state. The random variables $\xi_1, \ldots, \xi_k$ describe a random walk with k steps visiting n possible states. The probability of the event that the random point goes from state r to the state s is denoted by p_{rs} and is called the transition probability. This is independent of the number of steps that have already been taken, that is, we have for any i:

(8.8) $$p_{rs} = P\{\xi_{i+1} = s | \xi_i = r\}, \qquad r, s = 1, \ldots, n.$$

Then the set of indices I can be considered as a k step random walk, during which a random point starts from the state i and ends in state j_k through the states $j_1, \ldots, j_{k-1}$. From the other side, starting with an initial value $\xi_0 = i$, a realization of the random variables $\xi_1, \ldots, \xi_k$ (given by equations (8.8) and the matrix $\mathbf{P}$ of the transition probabilities) gives a feasible sequence of indices I, and the generation of I is done with the required probability $p = p_{ij_j} \cdots p_{j_{k-1}j_k}$ precisely.

Now we summarize the things said up to this point. A component of the approximate solution to equation (8.2) can be evaluated by the following Monte Carlo method. Choose a matrix $\mathbf{P}$ satisfying the conditions formulated in (8.6). Besides the index i of the component to be evaluated we need the specification of the iteration number k that determines the goodness of the approximation, the sample size N that is responsible for the error term in the Monte Carlo calculations, and finally we need the evaluation of the quantities d_{ij} from (8.7). Afterwards we sample N random walks. For the sequence of indices $I = (i, j_1, \ldots, j_k)$ obtained during the course of sampling the rth random walk compute the sum (we omit index r in the indices of I, though they are dependent on r).

(8.9) $$S^{(r)} = b_i + d_{ij_1} b_{j_1} + \ldots + d_{ij_1} d_{j_1 j_2} \ldots d_{j_{k-1} j_k} b_{j_k}.$$

The probability of generating this very sum is $p = p_{ij_1} \cdots p_{j_{k-1}j_k}$.

Having accomplished the generation of the N random walks, an approximation to the value $x_i^{(k)}$ is given by the arithmetic average

$$x_i^{(k)} \sim \frac{1}{N} \sum_{r=1}^{N} S^{(r)}.$$

In the formal algorithmic description of the method for evaluating the estimator, the values n, i, k, N and p_{ij}, d_{ij}, i, $j = 1, \ldots, n$, are assumed to be given.

A1 Solving a system of linear equations

1. Set $m \leftarrow 0, S \leftarrow 0$.
2. [Initializing.] Set $x_0 \leftarrow i$, $q \leftarrow 0$, $j_0 \leftarrow i$, $r \leftarrow 1$, increase the index of the iteration $m \leftarrow m + 1$ and put $S^{(m)} \leftarrow b_i$.
3. [Generation of the sequence of indices I.] Increase the index $q \leftarrow q + 1$,

sample an index j_q from the numbers $1, 2, \ldots, n$, according to the discrete distribution described by the probabilities $p_{j_{q-1}i}$ (that is, here $P\{j_q = i\} = p_{j_{q-1}i}$, $i = 1, \ldots, n$). Compute $r \leftarrow r d_{j_{q-1}j_q}$ and $S^{(m)} \leftarrow S^{(m)} + r b_{j_q}$. If now $q = k$ then go to Step 4, otherwise repeat this step from the beginning.
4. [Averaging.] Let $S \leftarrow S + S^{(m)}$. If $m < N$, then go back to Step 2 for generating one more random walk, otherwise deliver $S \leftarrow S/N$ as the required approximation to the value $x_i^{(k)}$.

Theorem 2. The above described algorithm A1 results in an unbiased estimator of the sum (8.5).

Proof. The expected value of the sample $S^{(r)}$ in (8.9) is

$$\sum_{j_1=1}^{n} \cdots \sum_{j_k=1}^{n} S^{(r)} \cdot p_{ij_1} p_{j_1 j_2} \cdots p_{j_{k-1}} p_{j_k} =$$

$$= \sum_{j_1=1}^{n} \cdots \sum_{j_k=1}^{n} b_i p_{ij_1} \cdots p_{j_{k-1}j_k} +$$

$$+ \sum_{j_1=1}^{n} \sum_{j_k=1}^{n} d_{ij_1} b_{j_1} \cdot p_{ij_1} \cdots p_{j_{k-1}j_k} + \ldots +$$

$$+ \sum_{j_1=1}^{n} \cdots \sum_{j_k=1}^{n} d_{ij_1} d_{j_1 j_2} \ldots d_{j_{k-1}} d_{j_k} b_{j_k} \cdot p_{ij_1} p_{j_1 j_2} \cdots p_{j_{k-1}j_k} =,$$

$$= b_i + \sum_{j_1=1}^{n} a_{ij_j} b_{j_1} + \sum_{j_1=1}^{n} \sum_{j_2=1}^{n} a_{ij_1} a_{j_1 j_2} b_{j_2} + \ldots +$$

$$+ \sum_{j_1=1}^{n} \cdots \sum_{j_k=1}^{n} a_{ij_1} a_{j_1 j_2} \ldots a_{j_{k-1}j_k} b_{j_k}$$

when one makes use of (8.7). The right-hand side of this equation is the same as of (8.5).

8.3. Variants and applications of solving linear equations

In this section some variants of the basic technique described in the previous section will be elaborated upon and adaptation to evaluation of an inverse matrix and to computations in integral equations is carried out.

A version of the basic method arises when the Markovian chain is allowed to have termination. The normalizing constraints $\sum_{j=1}^{n} p_{ij} = 1$, $i = 1, \ldots, n$, in the construction of the matrix $\mathbf{P}$ are replaced by the inequalities $\sum_{j_1=1}^{n} p_{ij} \leqq 1$. The probability $p_i = 1 - \sum_{j=1}^{n} p_{ij}$ denotes the probability of absorption, that is, the random point stops in the ith state (the random walk terminates); this can sometimes be conveniently considered as an $(n+1)$st state, from where we have no positive transition probabilities. The generation of indices $I = \{i, j_1, \ldots, j_k\}$ is to be conducted in the same way as in the basic version — but the occurrence of an absorption is possible before generating k indices. Another deviation from the basic version is that realizations of the sum

$$S^{(r)'} = b_i + d_{ij_1} \frac{b_{j_1}}{d_{ij_1'}} + \ldots + d_{ij_1} d_{j_1 j_2} \ldots d_{j_{k-1} j_k} \frac{b_{j_k}}{p_{j_k'}} \tag{8.10}$$

should be generated instead of $S^{(r)}$ in (8.9).

A different variant of the basic Algorithm A1 offering better computational performance — at least in some large-scale problems — is the following. Basically the method generates a sufficiently long sequence of indices $i, j_1, \ldots, j_T$ meaning by this that the indices $\ell = 1, \ldots, n$ occur at least once among the indices $j_1, \ldots, j_T$. Let us introduce the notation

$$T_\ell = \min \{k | j_k = \ell\}, \qquad \ell = 1, \ldots, n,$$

that is, T_j is the index of the sequence when the random point first hits state j. Consider now the sequence $i, j_1, \ldots, j_T, \ldots, j_{T+k}$ and the n — at least partly — different subsequences:

$$j_{T_\ell}, \ldots, j_{T_\ell + k}, \qquad \ell = 1, \ldots, n.$$

Using these n subsequences we can get an estimation of *all* the n components of the solution vector $\mathbf{x}$. This is going to be worthile if $T + k << nk$, that is, where very big common portions are in the different subsequences. The validity of the procedure is based on the ergodicity of the generated index sequence. More formally, the validity can be proved by that property of the random walk indices $j_1, \ldots, j_t, \ldots$ (which is a homogeneous Markov chain) that for any bounded function f and index set $j_t, \ldots, j_{t+k}$ we have

$$E\{f(j_t, \ldots, j_{t+k}) | j_t = i\} = E\{f(j_1, \ldots, j_k) | j_1 = i\}.$$

The formal algorithm can be given the same way as for the case of the basic method.

Finally, if we want to obtain the whole vector $\mathbf{x}$ we may use a probabilistic approach in determining the starting value i. In particular, consider a probability distribution $q_1, \ldots, q_n$, where $q_i > 0, i = 1, \ldots, n$ and $\sum_{i=1}^{n} q_i = 1$. This is the initial distribution of the Markov chain. The determination of an initial index i will be done by sampling the index i with probability q_i. Correspondingly, the weight $\frac{1}{q_i}$ has to be attached to the related term in the summation, that is, we will average the sums

$$S^{(r)''} = \frac{1}{q_i}(b_i + d_{ij_1} b_{ij_1} + \ldots + d_{ij_1} \ldots d_{j_{k-1}j_k} b_{j_k}).$$

Of course the deterministic approach to determine the whole vector $\mathbf{x}$ by computing x_i, $i = 1, 2, \ldots, n$ according to Algorithm A1 is also feasible.

Consider now a matrix $\mathbf{B}$. We would like to determine the inverse matrix $\mathbf{B}^{-1}$. Denote the columns of $\mathbf{B}^{-1}$ by $\mathbf{d}_1, \ldots, \mathbf{d}_n$ respectively, then we have

$$I = \mathbf{B}\,\mathbf{B}^{-1},$$

or

$$\mathbf{e}_i = \mathbf{B}\,\mathbf{d}_i, \qquad i = 1, \ldots, n$$

where $\mathbf{e}_i$ is the ith unit vector. After introducing the notation $\mathbf{A} = \mathbf{I} - \mathbf{B}$ ($\mathbf{I}$ is the identity matrix), we arrive at

$$\mathbf{d}_i = (\mathbf{I} - \mathbf{B})\mathbf{d}_i + \mathbf{e}_i = \mathbf{A}\mathbf{d}_i + \mathbf{e}_i,$$

which is of the same shape as the original equation (8.1). It means the columns and by the same token the elements of the inverse matrix $\mathbf{B}^{-1}$ can be determined by the above described Monte Carlo procedure (Algorithm A1).

The application of the Monte Carlo method to solve linear equations may be justified only in exceptional cases; for instance, when one or a few crucial elements of the solution (or of the inverse) are required. Otherwise the deterministic techniques are to be preferred.

There is, though, an area where analogous ideas — or one may say a continuous variant of the basic idea — secure better results than the deterministic ones, namely in the case of the solution of integral equations. Consider the integral transform K given by the equation

$$K\psi(\mathbf{x}) = \int K(\mathbf{x},\mathbf{y})\psi(\mathbf{y})d\mathbf{y},$$

where the points $\mathbf{x}, \mathbf{y}$ are in the two-dimensional domain D.

Define the scalar product of the functions $f(\mathbf{x})$ and $g(\mathbf{x})$ as $\int f(\mathbf{x})g(\mathbf{x})d\mathbf{x}$. We assume that all the integrals in the sequel are meaningful, that is, finite; to assure it the functions are to be L_2 integrable, that is, e.g. $\int \psi^2(\mathbf{x})d\mathbf{x} < \infty$, $\int\int K^2(\mathbf{x},\mathbf{y})d\mathbf{x}\, d\mathbf{y} < \infty$, etc.

Repeatedly applying the transformation K we obtain the kth iteration of ψ with respect to the kernel $K(\mathbf{x},\mathbf{y})$ as

$$K^k\psi(\mathbf{x}) = \int K(\mathbf{x},\mathbf{z})K^{k-1}\psi(\mathbf{z})d\mathbf{z}, \qquad k = 2,3,\ldots.$$

Introduce a probability distribution $p(\mathbf{x},\mathbf{y})$, where $\int p(\mathbf{x},\mathbf{y})d\mathbf{y} = 1$ and $p(\mathbf{x},\mathbf{y}) > 0$, if $K(\mathbf{x},\mathbf{y}) \neq 0$, which is actually the density function of the transition probabilities of a random walk from the point $\mathbf{x}$ to the point $\mathbf{y}$. Consider the random walk of k steps $\mathbf{x}_0, \mathbf{x}_1, \ldots, \mathbf{x}_k$ and the related random variable S, where $\mathbf{x}_0$ is fixed and $\mathbf{x}_i$ is sampled with density $p(\mathbf{x}_{i-1}, \mathbf{x}_i)$

$$S = \frac{K(\mathbf{x}_0,\mathbf{x}_1)K(\mathbf{x}_1,\mathbf{x}_2)\ldots K(\mathbf{x}_{k-1},\mathbf{x}_k)}{p(\mathbf{x}_0,\mathbf{x}_1)p(\mathbf{x}_1\mathbf{x}_2)\ldots p(\mathbf{x}_{k-1},\mathbf{x}_k)}\psi(\mathbf{x}_k).$$

The expected value of this random variable is clearly $K^k\psi(\mathbf{x}_0)$ since the density of the event of generating the walk $\mathbf{x}_0, \mathbf{x}_1, \ldots, \mathbf{x}_k$ is the product that stands in the denominator of the right-hand side.

The formal algorithm will be given for a slightly more general version: two additional features will be provided. First, assume that the starting point $\mathbf{x}_0$ is not fixed, but randomly chosen according to some density function $q(\mathbf{y}), \mathbf{y} \in D$. Second, we will look for the scalar product of $h(\mathbf{x})$ and $K^k\psi(\mathbf{x})$ (defined as $\int h(\mathbf{x})K^k\psi(\mathbf{x})d\mathbf{x}$) rather than $K^k\psi(\mathbf{x})$. Were the functions $h(\mathbf{x})$ and $q(\mathbf{y})$ replaced by the Dirac function we would be back to the estimation of $K^k\psi(\mathbf{x}_0)$.

The procedure to estimate the scalar product $\int h(\mathbf{x})K^k\psi(\mathbf{x})d\mathbf{x}$ is the following

A2 Solving an integral equation

1. [Initialize.] Set $i \leftarrow 0$, $S \leftarrow 0$.
2. Increase index $i \leftarrow i + 1$, if $i > N$, go to Step 5. Generate a starting point $\mathbf{x}_0$ in the domain D according to the density function $q(\mathbf{y})$. Set $m \leftarrow 0$.
3. [Start the generation of the ith random walk.] Increase $m \leftarrow m + 1$ if $m > k$, then go to Step 4, otherwise generate the next point $\mathbf{x}_m$, according to the density function $p(\mathbf{x}_{m-1}, \mathbf{y})$ and repeat this step from the beginning.

4. [Averaging.] Compute

$$S \leftarrow \frac{h(x_0)}{q(x_0)} \cdot \frac{K(\mathbf{x}_0,\mathbf{x}_1)K(\mathbf{x}_1,\mathbf{x}_2)\ldots K(\mathbf{x}_{k-1},\mathbf{x}_k)}{p(\mathbf{x}_0,\mathbf{x}_1)p(\mathbf{x}_1,\mathbf{x}_2)\ldots p(\mathbf{x}_{k-1},\mathbf{x}_k)}\psi(\mathbf{x}_k)$$

and add $S \leftarrow S + s$. Go back to Step 2.

5. Deliver $x \leftarrow S/N$ as an unbiased estimation of the sought after value.

Algorithm A2 and the underlying method can be applied to several problems. We outline two possible areas of application. One of them is the computation of eigenvalues and eigenfunctions; the other is the computation of Fourier coefficients. Consider the integral equation

$$g(\mathbf{x}) = \lambda \int K(\mathbf{x},\mathbf{y})g(\mathbf{y})dy.$$

We have to make two assumptions: the smallest eigenvalue λ_1 is positive and furthermore the kernel function is symmetric $K(\mathbf{x},\mathbf{y}) = K(\mathbf{y},\mathbf{x})$ and positive definite: $\int \psi(\mathbf{x})\ K\psi(\mathbf{x})d\mathbf{x} > 0$ if $\psi(\mathbf{x}) \neq 0$, $\mathbf{x} \in D$. Then for any two functions $f(\mathbf{x}) > 0$ and $h(\mathbf{x}) > 0, \mathbf{x} \in D$ we have

$$\lambda_1 = \lim_{n\to\infty} \frac{\int h(\mathbf{x})K^n f(\mathbf{x})d\mathbf{x}}{\int h(\mathbf{x})K^{n+1} f(\mathbf{x})d\mathbf{x}},$$

$$g_1(\mathbf{x}) = \lim_{n\to\infty} \frac{K^n f(\mathbf{x})}{\sqrt{\int K^n f(\mathbf{x})K^n f(\mathbf{x})d\mathbf{x}}}$$

where $g_1(\mathbf{x})$ is the eigenfunction corresponding to λ_1. Since both λ_1 and $g_i(\mathbf{x})$ are expressed with integral transforms, their evaluation by A2 is straightforward.

We turn now to the problem of the Fourier coefficients. Consider a system of orthonormalized functions $\psi_1(\mathbf{x}), \ldots, \psi_n(\mathbf{x})$, that is, $\int \psi_i(\mathbf{x})\psi_j(\mathbf{x})d\mathbf{x} = \delta_{ij}$. Then the Fourier coefficient c_{mi} of the function $K^m f(\mathbf{x})$ with regard to the system $\{\psi_i(\mathbf{x})\}$ can be obtained as

$$c_{mi} = \int \psi_i(x) \cdot K^m f(\mathbf{x})d\mathbf{x}$$

which can be evaluated by Algorithm A2 since it is the scalar product of $\psi_i(\mathbf{x})$ and $K^m f(\mathbf{x})$. Having obtained all the coefficients $c_{mi}, i = 1, \ldots, n$ one obtains the expansion of $K^m f(\mathbf{x})$ as

$$K^m f(\mathbf{x}) \sim \sum_{i=1}^{n} c_{mi}\psi_i(\mathbf{x})$$

which may serve as a suitable approximation to the original function f.

8.4. A differential equation

A simulation method is described in this section for solving Dirihlet's problem. The problem can be exposed as follows. Find a function $u(x, y)$ of two variables that (i) satisfies the following differential equation

$$\frac{\partial^2 u(x,y)}{\partial x^2} + \frac{\partial^2 u(x,y)}{\partial y^2} = 0, \tag{8.11}$$

inside a given connected domain T and (ii) on the boundary G of the domain T, $u(x, y)$ is equal to a given function $g(x, y)$, that is, $u(x, y) = g(x, y)$ if $(x, y) \in G$. (The notation of $u(P)$, where P is a point in T, will also be used.)

Usually the accomplishment of this task is done by the following procedure. The domain T is split up with two sets of parallel straight lines that are parallel to the coordinate axes; furthermore, the distance between the parallel lines is h (Figure 37).

The lattice points (the intersections of the perpendicular lines) thus produced will be called interior points or boundary points. A lattice point (ih, jh) is an interior point if all the four neighbouring points, in particular the points $((i-1)h, jh)$, $((i+1)h, jh)$, $(ih, (j-1)h)$, $(ih, (j+1)h)$ are inside the domain T. All the remaining lattice points inside T will be called boundary points. Lattice points outside T are of no interest for us now. The second-order derivatives in equation (8.11) will be replaced by an approximation to them. One may use, for example, the following formulae

$$\begin{aligned} \frac{\partial^2 u(x,y)}{\partial x^2} &\sim \frac{1}{h^2}(u(x-h,y) - 2u(x,y) + u(x+h,y)), \\ \frac{\partial^2 u(x,y)}{\partial y^2} &\sim \frac{1}{h^2}(u(x,y-h) - 2u(x,y) + u(x,y+h)). \end{aligned} \tag{8.12}$$

The approximation for sufficiently small values of h is satisfactory, generally; other approximation schemes instead of the ones in (8.12) may be applied (but if more "neighbours" are involved in them in addition to the four above-mentioned ones, then the definition of the interior point has to be modified correspondingly).

Denote the function value of $u(x, y)$ in the interior point (ih, jh) by u_{ij} and by g_{ij} in the boundary point (ih, jh). While the value g_{ij} equals the nearest value of $g(x, y)$, that is, if (x', y') is that point of the boundary G that is nearest to the boundary point (ih, jh), then $g_{ij} = g(x', y')$, the value of u_{ij} is to be determined, and these values u_{ij} eventually yield

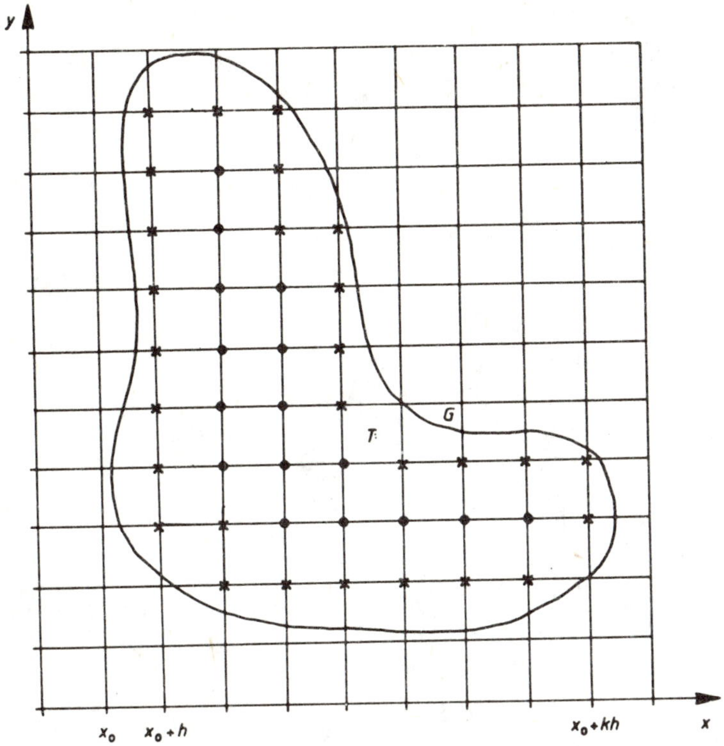

Fig. 37. Interior (•) points and boundary (×) points in the approximate solution of the Dirichlet problem

an approximation to the function $u(x, y)$. By the aid of equations (8.12) we can formulate the equality in (8.11) for every interior lattice point and the equations thus obtained provide us a large-scale system of linear equations in the unknown u_{ij}. By solving this system we obtain a set of values $\{u_{ij}\}$ approximating the function $u(x, y)$ looked for in the lattice points. By decreasing the distance h the size of the system may take on very large values (halving the distance h means quadrupling the number of unknowns) but on the other hand the solution of the system provides us with a better approximation to the function $u(x, y)$.

For solving this large-scale linear system the method of determining the solution of a system of linear equations (described in the previous section) may be deployed. However, there is another Monte Carlo method for solving the differential equation in (8.11) that has the name of "walk on circles" and it will subsequently be described and discussed.

Consider an interior point P_0 in the domain $T, P_0 \in T, P_0 \notin G$ and starting with this point P_0 construct a sequence of random points $P_0 = Q_1, Q_2, Q_3, \ldots, Q_n$. The way of constructing the point Q_k from the point Q_{k-1} is as follows (Figure 38).

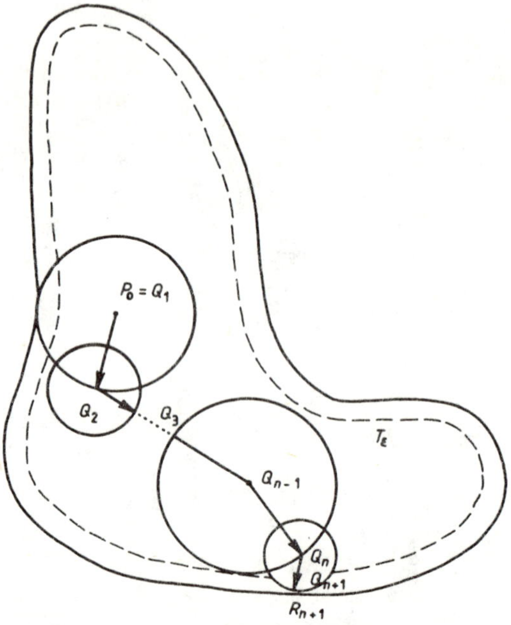

Fig. 38. Generating the point sequence $Q_1, \ldots, Q_{n+1}$

Draw a circle with radius r_{k-1} (that may be a random variable) around the point Q_{k-1} as a centre and furthermore the circle has the property that it is completely in the domain T. Choose a point according to the uniform distribution on the circle (on the surface of the circle); this will be the point Q_k. We now prove the following fundamental result concerning the function values $u(Q_k)$ that will be instrumental in developing the Monte Carlo method of solving (8.11).

Theorem. If the function $u(x, y) = u(\mathbf{P})$, $(x, y) = P \in T$ satisfies the Laplace equation

$$\frac{\partial^2 u(x,y)}{\partial x^2} + \frac{\partial^2 u(x,y)}{\partial y^2} = 0,$$

then for any k and any point sequence $Q_1, Q_2, \ldots, Q_n$ produced by the above described method we have

$$E(u(Q_k)) = u(\mathbf{P}_0),$$

that is, the expected value of the function u in every point of the point sequence is equal to the value of u in the starting point $\mathbf{P}_0$.

Proof. First, some notation will be introduced. Denote the density function of the radius r_k by $q_k(r)$, for which $q_k(r) = 0$ if $r < 0$ or if r is greater than the distance from the point Q_k to the nearest point of the boundary G; the choice of the density function $q_k(r)$ is in no other way restricted. The sampling of a uniformly distributed point on the circumference of the circle is equivalent to the generation of an angle ψ_k uniformly in the interval $[0, 2\pi)$ and the computation of the directional unit vector $\omega_k = (\cos\psi_k, \sin\psi_k)$. The point subsequent to Q_k can now be written as

$$Q_{k+1} = Q_k + r_k\omega_k, \qquad k = 1, 2, \ldots, n.$$

Denote the density function of the point Q_k by $p_k(P)$, then the expected value of interest can be expressed as

$$E(u(Q_{k+1})) = E(u(Q_k + r_k\omega_k)) =$$

$$= \int_T \int_0^{+\infty} \int_0^{2\pi} \frac{1}{2\pi} p_k(\mathbf{P}) q_k(r) u(Q_k + r\omega)\, \mathrm{d}\psi\, \mathrm{d}r\, \mathrm{d}\mathbf{P}, \tag{8.13}$$

since the density function of ψ is $1/2\pi$. It is well known from the theory of harmonic functions that the equality

$$\frac{1}{2\pi}\int_0^{2\pi} u(\mathbf{P} + r\omega)\, \mathrm{d}\psi = u(\mathbf{P}) \tag{8.14}$$

holds if u is harmonic, that is, satisfies equation (8.11) and $\omega = (\cos\psi, \sin\psi)$. Substituting this equality into (8.13) we easily obtain

$$E(u(\mathbf{Q}_{k+1})) = \int_T \int_0^{+\infty} p_k(\mathbf{P}) q_k(r) u(\mathbf{Q}_k)\, \mathrm{d}r\, \mathrm{d}\mathbf{P} =$$

$$= \int_T p_k(P) u(Q_k)\, \mathrm{d}\mathbf{P} = u(\mathbf{Q}_k),$$

since both $q_k(r)$ and $p_k(\mathbf{P})$ are density functions, thus their integral is equal to 1. Now we repeatedly apply this equation k times to arrive at

$$E(u(\mathbf{Q}_{k+1})) = E(u(\mathbf{Q}_k)) = \ldots = E(u(Q_1)) = E(u(P_0)) = u(P_0),$$

which was the equality looked for, thus the theorem is proved.

Based on the statement formulated in the theorem, the following simulational method can be developed for evaluating the value of the function u in any point $\mathbf{P}_0 \in T$. Let ε be a sufficiently small positive number and define the narrow strip T_ε around the boundary G defined by

$$T_\varepsilon = \{\mathbf{x} \mid d(\mathbf{x},\mathbf{y}) \leqq \varepsilon,\ \mathbf{x} \in T,\ \mathbf{y} \in G\}$$

where $d(\mathbf{x},\mathbf{y})$ is the Euclidean distance between the points $\mathbf{x}$ and $\mathbf{y}$. In other words, the domain T_ε is a narrow strip in the domain T around the boundary G and its "thickness" is ε. The sequence of points $\mathbf{Q}_1, \ldots,$ $\mathbf{Q}_k \ldots$ is generated until a point $\mathbf{Q}_n$ falls into the domain T_ε (that is sufficiently near to the boundary). Denote by $\mathbf{R}_n$ the point from G, which is the nearest to the point $\mathbf{Q}_n$, that is, $\mathbf{R}_n = \arg\min_{\mathbf{y} \in G} d(\mathbf{y}, \mathbf{Q}_n)$. In case of sufficiently small value of ε the value $u(\mathbf{Q}_n)$ will differ from the value $g(\mathbf{R}_n)$ with a negligible amount only assuming a reasonable smoothness of u so we can write the approximate equation $g(\mathbf{R}_n) \sim u(\mathbf{Q}_n)$. According to Theorem — it simultaneously ascertains

$$g(\mathbf{R}_n) \sim u(\mathbf{P}_0).$$

By generating N realizations of the point sequence $\mathbf{Q}_1, \ldots, \mathbf{Q}_n$, resulting in the values $g(\mathbf{R}_{n_i})$, $i = 1, \ldots, N$, we can derive the approximation

$$u(\mathbf{P}_0) \sim \frac{1}{N} \sum_{i=1}^{N} g(\mathbf{R}_{n_i}),$$

where $\mathbf{R}_{n_i}$ is the end point of the sequence near the boundary G, obtained in the simulation of the ith point sequence. The right-hand side part of the approximate equation is in fact an unbiased estimator of the function value $u(\mathbf{P}_0)$.

In the ensuing formal description of the algorithm we assume that the quantities $N, \varepsilon, T_\varepsilon$ and P_0 that control the solution procedure are given.

A3 Solving a differential equation

1. [Setting initial values.] Set $m \leftarrow 0$, $S \leftarrow 0$.
2. Increase $m \leftarrow m + 1$ and initialize $i \leftarrow 1$, $\mathbf{Q} \leftarrow \mathbf{P}_0$.
3. [Generation of the mth point sequence.] Generate u uniformly in $[0, 1)$, let $\psi \leftarrow 2\pi u$ and $\omega \leftarrow (\cos\psi, \sin\psi)$. Consider the circles drawn around the point $\mathbf{Q}$ with radius r and fully inside domain T and denote the greatest such radius by r_0.

4. [Determination of the next point.] Set $\mathbf{Q} \leftarrow \mathbf{Q} + r_0\omega$. If $\mathbf{Q} \in T_\varepsilon$ then go to Step 5, otherwise go back to Step 3 to continue the generation of the point sequence.
5. [Summation of the different realizations.] Let us denote the point of the border G nearest to the point $\mathbf{Q}$ by $\mathbf{R}$ and increase the total obtained so far $S \leftarrow S + g(\mathbf{R})$. If $m = N$ then deliver the value $S \leftarrow S/N$ as an approximation to the function value $u(\mathbf{P}_0)$, otherwise go back to Step 2 to generate the next sequence of points.

Notice here that the Monte Carlo procedure basically differs from the traditional deterministic one: by the simulational method we are able to determine the function value at one given point — by generating more and more point sequences an increasingly accurate approximation is supplied. By contrast the deterministic method simultaneously evaluates the function values at *all* of the grid points though at the expense of much more computational work. Henceforth if in an actual problem the overall shape of the function is to be assessed, then the deterministic approach is preferable, but if one function value (or a small number of function values, or values in a small neighbourhood) are required, then a simulational method (solving the system of linear equations for one component of the solution vector or using the random walk on circles) would be of greater help. The methods can actually complement each other: first by the aid of the deterministic method function values are approximated on a set of grid points produced by a rather coarse lattice, and then simulation is used to obtain more accurate approximations in points of interest (e.g. near to the maximum of the function).

The Monte Carlo method is not free of computational difficulties either. The determination of the greatest radius r_0, or checking the relation $\mathbf{Q} \in T_\varepsilon$ for a given ε might be quite a job, so a careful consideration of methods is advised.

8.5. Interpolation of a function of very many variables

Consider a one-dimensional function $g(x)$, $x \in [0,1]$. The function values $g(0)$ and $g(1)$ in the terminal points are assumed to be known. For approximating the function $g(x)$ one can use the equation

$$g(x) = (1-x)g(0) + xg(1) \tag{8.15}$$

of the linear interpolation (Figure 39).

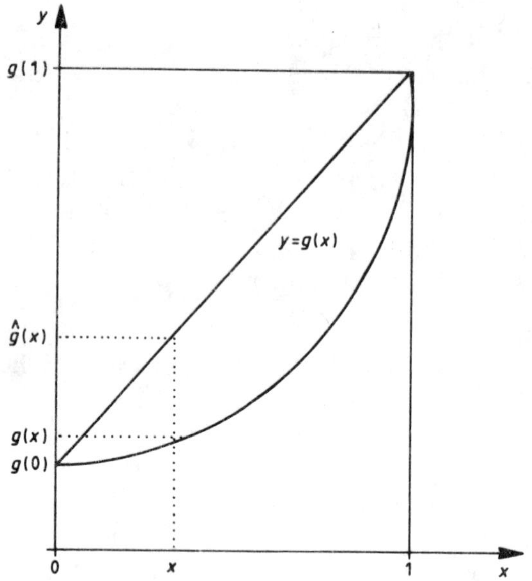

Fig. 39. Linear interpolation $\hat{g}(x)$ to the function $g(x)$ by the values $g(0)$ and $g(1)$

Consider now the function $f(\mathbf{x}) = f(x_1, \ldots, x_n)$ where the number n of the variables is very large. Assume that we know (or what is equivalent we can easily compute) the function $f(\mathbf{x})$ in the vertices $\mathbf{c}_i$ of the unit cube $C = \{\mathbf{x} | 0 \leqq x_i \leqq 1,\ i = 1, \ldots, n\}$, that is, the values $f(\mathbf{c}_i)$ where the components c_{ij} of the vector $\mathbf{c}_i$ may take on the value 0 or 1 are available. Our task then becomes to define a good approximation to the function value $f(\mathbf{x})$ for any $\mathbf{x} \in C$.

The index set of the 2^n vertices $\mathbf{c}_i$ of the cube C will be denoted by I. Also the notation $s_{ij} = s_{ij}(x_j, c_{ij}) = (1 - x_j)^{1-c_{ij}}(x)^{c_{ij}}$, $j = 1, \ldots, n$, $i \in I$ will be introduced (here $s_{ij} = x_j$ if $c_{ij} = 1$ and $s_{ij} = 1 - x_j$ if $c_{ij} = 0$). Then for approximating the function value $f(\mathbf{x})$ (in correspondence with equation (8.15)) by linear approximation we will use the following formula:

$$f(\mathbf{x}) = \sum_{i \in I} s_{i1} s_{i2} \ldots s_{in} f(c_{i1}, \ldots, c_{in}). \tag{8.16}$$

In the course of evaluating the sum in equation (8.16) one may encounter difficulties because of two reasons. On the one hand, in the case of large n the number of the terms to be totalled is prohibitive; on the other, the eventual result would probably be quite off the mark due to rounding errors. The procedure for computing the sum that emerges as a matter of course is that we sample some of the 2^n terms according some probability distribution and these will be added up. It is similar to the idea that has been employed in evaluating the sum during the calculation of an approximate solution to a system of equations.

Define the discrete random variables $\xi_j, j = 1, \ldots, n$ to be independent and distributed according to the equations

$$P\{\xi_j = 0\} = 1 - x_j \quad \text{and} \quad P\{\xi_j = 1\} = x_j. \tag{8.17}$$

The random vector variable $\boldsymbol{\xi} = (\xi_1, \ldots, \xi_n)$ takes on the value $\mathbf{c}_i$ with probability $s_{i1} \ldots s_{in}$, since the independence of the components ξ_i implies the equation

$$P\{\boldsymbol{\xi} = \mathbf{c}_i\} = P\{\xi_i = c_{i1}, \xi_2 = c_{i2} \ldots, \xi_n = c_{in}\} = s_{i1}s_{i2} \ldots s_{in}.$$

The expected value of the random variable $f(\boldsymbol{\xi})$ is just equal to $f(\mathbf{x})$ because of (8.16) and it means that an unbiased estimator of the interpolated values of $f(\mathbf{x})$ is

$$\frac{1}{N} \sum_{i=1}^{N} f(\boldsymbol{\xi}^{(i)}),$$

where $\boldsymbol{\xi}^{(i)}, i = 1, \ldots, N$ are the realizations of the random vector variable $\boldsymbol{\xi}$, given by equations in (8.15). The formal algorithm of computing the estimator is the following (the values $\mathbf{x}$ and N are given):

A4 Interpolation of a function of very many variables

1. [Initialize.] Let $i \leftarrow 0$, $S \leftarrow 0$.
2. [Sampling the ith random vertex.] Increase the index $i \leftarrow i + 1$. If $i > N$, then go to Step 5, otherwise set $j \leftarrow 0$.
3. [Generating the jth component of the ith vertex.] Increase $j \leftarrow j + 1$, if $j > n$, then go to Step 4, otherwise generate u uniformly in interval $[0,1)$. If $u > x_j$, then set $c_j \leftarrow 0$, otherwise set $c_j \leftarrow 1$. Repeat this step from the beginning.
4. [Summing up the function values.] Add up $S \leftarrow S + f(c_1, c_2, \ldots, c_n)$ and go back to generate the next term in the sum to Step 2.
5. Deliver the value $S \leftarrow S/N$ as an approximate value of the linear interpolation to the function value $f(\mathbf{x})$.

Take notice that the generation of a sample from the n-dimensional large scale discrete distribution of the random vector $\boldsymbol{\xi}$ is not a very simple task. In practical application it might be worthwhile to write a machine coded algorithm for generating the values $\boldsymbol{\xi}^{(i)}$.

8.6. Algorithms of optimization

One of the basic problems in operations research is determining extreme values of functions. It can be formulated the following way: determine a point $\mathbf{x}^*$ for which the function $f(\mathbf{x})$ attains its maximum in the domain D, that is,

$$\mathbf{x}^* = \arg\max_{\mathbf{x} \in D} f(\mathbf{x}). \tag{8.18}$$

The domain D is called the set of feasible solutions and generally it is determined by inequalities imposing constraints on the possible values of $\mathbf{x}$. If the domain D is the whole n-dimensional space R^n then problem (8.18) is called one of unconstrained optimization.

There is a great variety of algorithms developed for determining the optimum $f(\mathbf{x}^*)$ and an optimal point $\mathbf{x}^*$; the theoretical and the practical applicability of these methods depends on the properties of the function $f(\mathbf{x})$ and the domain D. One may, though, frequently encounter cases where our knowledge of the function $f(\mathbf{x})$ is very limited. We can, for example, determine the function value $f(\mathbf{x})$ for any $\mathbf{x} \in D$ only. Sometimes even this function value is imprecise. In these cases, and in some others as well, it is worthwhile to use some kind of simulational approach to obtain $\mathbf{x}^*$ and $f(\mathbf{x}^*)$. Here and in the next chapter we describe some of these methods.

Suppose the domain D be bounded; its volume will be denoted by V, that is, $V = \int_D d\mathbf{x} > 0$. Generate N uniformly distributed points $\mathbf{x}_i$, $i = 1, \ldots, N$ in domain D and consider the greatest value among $y_i = f(\mathbf{x}_i)$ — we assume the computability of $f(\mathbf{x})$ as well. The point $\mathbf{x}_j$ where the maximum is achieved — $\mathbf{x}_j = \arg\max_{i=1,\ldots,N} f(\mathbf{x}_i)$ will be called an approximation to the optimal solution $\mathbf{x}^*$. Consider D_ε, an ε-neighbourhood of the point $\mathbf{x}^*$ for a given radius $\varepsilon > 0$ and $D_\varepsilon D$, i.e. let $D_\varepsilon = \{\mathbf{x} | \sum_{i=1}^{n} (x_i - x_i^*)^2 \leqq \varepsilon\}$, the volume V_ε of the neighbourhood D_ε is positive. The probability that a point uniformly distributed in domain D will fall into D_ε is $V_\varepsilon / V > 0$. It means that having generated a sufficiently large number of uniformly distributed points in

D the approximation $\mathbf{x}_j$ will be in D_ε, sufficiently near to the optimal solution $\mathbf{x}^*$. In other words, denoting the point realizing the maximum in the case of generating N uniformly distributed points by $\mathbf{x}_j(N)$ the point sequence $\mathbf{x}_j(N)$ will converge to the optimal solution $\mathbf{x}^*$ with probability 1 if $N \to \infty$.

A5 Maximizing a function (crude method)

1. Generate the points $\mathbf{x}_i$, $i = 1, \dots, N$ uniformly in domain D and let j be an index for which the maximum value $\max_{i=1,\dots,N} f(\mathbf{x}_i)$ is attained.
2. Deliver the point $\mathbf{x}_j$ as an approximation to the optimal solution. If we want to have our approximate solution $\mathbf{x}_j$ to be in D_ε with a preassigned ε, then we have to repeat the algorithm from the beginning, until $\mathbf{x}_j \in D_\varepsilon$ is not fulfilled.

This algorithm is not really more than a possibility and a rather vague one to that . The practical application of the above outlined procedure is bound to fail in most cases because of several difficulties. Some of them are given below.

(i) The generation of uniformly distributed points in domain D can be very troublesome because of its "bad" shape (the acceptance-rejection ratio, that is, the efficiency of the generation procedure, may be small).

(ii) The domain D compared to domain D_ε with a sufficiently small ε is very large, then we have a very small probability V_ε/V and thus the sample size N is to take on a very great value.

(iii) The function $f(\mathbf{x})$ has more, maybe many local maxima, and they are very near to the global maximum. In this case the comparison between function values or termination of the procedure (sufficiently small ε) becomes difficult.

(iv) The same problem of comparing the different function values occurs when we can get inaccurate function values $f(\mathbf{x}_i)$ only.

(v) Testing the relation $\mathbf{x}_j \in D_\varepsilon$ may be difficult if possible at all since knowledge of D_ε assumes at least some knowledge of the set of optimal values.

The deficiencies of the basic procedure A5 can be remedied in several ways. One of the obvious ways of changing the algorithm is changing the distribution of random points: the points $\mathbf{x}_i$, $i = 1, \dots, N$ will be generated according to a density function $h(\mathbf{x})$ (in the way importance sampling is used) rather than uniformly. An expedient way of determining the function $h(\mathbf{x})$ would be to have great values of $h(\mathbf{x})$ near to the optimal solution (global maximum) and relatively small values elsewhere. It can be achieved by taking a density function $h_1(\mathbf{x}) = cg_1(\mathbf{x})$, proportional to the function $g_1(\mathbf{x}) = f(\mathbf{x}) + K$, where K and c are constants: K is large enough to ensure $f(\mathbf{x}) + K \geqq 0$, $\mathbf{x} \in D$ and c is determined by the requirement $1 = \int h_1(\mathbf{x})\, d\mathbf{x} = \int cg_1(\mathbf{x})\, d(\mathbf{x})$. (Of course, the integral $f(\mathbf{x})$ has to be finite over D.) The points to be generated are even better concentrated around the optimal solution by a suitable scaling we may use; for instance, we may take the density function $h_2(\mathbf{x}) = c_2 g_2(\mathbf{x})$ proportional to the function $g_2(\mathbf{x}) = \exp\{f(\mathbf{x})\}$ or the density function $h_3(\mathbf{x}) = c_3 g_3(\mathbf{x})$ obtained from the double exponential function $g_3(\mathbf{x}) = \exp\{\exp\{f(\mathbf{x})\}\}$. A point generated by density function $h_2(\mathbf{x})$ (or $h_3(\mathbf{x})$) will fall with greater probability into the domain D_ε than one with density $f(\mathbf{x})$.

Utilization of a well-chosen density function $h(\mathbf{x})$ can render the solution of difficulties (i) and (ii) possible. Furthermore, at the same time we can get rid of the restriction on the boundedness of the domain D — but we still need the finiteness of integrals $\int g_i(\mathbf{x})\, d\mathbf{x}$. The problems in this variant of the algorithm surface in giving an explicit form of the density function $h(\mathbf{x})$. Generally the determination of the integrals $\int g_i(\mathbf{x})\, d\mathbf{x}$ cannot be warranted; in practice one is inclined to use an approximation of $f(\mathbf{x})$ in deriving a suitable function $g_i(\mathbf{x})$. In actual optimization work this can give a reasonable approach. Still, the question of sampling with density function $h_i(\mathbf{x})$ remains a problem causing sometimes insurmountable obstacles.

The difficulty listed under (iii) can only be overcome utilizing similar methods to those used in deterministic procedures. One can for instance restart the procedure from the very beginning several times, and if we arrive in the same neighbourhood we can state more or less safely that it is the domain around the absolute maximum we have found.

As far as the problem of inaccurate function values is concerned (problem (iv)), there is an ample resource of methods developed for this type of problem that will be elaborated in the next section.

Another possible solution procedure for solving (8.18) is a random search technique. The essence of this iterative procedure is that we will proceed in the direction where the objective function increases, thus eliminating vast areas where the objective function is small.

Let $\mathbf{x}_1$ be the starting point of the algorithm and assume that we have already determined the points $\mathbf{x}_2, \ldots, \mathbf{x}_k$. Now we show how we can obtain the next point $\mathbf{x}_{k+1}$. Consider the sphere $S(\mathbf{x}_k, r_k)$ which is drawn around the point $\mathbf{x}_k$ with radius r_k and is inside D. Generate points $\mathbf{y}_i$ uniformly on the surface of the sphere $S(\mathbf{x}_k, r_k)$ until we get a point $\mathbf{y}_s$, satisfying the condition $f(\mathbf{y}_s) > f(\mathbf{x}_k)$. Then we proceed from the point $\mathbf{x}_k$ to the point $\mathbf{x}_{k+1} = \mathbf{y}_s$ and repeat the random search. If we have $f(\mathbf{y}_i) \leqq f(\mathbf{x}_k)$ for each point $\mathbf{y}_i$, $i = 1, \ldots, N$, where N denotes the maximum number of points to be generated on the sphere $S(\mathbf{x}_k, r_k)$ then the radius of the sphere is decreased by a multiplicator $0 < q < 1$, that is, we set $r_k \leftarrow qr_k$ and another set of N_k points will be generated on this smaller sphere around $\mathbf{x}_k$. This procedure of shrinking and proceeding spheres will be terminated if the radius r_k is small enough: the point $\mathbf{x}_k$ can be considered a local maximum.

A variant of the method may be the following: if a point $\mathbf{y}_s$ has been found, where the objective function increased compared to $f(\mathbf{x}_k)$, then maximization of the objective function is carried out along the halfline (or interval) $\mathbf{x} = \mathbf{x}_k + \lambda(\mathbf{y}_s - \mathbf{x}_k), \lambda > 0,\ \mathbf{x} \in D$ originating in $\mathbf{x}_k$. It calls for obtaining that multiplier λ_k for which $f(\mathbf{x})$ is the greatest, that is, $\lambda_k = \arg \max\limits_{\lambda>0} f(\mathbf{x}_k + \lambda(\mathbf{y}_s - \mathbf{x}_k))$ and then the next point in the iterative procedure will be $\mathbf{x}_{k+1} = \mathbf{x}_k + \lambda_k(\mathbf{y}_s - \mathbf{x}_k)$. This line search technique following the direction finding can successfully counterbalance the misdeeds of an inadequate choice of the initial radius r_1 especially if we couple it with correspondingly changing the radius. That is, if $\lambda_k > r_k$, then r_{k+1} should be greater than r_k and vice versa. We can leave r_k unchanged if for instance $1/\mu < \lambda_k/r_k < \mu$, otherwise the value $r_{k+1} = \lambda_k/\mu$ if $\lambda_k > r_k$ (or $r_{k+1} = \mu\lambda_k$, if $\lambda_k < r_k$) may be set, where $1 \leq \mu$ is a constant, $\mu = 2$, say.

The main advantage of this and other algorithms that can be built along similar reasonings is the great number of arbitrarily adjustable parameters — and it proves to be its main drawback — the starting point $\mathbf{x}_1$, the initial radius r_1, the number of the points N_k to be generated in the kth step, the multiplicator q for decreasing the radius and the quantity ε serving as terminating criterion. If the structure of the problem is known or the previous experimentation has already given an insight into the solution (for example, we have a starting point x_1 that is known to be nearly optimal), then by a good choice of parameters we can obtain good results, in very few iterations, otherwise in general there is not much that can be said about the speed of convergence. Convergence with probability 1 is ensured by assuming very mild conditions.

It is highly advisable to determine the number of the possible iterations M in advance in order to make available the interactive option so that the user can see the progress of optimization and modify the parameters whenever it seems proper, that is, if the domain of feasible solution is not bounded or, compared to r_k, is disproportionately large.

Assume that the values of $\mathbf{x}_1$, r_1, q, N_k, ε, M are fixed, then the formal algorithm for optimization is the following:

A6 Maximizing a function (random search)

1. Initialize $k \leftarrow 0$, $r \leftarrow r_1$.
2. Increase the index $k \leftarrow k+1$. If $k > M$ then go to Step 5. Set $i \leftarrow 1$ and $c \leftarrow f(\mathbf{x}_k)$.
3. Generate $\mathbf{y}_i$ uniformly on the surface of the sphere $S(\mathbf{x}_k, r)$. If $f(\mathbf{y}_i) > c$ then set $\mathbf{x}_{k+1} \leftarrow \mathbf{y}_i$ and go back to Step 2.
4. If $i = N_k$, then decrease the radius determining $r \leftarrow qr$, and furthermore if $r < \varepsilon$ now, then proceed to Step 5, otherwise set $i \leftarrow 0$. If $i < N_k$, then increase the index $i \leftarrow i+1$ and go back to Step 3 to generate another uniformly distributed point.
5. Deliver the point $\mathbf{x}_{k-1}$ as an approximation of the optimal solution.

Since the requirement for the sphere to be in the feasible domain, that is, the condition $S(\mathbf{x}_k, r) \subset D$ cannot be checked in a number of cases, another little weaker condition should then be examined: in Step 3, after generation of $\mathbf{y}_i$ one should check whether $\mathbf{y}_i$ is in D (reject and repeat the generation if no).

The line search can be inserted into Step 3: the expression $\mathbf{x}_{k+1} \leftarrow \mathbf{y}_i$ should be replaced by $\mathbf{x}_{k+1} \leftarrow \mathbf{x}_k + \lambda_k(\mathbf{y}_i - \mathbf{x}_k)$, where λ_k is given by

$$\lambda_k = \arg \max_{\substack{\lambda > 0 \\ x_k + \lambda(y_i - x_k) \in D}} f(\mathbf{x}_k + \lambda(\mathbf{y}_i - \mathbf{x}_k)).$$

In practice simulational procedures are used alternatively with deterministic ones. Generally, simulational algorithms yield a solution near to the optimum fast enough — this may be checked by measuring the distance between consecutive points $\mathbf{x}_k, \mathbf{x}_{k+1}, \ldots$ by some measure (e.g. Euclidean $\sum_{i=1}^{n}(\mathbf{x}_{ki} - \mathbf{x}_{k+1,i})^2$) — and then a deterministic fast (e.g. quadratically convergent) procedure is applied.

8.7. Stochastic quasi-gradient algorithms

A procedure for evaluating an approximation of an optimization problem will be described, that may be used in case of random noise as well; that is, if the computation of the function value or the gradient of the objective function, or both, cannot be done accurately.

Consider the problem of convex optimization, that is, minimization of a convex function $f(\mathbf{x})$ over a convex domain D, given as follows:

$$
(8.19) \qquad \begin{aligned} &\min f(\mathbf{x}), \\ &g_1(\mathbf{x}) \leqq 0, \\ &\quad\vdots \\ &g_m(\mathbf{x}) \leqq 0, \end{aligned}
$$

where $g_1(\mathbf{x}), \dots, g_m(\mathbf{x})$ are convex functions. There is a family of iterative deterministic optimization procedures, using the gradient $f_x(\mathbf{x})$ of the objective function $f(\mathbf{x})$ for solving (8.19). If $f(\mathbf{x})$ and/or $f_x(\mathbf{x})$, $g_i(\mathbf{x})$ cannot be determined accurately, or the computation to be carried out is too laborious, we may rely on stochastic quasi-gradient methods where we construct statistical estimates of the function values and the gradient.

Assume we construct a sequence of approximate solutions $\mathbf{x}_1, \dots, \mathbf{x}_k \cdots$ to problem (8.19) by the aid of the sequence of random variables η_{ik} and ξ_k, $k = 1, 2, \dots$ for which we have

$$
E\{\eta_{ik} | \mathbf{x}_1, \dots, \mathbf{x}_k\} = g_i(\mathbf{x}_k) + a_{ik}
$$

$$
E\{\xi_k | \mathbf{x}_1, \dots, \mathbf{x}_k\} = f_x(\mathbf{x}_k) + \mathbf{b}_k.
$$

In order to ensure convergence of $\mathbf{x}_1, \dots, \mathbf{x}_k$ we have to assume that the sequences $\{a_{ik}\}_{k=1}^{\infty}$ and $\{\mathbf{b}_k\}_{k=1}^{\infty}$ are diminishing in a sense but not too rapidly. The rigorous theory is developed on the basis of the stochastic quasi-Feyer sequences and proves the convergence with probability 1 under mild restrictions. (A sequence of random vectors $\eta_0, \eta_1, \dots, \eta_k, \dots$ is quasi-Feyer with respect to a set X if for any $\mathbf{x} \in X$ we have $E\{\|\mathbf{x} - \eta_{k+1}\|^2 | \eta_0, \eta_1, \dots, n_k\} \leqq \|\mathbf{x} - \eta_k\|^2 + \delta_k$, $k = 0, 1, \dots$, where $\delta_k \geqq 0$ and $\sum_{k=0}^{\infty} E\{\delta_k\} < \infty$. For a quasi-Feyer sequence η_k, $\|\mathbf{x} - \eta_k\|^2$ converges with probability 1 for any $\mathbf{x} \in X$, $E\{\|\mathbf{x} - \eta^k\|^2\} < C$).

In what follows we describe two specific forms of constructing the random variables η_{ik} and ξ_k, $k = 1, 2, \dots$ and by the same token stochastic quasi-gradient methods.

Consider problem (8.19). A deterministic method for obtaining an optimal solution is if we start from the initial point $\mathbf{x}_1 \in D$ and proceed from $\mathbf{x}_k$ to $\mathbf{x}_{k+1}$ according to the equation

$$
(8.20) \qquad \mathbf{x}_{k+1} = \prod_B (\mathbf{x}_k + r_k f_x(\mathbf{x}_k)), \qquad k = 1, 2, \dots,
$$

where $B \subset D$ is a convex compact set and $\prod_B(\mathbf{x})$ is the projection of $\mathbf{x}$ to B, r_k is the step length. The stochastic variant of this method is given by

$$\mathbf{x}_{k+1} = \prod_B(\mathbf{x}_k + \varsigma_k \mathbf{x}_k), \qquad k = 1, 2, \ldots$$

$$E\{\xi_k | \mathbf{x}_1, \ldots, \mathbf{x}_k) = f_x(\mathbf{x}_k) + \mathbf{b}_k. \tag{8.21}$$

The validity of this stochastic quasi-gradient method is established in the following

Theorem. Assume that $f(\mathbf{x})$ is a convex continuous function, B is convex and compact, the length of the gradient is bounded, i.e. $E\{\|\xi_k\|^2 | \mathbf{x}_1, \ldots, \mathbf{x}_k\} \leqq K$ and with probability 1 the following conditions are satisfied:

$$\sum_{k=1}^{\infty} \varsigma_k = \infty, \qquad \varsigma_k \geqq 0,$$

$$\sum_{k=1}^{\infty} E\{\varsigma_k^2\} < \infty, \tag{8.22}$$

$$\sum_{k=1}^{\infty} E\{\varsigma_k \mathbf{b}_k\} < \infty,$$

then $\mathbf{x}^* = \lim_{k \to \infty} \mathbf{x}_k$ exists and is an optimal solution to problem (8.19) with probability 1.

Generally, one of the following two cases takes place: (i) the gradient $f_x(\mathbf{x})$ has an explicit form (one may evaluate it but only with some error — in this case one can use this noisy gradient), (ii) if the gradient is not available then one can resort to some approximation such as

$$\xi_k = \frac{1}{2\delta_k} \sum_{k=1}^{n} [f(\mathbf{x}_k + \delta_k \mathbf{e}_i) - f(\mathbf{x}_k - \delta_k \mathbf{e}_i)] \mathbf{e}_i, \tag{8.23}$$

where $\mathbf{e}_i$ is the ith unit vector. Even here one can allow an error in evaluating $f(\mathbf{x})$ provided ξ_k still satisfies (8.21) and (8.22).

The algorithm is simple; the only addition that needs to be made is a terminating criterion, which may require that subsequent points $\mathbf{x}_k$ are near enough, i.e. $\|\mathbf{x}_k - \mathbf{x}_{k+1}\|^2 < \varepsilon$ should hold for a small ε.

A7 Stochastic quasi-gradient algorithm

1. Initialize $k \leftarrow 1$.
2. Increase $k \leftarrow k+1$, compute $\mathbf{y}_k$ as an estimation of the gradient (either by directly computing $f_x(\mathbf{x})$ or by the form (8.23)). Evaluate $\mathbf{x}_{k+1} = \prod_B(\mathbf{x}_k + \varsigma_k \mathbf{y}_k)$.
3. If $\|\mathbf{x}_k - \mathbf{x}_{k+1}\|^2 < \varepsilon$, then deliver $\mathbf{x}_{k+1}$ as an optimal solution, otherwise go back to Step 2.

Sometimes, because of scaling problems or other reasons, the sequence $\{\mathbf{x}_k\}$ may rapidly oscillate. A variant of the above method will be described next. This method uses a filter that sums up the previous gradients and the point sequence produced moves into the slowly changing direction thus obtained.

Assume $E(\xi_k|\mathbf{x}_1,\ldots,\mathbf{x}_k) = f_x(\mathbf{x}_k) + \mathbf{b}_k$ and $\mathbf{x}_1 \in D$, furthermore we have a direction $\mathbf{d}_1$ that is at least partially in D. The transition from the point $\mathbf{x}_k$ to $\mathbf{x}_{k+1}$ is described by the following equations:

$$\mathbf{x}_{k+1} = \mathbf{x}_k + \tau_k \mathbf{d}_k, \qquad k = 1,2,\ldots$$

$$\mathbf{d}_{k+1} = \prod_B(\mathbf{d}_k + \varsigma_k(\xi_k - \mathbf{d}_k)).$$

The step lengths ς_k and τ_k are random variables, for which we assume that

(i) $\sum_{k=1}^{\infty} E\{\varsigma_k^2\} < \infty, \qquad \sum_{k=1}^{\infty} E\{\tau_k^2\} < \infty,$

(ii) $\sum_{k=1}^{\infty} \varsigma_k = \infty, \qquad \sum_{k=1}^{\infty} \tau_k = \infty$ with probability 1,

(iii) $\tau_k/\varsigma_k \to 0$ with probability 1, and

(iv) $\sum_{k=1}^{\infty} \|\xi_k\|^2 < \infty$ and $\sum_{k=1}^{\infty} E\{\varsigma_k \mathbf{b}_k\} < \infty$.

Under assumptions (i)–(iv) the point sequence $\{\mathbf{x}_k\}$ converges with probability 1 to an optimal solution.

Similarly to algorithm A7, applicability of this method for practical problems depends on the choice of τ_k, ς_k, $\prod_B$, B.

Finally, one possibility should be noted. An estimation of the gradient can be obtained if we generate s_k uniform directions emanating from $\mathbf{x}_k$ and the average of the finite differences computed from these function values may serve as ξ_k. The remarkable feature of this idea is that s_k is independent of n, the number of dimensions.

8.8. Equation of state calculations and optimization

A heuristic approach is outlined in this section that was originally devised and used for calculating the properties of any substance which may be considered as a set of individual particles interacting with each other. Subsequently it is shown how the same can be applied to derive optimal or near-optimal solutions to several classical problems in optimization that are not prone to be investigated by exact optimization procedures.

Consider a system of particles, each of them having a state characterized by its physical location. There is a potential function dependent on the distance of the particles from each other. From the locations of the particles through the potential we are able to evaluate the overall energy E of the system whatever it may be, dependent on the physical meaning of the particles and the potential. The temperature T of the system is also given. Random perturbations are now simulated in order to move the system into the most likely state. The simulation of a possible change in the system is done in the following way. Chose randomly a particle; generate a random displacement of same. (This random displacement should be done in correspondence with the properties of the physical system, i.e. maybe a uniformly distributed random point in a square, on the surface of a sphere, jumping to one of the adjacent unoccupied places in a crystalline structure, etc.). The energy of the modified system is computed and denoted by E^*. If the energy of the system is decreased, that is, $E > E^*$, then the change (the random displacement of a particle) is accepted without further investigation. However, the random disturbances will not be rejected outright even in other cases, but with some probability will still be accepted. Boltzmann's law gives the probability of the occurrence of a change in the level of energy by the expression

$$p = \exp\{-(E^* - E)/kT\}, \tag{8.24}$$

where k is the Boltzmann constant. Thus we generate u uniformly and if $u < p$, then the change is accepted, otherwise the particle is put back into its original position.

Repeating many times this procedure of generating random perturbations and accepting or rejecting them will yield a more or less steady state. Now decrease the temperature T and repeat the displacement generation-acceptance-rejection procedures, using this new lower temperature T^*. In this case the probability of accepting the same positive change in the level of energy will be less likely than it was in the case of temperature T

because of (8.24). Having completed several (or many) consecutive decreases in the temperature, the whole system will be "frozen", that is, no positive change in the energy may be accepted because of the low temperature and there are no more possibilities for decreasing the energy. The system is in a quasi-stable state, where the energy attained a (local) minimum.

T_0 is a prescribed temperature and $\triangle T > 0$ will be the amount of the decrease in one iteration.

A8 Thermodynamical system stabilization

1. Initialize the state and location of particles, temperature T and compute energy E.
2. Repeat this step N times. Choose randomly a particle, generate a random displacement and compute the new energy level E^*. If $E > E^*$, then accept the displacement, otherwise generate u, and if $u > p = \exp\{-(E^* - E)/kT\}$ then reject, if $u < p$ then accept the displacement.
3. [Termination.] If $T < T_0$, then stop the procedure, otherwise set $T \leftarrow T - \triangle T$ and go back to Step 2.

The criterion for stopping the simulation may be related to the energy level as well, if for instance the energy level reaches a prescribed value, or if there is no decrease in the energy during the last N displacements, etc.

The above described idea and algorithm A8 can be applied to many problems. The area that springs to mind for which this application is best suited is optimization. The set X of the feasible solutions will be the set of possible states of the system, the energy of the system will be the objective function (to be minimized). The temperature T of the system will be interpreted as a mobility parameter of the system, that is, a formal parameter showing the willingness of the system to accept changes contradicting the main aim, the minimization. Furthermore, there should be a random displacement generator $\mathbf{y} = G(\mathbf{x}, \xi)$ that from a state $\mathbf{x}$ by some random numbers ξ produces another feasible state $\mathbf{y}$. Consecutive applications of the transformation G should result in achieving any feasible state with a positive probability, that is, the set X should be connected with regard to the transformation G and random variable ξ.

Consider the quadratic assignment problem. Let A and B be two $n \times n$ matrices and A will be called the distance matrix and B the connection matrix. Denote $I = \{1, 2, \ldots, n\}$. The quadratic assignment problem consists of finding a permutation function t of the index set I for which the function

$$f(t) = \sum_{i \in I} \sum_{j \in I} a_{ij} b_{t(i)t(j)} \tag{8.25}$$

attains its minimum. The set X of feasible solutions is any of the permutations of the indices $1, 2, \ldots, n$, the objective function (energy) is given in (8.25), $G(\mathbf{x}, \xi)$ may be the exchange of two elements in a given permutation. Since any feasible solution may be produced by repeatedly applying the exchange of two elements, thus $G(\mathbf{x}, \xi)$ is appropriate. The change in the objective function if i and j are exchanged in a given permutation can be worked out as

$$\begin{aligned} \Delta E = E - E^* = \sum_{k \in I\{i,j\}} & (a_{jk} - a_{ik})(b_{t(i)t(k)} - b_{t(j)t(k)}) + \\ & + (a_{kj} - a_{ki})(b_{t(k)t(i)} - b_{t(k)t(j)}) + \\ & + (a_{jj} - a_{ii})(b_{t(i)t(i)} - b_{t(j)t(j)}) + \\ & + (a_{ji} - a_{ij})(b_{t(i)t(j)} - b_{t(j)t(i)}). \end{aligned} \tag{8.26}$$

Maybe one point would be needed: how to generate random displacements. It can be done by uniformly choosing two indices from the actual permutation and exchanging them. Now the application of algorithm A8 for the quadratic assignment problem is obvious. One more example of this type will be mentioned in the last section.

The method looks to be especially suitable for solving problems with more local minima, since there is always a chance the algorithm can jump out of a local pit.

8.9. The STABIL stochastic programming model

A stochastic programming model will be described in this section, the solution procedure of which requires simulation.

Consider a deterministic optimization problem given by the following inequalities:

$$\begin{aligned} & \min f(\mathbf{x}) \\ & g_i(\mathbf{x}) \geq b_i, \qquad i = 1, \ldots, m + M. \end{aligned} \tag{8.27}$$

In order to ensure solvability we have to assume concavity of the functions $g_i(\mathbf{x})$, $i = 1, \ldots, m + M$ and convexity of $f(\mathbf{x})$. If the first m right-hand side values are random variables and not deterministic values, then (8.27) as a model becomes meaningless. In this case it is better to construct a probability constrained model as follows:

$$\begin{aligned} & \min f(\mathbf{x}) \\ (8.28) \qquad & G(x) = P\{g_i(\mathbf{x}) \geqq \xi_i,\ i = 1,\dots,m\} \geqq p, \\ & g_i(\mathbf{x}) \geqq b, \qquad i = m+1,\dots,m+M, \end{aligned}$$

where the probability of fulfilment of the random inequalities is required to be greater than a prescribed value p $(0 < p < 1)$.

Suppose now that all of the functions $f(\mathbf{x})$, $g_i(x)$, $i = 1,\dots,m+M$ in (8.28) are linear, the resulting stochastic programming model is then called STABIL. We will need the next theorem, that we present without proof.

Theorem 1. If the joint distribution of the random variables ξ_i, $i = 1,\dots,m$ is continuous with density $\exp\{-Q(\mathbf{z})\}$, where $Q(\mathbf{z})$ is convex, then $G(\mathbf{x})$ is a logarithmic concave function.

We may assume that $E\{\xi_i\} = 0$, $D^2\{\xi_i\} = 1$, since if otherwise, then subtracting the expected value from both sides of the inequality $g_i(\mathbf{x}) \geqq \xi_i$ and dividing the same with the variance, it can be standardized. If the joint distribution of the random variables ξ_i is multidimensional normal, that is, its density is of the form

$$\varphi(\mathbf{x}) = (2\pi)^{-m/2}|R|^{-1/2}\exp\left\{-\frac{1}{2}\mathbf{x}R^{-1}\mathbf{x}\right\}$$

where R is a positive definite symmetric matrix (the correlation matrix), then it complies with the requirements of Theorem 1. We will assume hereinafter that the joint density function of $\xi_1,\dots,\xi_m$ is $\varphi(\mathbf{x})$. It should be noted that the gradient of the distribution function ϕ may be evaluated by computing $(m-1)$-dimensional distribution function values.

For solving problem (8.28) we will apply the feasible directions method of Zoutendijk. Basically it is an iterative procedure that may be used for solving the problem

$$\begin{aligned} & \min f(\mathbf{x}), \\ (8.29) \qquad & G(\mathbf{x}) \geqq p, \\ & \mathbf{a}_i'\mathbf{x} \geqq b_i, \qquad i \in I, \end{aligned}$$

where G and f are twice differentiable. Assume that the set of feasible solution is nonempty and bounded, furthermore there exist $\mathbf{x}_1$ an initial solution satisfying the constraints in (8.29) and there is at least one solution, for which $G(\mathbf{z}) > p$. An iteration step from $\mathbf{x}_k$ to $\mathbf{x}_{k+1}$ will be described in two parts: first we have to find a direction and second we optimize along this line. More specifically, the direction finding consists of solving the following problem for the unknowns $(\mathbf{x}, y)$:

$$
\begin{aligned}
&\min y, \\
&G(\mathbf{x}_k) + \nabla G(\mathbf{x}_k)(\mathbf{x} - \mathbf{x}_k) + \theta y \geqq p, \\
&\qquad\qquad a_i'\mathbf{x} \geqq b_i, \qquad i \in I, \\
&\quad \nabla f(\mathbf{x}_k)(\mathbf{x} - \mathbf{x}_k) \leqq y,
\end{aligned}
\tag{8.30}
$$

where θ is a positive (generally we choose a small value for it), fixed number. The number of unknowns in (8.30) is $n+1$ since it is $(\mathbf{x}, y)$. An immediate feasible solution can be given as $\mathbf{x} = \mathbf{x}_k$, $y = 0$. Having solved the direction finding problem, its optimal solution will be denoted by $(\mathbf{x}_k^*, y_k^*)$. If $y^* = 0$ then the procedure terminates and $\mathbf{x}_k^*$ is optimal to (8.29) as well. If $y^* \neq 0$, which in this case means $y^* < 0$, then we may proceed to determine the step length and this is done by a line search. We need that value λ_k of the parameter λ for which the function value $f(x_k + \lambda(x_k^* - x_k))$ is the smallest possible, that is,

$$
\lambda_k = \arg \min_{\lambda > 0} f(x_k + \lambda(\mathbf{x}_k^* - \mathbf{x}_k)). \tag{8.31}
$$

This λ_k helps us to produce the next point in the iteration procedure as

$$
\mathbf{x}_{k+1} = \mathbf{x}_k + \lambda_k(\mathbf{x}_k^* - \mathbf{x}_k).
$$

This feasible direction method, under the already posed assumptions (linear functions, normal distribution, inner point) can be and will be applied to solve (8.29) and the STABIL stochastic programming model.

Theorem 2. Under the above made assumptions the feasible directions method either terminates after a finite number of steps, giving an optimal solution, or produces a sequence $\mathbf{x}_1, \mathbf{x}_2, \ldots$ converging to an optimal solution.

The question of initial solution must not be neglected. If we do not have an initial solution $\mathbf{x}_1$, the same feasible directions method can be used to obtain one. More precisely, assume that $\mathbf{z}_0$ is a feasible solution to the linear constraints and consider the problem

$$
\begin{aligned}
&\min\,(p - G(\mathbf{x})) \\
&\quad \mathbf{a}_i'\mathbf{x} > b_i, \qquad i \in I.
\end{aligned}
$$

The feasible directions method can be used to obtain an optimal solution of this problem, using $\mathbf{z}_0$ as initial solution. An optimal solution of this problem is a suitable feasible solution $\mathbf{x}_1$ (or it will show that there is no such if the optimal function value is positive).

Examine in a little more detail the numerical problems in implementing the optimization procedure for the STABIL. Consider the function $G(\mathbf{x})$ in (8.28), which is

$$G(\mathbf{x}) = P\{\mathbf{c}_i'\mathbf{x} \geqq \xi_i,\ i = 1,\dots,m\} = \int_{-\infty}^{\mathbf{c}_1'\mathbf{x}} \cdots \int_{-\infty}^{\mathbf{c}_m'\mathbf{x}} \varphi(\mathbf{z})\, d\mathbf{z},$$

that is, the normal distribution function, the computation of which was dealt with in Chapter 7 ($g_i(\mathbf{x})$ in (8.28) was replaced by a linear function to obtain STABIL). Since it cannot be evaluated very precisely, we have to devise some countermeasures against failure. Generally during the optimization the function value $G(\mathbf{x})$ will be evaluated with a prescribed precision, say 5%. However, computing the step length cannot be done this way since we are faced with two kinds of error. Having linear objective function the determination of λ_k in (8.31) is basically reduced to the computation of the intersection of the ray $\mathbf{x}_k + \lambda(\mathbf{x}_k^* - \mathbf{x}_k)$ with the boundary facets of the domain of feasible solutions. If the facet is one given by a linear constraint, i.e. the facet is a hyperplane, we have only straightforward calculations.

The other case of concern for us is when the surface given by $G(\mathbf{x}) = p$ is involved. Here in determining λ_k we may make an error if we choose too small λ_k — it will mean that in the next iteration step the algorithm would move in the same direction once more, wasting time and computing effort. The other type of error we may commit is that we may choose too large λ_k, yielding a point outside of the domain of feasible solutions — this bears fatal consequences. Due to inaccuracy in evaluating the normal distribution functions, both errors may occur. So a special adaptive stepping algorithm was developed for obtaining the intersection of $\mathbf{x}_k + \lambda(\mathbf{x}_k^* - \mathbf{x}_k)$ with $G(\mathbf{x}) = p$. The main features are that the nearer we are to the borderline, the more accurately will calculations be carried out, the smaller will be the steps we take, and we safeguard against leaving the domain of feasible solutions.

One last detail in the implementation is the choice of optimality criterion. Because of the errors in the function evaluation the procedure is whimsical and multiple criteria have to be applied. The criteria that the difference between three consecutive solutions $(\mathbf{x}_k^*, y_k^*)$ should be small enough in each component proved to be restrictive enough if accompanied by the requirement that the objective function does not change greatly in these three points.

Actual work carried out using STABIL was done in designing a five year plan for the electricity industry in Hungary. As a very peculiar feature of the actual data we mention that the original optimal solution of the deterministic problem was of very low (0.09) probability, whilst the optimal solution created for the probability constrained model was of probability 0.90. And at the same time the cost function value was almost the same. It means that to achieve the same objective we can choose a solution with great realiability instead of one with small probability.

8.10. Optimal water level regulation of a lake

A system of stochastic programming models is presented here for the optimal control of a storage level. Each model in this system embraces only a few earlier periods and gives a policy of control only for the next or next two periods, but the time horizon is shifted continuously by one in the subsequent model. This idea is known as the *rolling horizon.*

The lake in consideration is Lake Balaton in the western part of Hungary. Monthly random inputs of rain and small rivers will add ξ_i to the initial content ς_0 of the lake. We have at our will the decision variables z_i, indicating the amount of water that is released from the lake via a channel in the ith month. Introduce furthermore the notation

$$\varsigma_k = \varsigma_0 + \xi_1 + \ldots + \xi_k, \qquad k = 1, 2, \ldots,$$

$$Z_k = z_1 + z_2 + \ldots + z_k, \qquad k = 1, 2, \ldots,$$

then the actual content $\varsigma_k - Z_k$ should be between two bounds in order to ensure environmental and recreational targets

$$a_k \leqq \varsigma_k - Z_k \leqq b_k, \qquad k = 1, 2, \ldots.$$

We have a restriction on the amount of water that might be released given by

$$0 \leqq z_k \leqq K, \qquad k = 1, 2, \ldots.$$

We formulate now a dynamic type decision methodology. In each month we take into account the previous two months and we decide upon the release in the next two months, that is, our first model will be

$$\max P\{a_1 \leqq \varsigma_1 - Z_1 \leqq b_1,\ a_2 \leqq \varsigma_2 - Z_2 \leqq b_2 | \varsigma_0\}.$$

Out of the optimal solution z_1^*, z_2^* we accept only z_1^*, we release z_1^* amount, encounter the realization of ξ_1, and formulate the next model, beginning with the next month, etc. So the kth model to be solved is

(8.32)

$$\max P\{a_k \leqq \varsigma_k - Z_k \leqq b_k,\ a_{k+1} \leqq \varsigma_{k+1} - Z_{k+1} \leqq b_{k+1} | \varsigma_{k-3},\ \xi_{k-2},\ \xi_{k-1}\}$$
$$\text{subject to } 0 \leqq z_k \leqq K,\ 0 \leqq z_{k+1} \leqq K, \qquad k = 1,2,\ldots.$$

Of course in other actual applications or models the time horizon can be extended — both ways — if need be, but in the Lake Balaton application it was completely sufficient to deal with two previous and two forthcoming months.

It can be shown — using the same fundamental results on log-concave measures as in proving the theorems of the previous section — that the objective function in (8.32) is a log-concave function of the variables z_k, z_{k+1}.

Several optimization techniques may be applied to solve problem (8.32) though there is a considerable help in studying the probability function: in a number of cases optimization can be reduced to simple equation-solving. In the actual computation (i) it was assumed that $\xi_i - s$ are normally distributed, (ii) evaluation of the objective function (two-dimensional normal distribution function) was done by utilizing a reduction formula, where numerical integration was applied. Nevertheless, the method for evaluating the objective function is a Monte Carlo method, if we want to keep our method competitive with more months to look ahead than two or if the random variables are of a different distribution.

We are especially forced to use simulation if the distribution of $\xi_i - s$ is not analytically given, only samples can be generated with the distribution of interest.

8.11. System simulation

Up to this section we have dealt with problems where one specific answer was looked for, a numerical solution to some mathematical problem: computing values of a special function, solving a system of equations or optimizing a function. Now we enter an area that may be termed system simulation or simulational case studies, where what we are interested in is not one specific answer but rather the behaviour of a system.

A typical example of such system simulations is analysis of processes in sophisticated systems, where the parts are connected by relations difficult to describe analytically. A computerized model of the actual system makes experimentation with it possible and the likely impact of future changes in the system can be investigated without actually changing it. It is very desirable in cases where the cost of carrying out the actual change would be very great (replacing a computer system with another configuration, for instance, or choosing among several different car production lines, etc.) or experimentation with the system is absolutely impossible (the whole economy of a country, the oil market, acid rain problems). There is an immense literature on this subject, part of which is classified under the headings of the related areas, economics, ecology, psychology, and the engineering sciences, to name a few.

System simulation involves model creation and simulation. Of course the former is subject to all problems and techniques relevant to this field of applied mathematics, that is:

(i) First we study the physical system we want to model.

(ii) The most significant components of the system and the main relationships and interactions among them are specified; all other characteristics and components are neglected for the time being.

(iii) The specified factors and interactions are described by a computer program; this yields a computer model of the reality.

(iv) The model is verified comparing its performance to that of the physical system. The verification is generally done the following way. Input data are gathered for a previous time period, fed into the computer model and the output data of the model are compared to the known outcomes of the previous period. If the discrepancy between the computer model's output and the actual outcomes of the physical system is small enough in some sense, then the model is considered to represent the physical system reliably.

(v) If the model fails verification then we have to refine the model by incorporating more components and interactions or amend the model by changing the parameters of interactions or the relationships among the factors already considered. The modified model is to be verified once more (go back to step iv). This kind of tuning up of the model may require several iterations and is to be continued until the modified model performs satisfactorily.

The computer program may be written in any of the high level languages (FORTRAN, Algol, PL1, etc.) occasionally with some routines in machine code or C or in one of the simulational languages (SIMSCRIPT, GPSS, SIMULA, CSL, DYNAMO). Simulational languages were developed to enhance program writing and performance in system simulation. It has been achieved by providing special features in the simulational language, for example, the "class" concept in SIMULA, that make the description of some simulational task very simple. The same things may be written in other high level languages only in a cumbersome way, if at all (consider for comparison nonstructured FORTRAN). Thus it is highly advisable that great system simulational models be developed in a simulational language.

8.12. A queuing system

Some of the most frequently encountered types in system simulation are the queuing models. The common feature of these is that some customers arrive, join a queue, wait until they can be served if the server is busy, then they are served and eventually depart from the system. The behaviour of the system is determined by the specification of the rules governing the system: what is the rule (random or deterministic) and the frequency of the arrivals, how long does one have to wait in the queue, who is served first, etc. The mathematical study of such systems can be performed in special cases only, so simulational computations are fully justified especially if several queuing systems are combined. Generally we are interested in gathering some statistics on the behaviour of the system: what is the average length of the queue, how much time does a customer spend in the system, in what percent of the time does the length of the queue exceed a given bound, etc. In this section we develop a simple queuing model hinting at possible developments and outlining an application.

In describing the queuing system we will use the terminology of a checkout counter in a supermarket. First, a customer enters the supermarket. Then, after having chosen the article(s) he wants to buy, the customer arrives at the checkout, lines up in the queue, waits until he comes up to the counter and after being served (made payment, obtained receipt, etc.), departs. Basically we will deal with arrival at the counter, waiting in the queue and being served — arrival at the supermarket and the time spent selecting goods may be combined in one singular time item at this point of our investigation and we need only the time of arrival at the queue. So this is a single-queue single-server system with an FIFO rule (the one First In will be First Out of the queue).

Simulation of the queue should begin with scheduling the arrivals. This is best done labeling the $1, 2, \ldots, n$th customers with their respective arrival times a_i and service times s_i, $i = 1, 2, \ldots, n$ (the ith customer arrives at the queue at time moment a_i and will spend s_i time at the counter, occupying the server). In accordance with the remark in the previous paragraph, we disregard for now the time t_i spent in the supermarket before joining the queue. The quantities a_i and s_i may be deterministic values handed down to us by someone or may be samples from a probability distribution. In the latter case, using standard random number generation techniques, we can generate arrival times a_i and service times s_i, $i = 1, \ldots, n$ according to this distribution in advance in a separately set up part of the computer program. The arrival and service times together with the FIFO rule govern the system. In the following formal description of the queuing system the variable t denotes the simulation clock, x is the cumulative time spent in the system by the customers and c is the cumulative waiting time spent in the queue.

A9 Single-server queue

1. [Set up.] Generate arrival times a_i and service times s_i, $i = 1, \ldots, n$.
2. Initialize $t \leftarrow s_1$, $x \leftarrow t$, $i \leftarrow t$, $c \leftarrow 0$.
3. [Proceed to the next arrival.] Increase $i \leftarrow i + 1$. If $t < a_i$, then change $t \leftarrow a_i$.
4. [Accumulation of waiting time.] Let $c \leftarrow c + t - a_i$, compute the time when the server becomes free $t \leftarrow t + s_i$ and sum up the transit time $x \leftarrow x + t - a_i$. If $i < n$ then go back to Step 3.
5. [Computation of statistics.] Deliver t when the whole system will become empty, $c \leftarrow c/n$ as the average waiting time and $x \leftarrow x/n$ as the average transit time of a customer.

A bit of explanation is in order here. The first step is self-explanatory. When the first customer arrives (Step 2) the server is free so he will immediately be served and will occupy the server for s_1 time. The next arrival is at moment a_2, and we have here two possibilities: (i) either the server will be free by this time — this occurs if $t = s_1 < a_2$, (ii) or the counter is busy so the second customer has to wait until $t = s_1$.

In the third and fourth steps of Algorithm A9 the distinction is made between the cases, and the actual computation is carried out.

In case (i) the system will be empty for $a_2 - s_1$ time, so when the second customer arrives we are back to the original situation with an empty queue, idle server. With the arrival of the second customer, we set the simulation clock time $t = a_2$, waiting time $t - a_2 = 0$, and the server will be free at time $t + s_2$. In case (ii) the customer waits $w_2 = t - a_1$ time, adds his own serving time to the simulation clock $t = t + s_2$ to obtain the time when the server will be idle and the overall time spent in the system is $w_2 + s_2 = s_1 + s_2 - a_2 = t - a_2$.

In Step 5 the necessary statistics, of course sometimes much more comprehensive than given here, are evaluated.

Several other rules may be used in the queuing models, for example there may be an LIFO (Last In, First Out) or a random choice. We may attach priorities to the customers as well. There is a possibility of incorporating frustrated behaviour – customers standing in line may become irritated because of lengthy waiting and with a certain probability may leave the queue without being served. This probability should increase with the lapse of time.

Using several realizations of the above described single-server queue with different rules, very intricate situations can be simulated. One example is given here. Consider a cold rolling mill, containing two (or more) train rolls the input of which is roughly rolled steel spools and the output is thin rolled steel spools or sheets of different sizes and qualities. An incoming steel spool enters first a heat treatment unit where it stays for several hours, even a couple of days, to be heated up. The duration of heat treatment is dependent on the temperature of the incoming spools, the weather, and partly on the relative location of the spools inside the heat treatment unit. A spool will proceed to either of the train rolls, that have different technological restrictions as to the spools they can handle (sizes, quality, etc.). A spool comes then to one of the three possible piece cutters that cut the spool into sheets or narrower spools which are in turn eventually packed into bundles on one of the wrapping machines. During this technological process several queues are formed in front of the different servers and the rules governing the queues (greatest possible size of the queue, what leaves the queue first, for which machine does it leaves) are greatly different from each other because of the technological constraints. Consider, for example, the transition from the heat treatment unit to the train rolls. The choice of the spool to be processed is influenced by several technological restraints also, maybe in an even greater degree, by a stack of orders to be filled; the orders have different priorities, and some of them can be combined into one order that can be satisfied by processing one single spool.

A perfect device to study the behaviour of the rolling mill, the bottleneck effects and impacts of would-be-facilities, is a simulational model. To this aim we need to gather the statistics of the incoming spools, the technological constraints (at least the most importance ones) and a sample of the orders to be filled. The final model will consist of about ten interconnected and interacting single-server queues.

8.13. A SIMSCRIPT example

The basic features of the simulational language SIMSCRIPT will be described and their use examplified on the single-server queue of the previous section. The SIMSCRIPT evolved from FORTRAN, uses most of its notation, sometimes in a slightly changed form, and has some additional features useful in simulation.

In presenting the system to be simulated we have entities, attributes, states of the system, activities, events, and a simulational clock. The components of the system are the *entities* (customers, checkout counter); their properties are called the *attributes*. The *state of the system* is characterized by the set of the attributes and the actual relations between the entities. Entities are engaged in activities (arrival, being busy) and an occurrence of an activity is called an *event*. An event changes the state of the system. The time elapsing in the simulated system is measured by a *simulational clock*; all the events are influenced by the flow of time, so all events have to be connected to the simulational clock. The simulational clock and the time displayed on it are independent of the real time. The state of the system does not change between events, so we can restrict ourselves to examining the particular time points when an event takes place. In our approach the computer program will leap from one event to the next — this is called the critical-event approach, used in the previous section. (The other possibility of dealing with the lapse of time is the time-slicing approach: the simulational time is divided into small time intervals and the system is examined thoroughly in each time interval before moving to the next time interval.)

In the SIMSCRIPT language the activities are described by subroutines called here event routines. In a SIMSCRIPT program five major parts are to be distinguished:

(i) In the *definition* section rules of the queues, system variables: sets, temporary or permanent entities and their attributes are specified.

(ii) For each event we have to write an *event routine* that describes the effects and interaction of the event on the other entities and attributes of the system.

(iii) At the end of the definition part one should place a *list of event routines*, distinguishing the exogenous routines (that are influenced by outside variables, i.e. input only) from the endogenous event routines (that are scheduled only from event routines within the simulation). An arrival is an endogenous event, whilst the initialization procedure is exogenous.

(iv) An *initialization section* produces the starting value of the system variables, at which the simulation model starts its work.

(v) The final part of the program is the exogenous event list, the records of which contain an ID number of an exogenous event routine and the time when the event routine will be executed.

To describe the queuing system of the previous section four event routines are necessary — two exogenous event routines: START, that triggers the system and ANALIS, that computes the statistics; and two endogenous event routines: ARRIV, scheduling arrivals and DEPART, simulating the departure of the consumers.

Some of the language elements of interest are the following: "CREATE entity", that creates a new element (its opposite is DESTROY, that is, freeing the space occupied by the entity). The "CAUSE event" instruction starts the event given, inserting it into its proper place in the event queue. A frustrated consumer after a long waiting time can be removed by the statement CANCEL. The FILE statement is one of the most important ones: it makes possible filing the entities into a set, with priorities even. The ACCUMULATE statement collects subtotals and the COMPUTE statement may be used to obtain statistics of the system variables.

The main steps in the endogenous event ARRIV can be outlined as follows:

```
   ENDOGENOUS EVENT ARRIV
   DESTROY ARRIV
   CREATE CUSTR
   Produce service time s_i
   IF CKOUT IS NOT EMPTY GOTO 1
   CREATE DEPART
   CAUSE DEPART AT TIME +s_i
   FILE CUSTR IN CKOUT
   GO TO 2
 1 ACCUMULATE LQ INTO CULQ SINCE LAST
   LET LQ = LQ + 1
   FILE IN
```

2 Generate arrival time a_i
CREATE ARRIV
CAUSE ARRIV AT TIME $+s_i$
RETURN
END.

The SIMSCRIPT simulational language has an automatic simulation clock, that is, the system after executing an event automatically proceeds to the event to be executed soonest.

The ARRIV routine executes the following things: destroys the event notice bringing it into action, creates a new customer. If the checkout is empty then computes the departure of the customer much in the way of algorithm A9 and produces a new arrival. If the checkout is busy then the customer is filed into the queue (upon corresponding specification it will be an FIFO filing in), and proceeds to the generation of the next arrival. The creating and causing of a new arrival is sustaining the system. Meantime necessary statistics are gathered by ACCUMULATE, where the length of queue LQ is added to the cumulative length. The generation of necessary time (for example s_i) can be done by the standard random number generator RAND, using the statement LET S =RAND I (MIN,MAX) producing S uniformly between MIN and MAX.

Though this system does not look simpler than the FORTRAN program, nevertheless it offers a much more flexible way of generalizing and expanding it. Possibilities may include cases with more servers, different queue rules, combined queues, etc. On the longer run the greater initial effort is well worth its price.

8.14. System simulational examples

Some applications of simulation are sketched in this final section to give a glimpse of the areas of possible application.

Consider a fjord that will be investigated as an ecological system with emphasis on the pollution of the water. There are several points (towns) on the shore where pollution can enter the water through sewers. The water content of the fjord is divided into several imaginary parts, mainly according to the physical limitations (depths, lengths, bends) of the fjord. The water of the neighbouring parts may be interchanged according to specific rules: the speed of change is mainly effected by the tide and the wind. A quantitative description of the interchange may be obtained by measuring the salinity of the water. Fresh water enters the fjord solely through the outermost part of the fjord that neighbours the open sea. The quality of the water is represented by two characteristics: the oxygen content in the water and the transparency of the water. The quality of the water is a function of several other things: the quantity of the algae growing in the water, the quantity of the phosphor and nitrogen impouring into the fjord, the temperature of the water, the amount of sunny hours, wind speed, tide and the speed of water exchange. Modelling the behaviour of the fjord involves the programming of the interactions of several components of the system. Such interactions are: chain of food (who is eating whom) consisting of different kinds of microbes, algae, fishes, etc., the decomposition of organic substances into some other materials under specific conditions in temperature, salinity, etc., their casual relationships, the mechanism of the water exchange, the interrelation of the sunshine and water temperature (plus the transparency) and their trend of changing with the seasons. The model gives an excellent device for studying the impact of some unexpected event on the ecology of the fjord, for instance of a longer windless period, sudden rise in the amount of impouring pollution, change in the composition of the pollution (new factory or technology), sharp decrease in the temperature, implementation of a new sewage purification plant.

Consider the following problem. We are given a system of rivers and creeks where many possible sites exist for building flood preventing dams. The problem is one of choosing the sizes of dams large enough to minimize the probability of flood but at minimum cost. First, a mathematical model of the system of dams is built up and as starting size of the dams a medium value is assigned to each dam. Using random rainfall and other input data derived from observations we can simulate several years, and by this evaluate an estimate of the probability of flood. Our task is to decrease the probability of flood simultaneously decreasing the sizes of the dams (the cost of building). The crude technique for solving the problem can be the following iterative procedure. In one iteration the sizes of all dams are simultaneously decreased or increased (or may remain unchanged) with equal probability with a given quantity (e.g. one-tenth of the average size). The probability of flood in the case of this perturbed system can be estimated by some repeated simulations. If the probability is decreased then we accept the change in the system, otherwise retain the original system. Then the system is randomly perturbed, etc. The procedure is repeated until the probability of flood reaches a predetermined level. Then we can repeat the generation of random disturbances to the dam system with the aim of decreasing the size of the dams (the overall cost of construction) with the additional constraint that the probability of flood does not increase.

The strategy of stabilizing thermodynamical systems can be applied here: the energy level is the probability of flood, random changes in the sizes of dams are the random displacements of the system, etc.

This kind of simulational computation, which is sometimes called evolutional strategy (or simulated annealing), is very useful when the state of the system can be characterized by one parameter (energy level or probability of flood). This strategy can be applied when a rocket's combustion chamber is being planned, a set of parameters in an approximate solution of differential equations is to be determined, etc.

A simulational model for genetic populations with biochemical properties can be constructed along the following lines. The population consists of a number of single cell individuals, each of them having a single pair of chromosomes. At regular intervals sexual mating takes place among some individual cells; the nature of the mating can be random or deterministic, governed by some properties of the cells. Two properties refine the model: (i) the biochemistry of the cells is under the control of a set of enzymes that in turn are defined by the genotype, and (ii) a cell membrane that regulates the flow of the chemicals into and out of the cell. The phenotype maybe defined as a simple numeric transformation of the genotype, which is given as a set of binary digits or, more realistically, the phenotype may be a product of interactions among various chemicals, whose quantities and interactions are controlled by the cell enzymes. This gives a time dependent nature to the phenotype. The presence or absence of a particular enzyme corresponds to the presence or absence of an allele. The cells are subjected to a birth and death process as well, that is, on the basis of their age or ability to survive they may be cancelled from the population (either random or deterministic selection rule). The model can be used to study the process of selection on the crossover probabilities of a small population.

VLSI (that is, very large scale integrated) semiconductor device simulation is widely used in developing new semiconductor technologies. Development of a new technology might involve several lithography steps, ion implantations, diffusion, annealing and oxidation steps. One can easily simulate the whole process in at least a couple of hours having at hand a suitable computer model for imitating the device, contrary to a real experiment that may need a month or months. The suitable computer model should consist of device simulation and process simulation as well.

Nonlinear partial differential equations are used to describe the static and dynamic behaviour of carriers in semiconductors; these equations can be derived easily from the Boltzmann transport equations. Of course, for solving the partial differential equations one has to implement special techniques, involving nonlinear iterations, discretization and sparse matrix procedures.

The descriptional model and stochastic experimentation outlined in the previous four applications may serve as a reliable tool in investigating many problems. We mention a few more. What is the load distribution like on a telephone network geographically and as a function of time? Necessary data include the topography of the network, the technical characteristics and the distribution of incoming calls and their respective service time. What will be the purchasing power of a country in the forthcoming years, given some constraints on the development of the economy? The model requires a sample of the population, the birth and death rate of each population stratum (white, black, hispanic, etc.), full details on probabilities of mobility and income possibilities, percentages of income spent on specific targets (housing, food, etc.) and the changes in percentages.

One can also simulate the business policy of concurrent firms using models or provide some training for military personnel by warfare simulations, but these approaches fall into a different branch of the simulational model — what may be called simulational games. The gaming approach to solve problems is an educational tool rather than a straightforward simulation. A model is built up with plenty of details, that may include stochastic behaviour as well (change in the interest rate or landing of paratroopers in an unexpected spot and time). Then generally more than one player participates in the game, that is, two or more people independently give instructions to the model to be carried out. These may be contradictory so a time priority of these instructions is to be maintained. The main asset of gaming approach is educational; the participating players may better understand the effect of their decisions on the whole system. As examples we mention one on acid rain problems and another on production management.

A model may be built for studying the effect of energy production increase on acid rain, if the neighbouring countries have only partial information on the policies of the other countries — but of course the impact of the decisions in one country is felt outside too. The players are the leaders of countries, who are making decisions on increases in energy production, oil and coal rates, and investment in pollution abatement.

A training model for production managers may enhance their decision making abilities. In order to get a realistic model one has to consider the materials system, the manpower system, the financial system and the capital equipment system. The materials system, for instance, may include a raw material subsystem, a production subsystem and a finished goods subsystem, everywhere plenty of inventory details and technological restrictions. In order to obtain a practical answer for a decision in production (changing a set of tools or switching to a new product) the whole system has to be sufficiently detailed.

8.15. Notes

8.1 Readers interested in more applications of simulation may also want to consult other books, such as Fishman (1979), Knuth (1969), Hammersley & Handscomb (1964), Maisel & Gnugnoli (1972), Rubinstein (1981) and Szobol (1981).

8.2 On the method and its applicability consult Hammersley and Handscomb (1964) and Rubinstein (1981).

8.3 The application of subsequences was proposed by Rubinstein (1981). On the integral equations see Szobol (1981).

8.4 The material here is mainly from Szobol (1981).

8.5 The idea was presented in Hammersley and Handscomb (1964).

8.6 The methods described here do not seem to be attributed to anyone in particular, but are common knowledge.

8.7 We relied here on Ermoliev (1983), who started the research in this field and stated the first results (Ermoliev (1969)). The filter technique we gave was presented and proved by Ruszczynski (1982).

8.8 The basic method was presented by Metropolis et al. (1953), the variant and its application is taken from Burkard and Rendl (1983).

8.9 These results were summarized in Prékopa et al (1980). The theoretical background was developed by Prékopa (1973).

8.10 The model and the computations with numerical data were presented in Prékopa & Szántai (1978).

8.11 Interested readers should consult the books in Note 8.1.

8.12 and 8.13 See Maisel and Gnugnoli (1972) and Wyman (1970). With regard to the cold rolling mill see Deák (1976).

8.13 The simulation of a fjord was given in Beyer & Madsen (1979). A similar study was carried out in the Computer and Automation Institute HAS with regard to Lake Balaton. A method using stochastic programming tools for a method using stochastic programming tools for optimal determination of the dams' sites and sizes was presented by Prékopa et al. (1976). Genetic populations were simulated in Rosenberg (1970), semiconductor simulations in Fichtner et al. (1983) and Bank et al. (1983). See also the paper of Choi (1983). For gaming approaches see Baba et al. (1984), Partridge et al. (1984) and Hamidi-Noori (1984).

APPENDICES

A1. The Buffon's needle problem

Assume that we have a big sheet of paper where parallel straight lines are drawn d distance apart. Let a needle of length r, be given; for the sake of simplicity we assume $r < d, r > 0$. Consider the following random experiment. The needle is thrown randomly onto the sheet of paper and the question now is what is the probability of the event that the needle crossects one of the lines drawn on the paper.

Assume we have a coordinate system positioned in such a way that the x axis falls unto one line. The position of the needle is essentially determined in this case by two data: the angle φ between the x axis and the needle and the y coordinate from the coordinates (x, y) of the center of the needle. The x coordinate of the center of the needle is of no concern for us, because the fact of the intersection which is under examination does not change under a translation, parallel to the x axis, thus we can safely assume $x = 0$. Similarly it can be shown that only the cases with $0 \leqq y \leqq d$ are to be dealt with. The angle φ may change in the interval $[0, \pi)$. Our assumption about the needle being randomly thrown onto the sheet, that is its each and every position is equally probable has to be described in mathematical terms, too. This random throw can be formulated by assuming that the random variables y and φ are independent of each other and uniformly distributed in their domain.

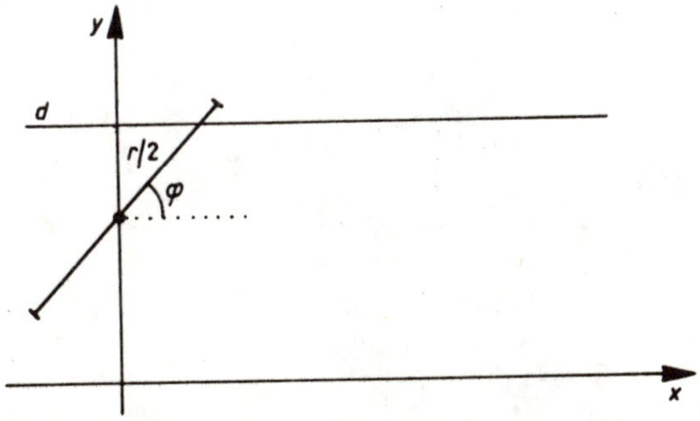

Fig. 40. A needle of lenght r thrown onto a set of parallel lines d distance apart

The distance of the needle's end from the straight line drawn parallel to the x axis through the center of the needle is $(r/2)\sin\varphi$ (see Fig. 40).

Consequently the needle intersects a line on the sheet if one of the inequalities

$$0 \leqq y \leqq (r/2)\sin\varphi,$$
$$d-(r/2)\sin\varphi \leqq y \leqq d$$

is satisfied (see Fig. 41).

Integrating over the domain of the random variables y and φ we can obtain for instance the probability of the first inequality's occurrence

$$P(0 \leqq y \leqq (r/2)\sin\varphi) = \frac{1}{\pi d}\int_0^\pi \int_0^{(r/2)\sin\varphi} \mathrm{d}y\,\mathrm{d}\varphi =$$

$$\frac{1}{\pi d}\int_0^\pi (r/2)\sin\varphi\,\mathrm{d}\varphi = \frac{r}{\pi d},$$

where $\frac{1}{\pi d}$ is the joint density function of the uniformly distributed random variables y and φ. Obviously the probability of the second inequality's occurrence is the same, that is the needle thrown onto the paper intersects a line with probability $\frac{2r}{\pi d}$.

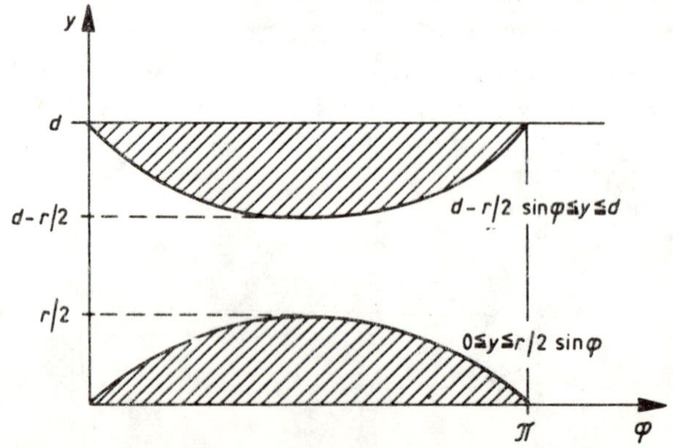

Fig. 41. The probability of the intersecting needle determined by (φ, y) is proportional to the two shaded areas, that are determined by the inequalities

If now this above-described experiment is actually executed N times and in M cases the needle intersects a line the relative frequency of the intersection can be given as $\frac{M}{N}$, and this is an approximation to the probability $\frac{2r}{\pi d}$. From the approximate equation $\frac{2r}{\pi d} \approx \frac{M}{N}$ an unbiased estimator

$$(\pi \approx) \frac{2rN}{dM}$$

can be derived for the value of π.

This experiment can be regarded as a characteristic example of the Monte Carlo methods:

(i) we need an approximate value of π,

(ii) we describe a random experiment in which the expected value of a random variable is π or a function of π,

(iii) actually executing the experiment several times the value of π can be expressed from the function, that yields the approximate value.

A2. Fields and polynomial rings

Consider a set K of numbers, where two operations are defined, one is called addition and denoted by $+$, while the other is called multiplication and denoted by $\circ$. In other words it means, that for any two elements a and b from the set K (they are not necessarily different) a unique element of the set K is defined which is denoted by $a + b$ (or $a \circ b$ respectively). We assume the following be true for these operations.

Assumptions

1. Addition is commutative: $a + b = b + a$
2. Multiplication is commutative: $a \circ b = b \circ a$.
3. Addition is associative: $a + (b + c) = (a + b) + c$.
4. Multiplication is associative: $a \circ (b \circ c) = (a \circ b) \circ c$.
5. Multiplication is distributive with respect to addition: $a \circ (b + c) =$ $= a \circ b + a \circ c$.
6. There exists a unique operation substraction (addition is unique), that is for any $a, b \in K$ we have exactly one element $c \in K$, for which $a + c = b$.
7. There is a unique zero element 0, that is for any element $b \in K$ we have $b + 0 = b$.
8. There exists an operation division (the unique inverse of multiplication), that is for any $a, b \in K$ (but $g \neq \emptyset$) we have exactly one element c in the set K, for which $a \circ c = b$ holds.

Definition. The set K with operations $+$ and $\circ$ is called a field if there are more than one element in K and Assumptions 1–8 are satisfied.

When dealing with fields in context to random number generation some more definitions are needed for fields with a finite number of elements, that are called finite fields.

Definition. The number m of the elements in a field K is also called the order of the field.

Definition. The order of an element $b \in K$ is the smallest integer r for which $b^r = 1$.

Definition. An element b of the field K of order m is called a primitive element if its order is $m-1$. (It is actually a generator of the multiplicative cyclic group of K.) Alternatively we can say that b is a primitive element if the powers b^1, $b^2, \ldots, b^{m-1}$ produce all the nonzero elements of K.

In context of random number generation we are looking for a "suitable" primitive element of the congruence classes mod m, that is a multiplier b for the congruence generator mod m.

Definition. A set R with operations $+$ and $\circ$ is called a ring if Assumptions 1–7 hold.

Examples

1. Consider the set A of rational numbers, it will evidently be a field with the usual addition and multiplication. Similarwise the set R of real numbers is also a field.

2. Consider now numbers of the form $a+b\sqrt{x}$, where a, b, x are rational numbers and $\sqrt{x}$ is irrational. Then the set W of the numbers in the form $a + b\sqrt{x}$ is a field (x is fixed, a and b are variables). To see this we have to examine the assumptions: addition and multiplication can be carried out in a usual way, they have the properties of commutativity, associativity and distributivity, subtraction can be defined and there is a unique null-element, the number $0 + 0\sqrt{x}$. We have to consider the division. Let $a + b\sqrt{x}$ and $c + d\sqrt{x}$ to be two elements of the set W, assuming $a + b\sqrt{x} \neq 0$, now we have to prove that

$$\frac{c + d\sqrt{x}}{a + b\sqrt{x}}$$

is an element of W. This can be seen from the equation

$$\frac{c+d\sqrt{x}}{a+b\sqrt{x}} = \frac{(c+d\sqrt{x})(a-b\sqrt{x})}{a^2-b^2x} =$$
$$= \frac{ac-bdx+(ad-bc)\sqrt{x}}{a^2-b^2x} = \frac{ac-bdx}{a^2-b^2x} + \frac{ad-bc}{a^2-b^2x}\sqrt{x},$$

and from the fact that $a^2-b^2x \neq 0$ (otherwise we would have $x = a^2/b^2 = = (a/b)^2$, that is x would be the square of a rational number contrary to our original assumptions).

3. Consider the polynomials with real coefficients. The set of these polynomials does not constitute a field since division cannot be executed among the polynomials always, for instance $1/x$ is not a polynomial.

4. Up to this point we considered fields with infinitely many elements, but we do have fields with finite number of elements as well. Define the set K with two elements only 0 and 1, the addition + and the multiplication $\circ$ is defined by the following tableaus:

+	0	1
0	0	1
1	1	0

$\circ$	0	1
0	0	0
1	0	1

Proving the fact that the set K is a field is left to the Reader. As a hint we note, that for the substraction $0-1=1$ and the null element is 0.

5. Consider the set K of the natural numbers $1, 2, 3, \ldots, m-1$ and the 0, where m is a prime number, let the addition and multiplication of two elements i and j be defined by the relations

$$i + j \pmod m \quad \text{and} \quad i \cdot j \pmod m$$

respectively. Is the set K a field? For the case $m = 1$ the answer obviously no, for the case $m = 2$ we are back to Example 4. In the general case $(m > 1)$ Assumptions $1-5$ are automatically satisfied. The subtraction is to be interpreted for the residue classes $c \equiv b - a \pmod m$, that is $c = 2 - 3 \equiv m - 1 \pmod m$, similarwise the inverse of the multiplication $a \cdot c = b$ is defined as $c \equiv \frac{b}{a} = \frac{km+b}{a} \pmod m$ with a suitable integer k, which makes the quotient an integer.

Finite fields are called Galois fields after their discoverer, for a field K with m elements the notation $GF(m)$ is used. (Finite fields with the same number of elements are basically the same.) Thus Example 4 presents $GF(2)$. If p is a prime then the classes mod p (Example 5) form $GF(p)$. If the modulus m is not a prime the residue classes mod m do not produce $GF(m)$ since division is not unique. (Consider as an example residue classes mod 4 where $2 \cdot \alpha = 1$ mod 4 does not hold for any α mod 4 and $2 \cdot \alpha = 2$ mod 4 holds for $\alpha_1 = 1$ and $\alpha_2 = 3$.)

Example. Consider residue classes mod 7, they give us $GF(7)$, where 3 is a primitive element with its successive powers being $1, 3, 2, 6, 4, 5$ (5 is a primitive element, too). The order of 2 mod 7 is 3 since $2^1 = 2$, $2^2 = 4$, $2^3 = 1$.

We are now ready to construct a field of polynomials. Let K be a finite field with m elements. The polynomial of the form

$$h(x) = a_0 + a_1 x + \ldots + a_n x^n, \quad a_i \in K, \quad a_n \neq 0$$

is called a polynomial of degree n over the field K. If addition and multiplication is carried out in the usual way the sum, difference and the product of two polynomials of this type will be polynomials over K, too. Thus we may have the following

Theorem. The set of polynomials over the field K is the polynomial ring over the field K, it is denoted by $K(x)$.

Definition. A polynomial $h(x)$ of degree n in $K(x)$ is called irreducible if it is not the product of two polynomials of lower degree in $K(x)$. (Basically an irreducible polynomial plays the same role in $K(x)$ as a prime number for integers.)

Fixing a polynomial $f(x)$ of degree n from $K(x)$ and defining multiplication of polynomials in $K(x)$ mod $f(x)$ we obtain a finite set of polynomials. If $f(x)$ is an irreducible polynomial in $K(x)$ then this finite set becomes a finite field denoted by $K[x]$ due to the following

Theorem. If $f(x)$ is irreducible, then every nonzero polynomial $h(x)$ of degree $n-1$ from $K(x)$ has a unique inverse, that is the polynomial $g(x)$ defined by $g(x) \cdot h(x) \equiv 1 \bmod f(x)$ is unique.

Example. Let K be the finite field $GF(2)$ consisting of two elements 0 and 1 (see previous Example 4). The polynomials

$$h(x) = a_0 + a_1 x \ldots + a_n x^n, \quad a_i = 0 \quad \text{or} \quad 1, \; i = 0, 1, \ldots, n-1$$

form the polynomial ring $K(x)$. Consider now the polynomial $f_1(x) = x^3 + x$ and define the addition and multiplication among the polynomials with coefficients 0 and 1 mod $f_1(x)$. The multiplication of $g_1(x) = x^2 + 1$ and $g_2(x) = x^2 + x + 1$ gives

$$g_1(x) \cdot g_2(x) \equiv x^2 + 1 \bmod f_1(x).$$

Since $f_1(x) = x(x^2 + 1)$ is not irreducible, this set of polynomials is not a field (for example the equation $g_1(x) \cdot g_3(x) \equiv 1 \mod f_1(x)$ does not have a solution for $g_3(x)$).

Take now the polynomial $f_2(x) = x^3 + x + 1$. Let multiplication now be defined mod $f_2(x)$ so

$$g_1(x) \cdot g_2(x) = x^2 + x \mod f_2(x).$$

This set of polynomials is a field since $f_2(x)$ is irreducible in $K(x)$.

Finally we arrive at the important concept of primitive polynomials which is given in

Definition. An irreducible polynomial having a primitive element as its root is called a primitive polynomial.

Note here that not all irreducible polynomials are primitive, for example $h(x) = x^4 + x^3 + x^2 + x + 1$ is irreducible over $GF(2)$ but it is not primitive (x, which is a zero of $h(x)$ is not a primitive element here). On the other side the congruence class α mod $f(x)$ containing the polynomial x is one of the roots of $f(x)$ trivially since $f(\alpha) = f(x) \equiv 0 \mod f(x)$ (though x is not necessarily a primitive element).

Determining primitive polynomials is not an easy computational job, but it can be done in various ways.

If $f(x)$ is a primitive polynomial then x is a primitive element. Using a primitive polynomial $f(x)$ of degree n we can produce a polynomial field $K[x]$ over the field $GF(p)$ and it has $p^n - 1$ nonzero elements.

Consider the field $GF(2)$ and the generated polynomial field $GF(2^n)$ associated to the primitive trinomial $f(x) = x^n + x^q + 1$. There is a one-to-one mapping of the n-tuples of binary digits $(x_k,\ x_{k+1}, \dots, x_{k+n-1})$ to the polynomials in $GF(2^n)$, that can be described as follows. Let the polynomials $g_k(x), \dots, g_{k+n-1}(x)$ be given by

$$\begin{aligned}
g_k(x) &= a_0 + a_1 x + \dots + a_{n-1} x^{n-1} \mod f(x) \\
g_{k+1}(x) &= g_0(x) x \mod f(x) \\
g_{k+2}(x) &= g_1(x) x = g_0(x) x^2 \mod f(x) \\
&\vdots \\
g_{k+n-1}(x) &= g_0(x) x^{n-1} \mod f(x)
\end{aligned}$$

the corresponding n-tuple $(x_k, \dots, x_{k+n-1})$ is produced as $x_k = g_k(1)$, $\dots x_{k+n-1} = g_{k+n-1}(1)$.

The inverse relationship between $(x_k, \ldots, x_{k+n-1})$ and the polynomials $g_k(x), \ldots, g_{k+n-1}(x)$ can be worked out the following way:

Given a sequence x_k, $x_{k+1}, \ldots, x_{k+n-1}$ of 0's and 1's the corresponding polynomials $g_k(x), \ldots, g_{k+n-1}(x)$ can be determined from the set of equations

$$g_k(1) = x_k$$

$$\vdots$$

$$g_{k+n-1}^{(1)} = x_{k+n-1}.$$

Here determination of $g_k(x) = a_0 + a_1x + \ldots + a_{n-1}x^{n-1}$ is equivalent to evaluating the coefficients $a_0, \ldots, a_{n-1}$. It can be done considering $g_k(1) = a_0 + \ldots + a_{n-1} = \sum_{i=0}^{n-1} a_i = x_k$, furthermore $g_{k+1}(x) =$ $= a_0x + a_1x^2 + \ldots + a_{n-1}x^n$. If $a_{n-1} = 0$ then $g_{k+1}(1) = \sum_{i=0}^{n-2} a_i$ and if $a_{n-1} = 1$ then $g_{k+1}(x) = a_0x + \ldots + a_{n-2}x^{n-1} + x^q + 1$ and $g_{k+1}(1) =$ $= \sum_{i=0}^{n-2} a_i + 1 + 1 = \sum_{i=0}^{n-2} a_i \bmod 2$. Thus the following system of equations emerge

$$g_k(1) = \sum_{i=0}^{n-1} a_i = x_k$$

$$g_{k+1}(1) = \sum_{i=0}^{n-2} a_i = x_{k+1}$$

$$\vdots$$

$$g_{k+n-1}(1) = \sum_{i=0}^{0} a_i = x_{k+n-1}$$

from which we finally obtain

$$a_0 = x_{k+n-1}$$

$$a_1 = x_{k+n-2} - x_{k+n-1}$$

$$\vdots$$

$$a_i = x_{k+n-1-i} - x_{k+n-i}$$

$$\vdots$$

$$a_{n-1} = x_k - x_{k+1}.$$

Example. Consider the initial n tuple $(1, 1, \ldots, 1)$, the corresponding initial polynomial $g_0(x) = a_0 + a_1 x + \ldots + a_{n-1} x^{n-1}$ has coefficients

$$a_0 = 1, \ a_1 = \ldots = a_{n-1} = 0,$$

thus the initial set of polynomials is

$$g_0(x) = 1, \ g_1(x) = x, \ldots, \ g_{n-1}(x) = x^{n-1}.$$

It can easily be seen that the generation of the next bit x_{k+n} by the Tausworthe generator based on $f(x)$ and given by $x_{k+n} \equiv x_k + x_{k+q} \bmod 2$ strictly corresponds to the same congruence relation among polynomials

$$g_{k+n}(x) \equiv g_k(x) + g_{k+q}(x) \mod f(x)$$

since we have

$$g_{k+n}(x) = g_k(x) + g_k(x) \cdot x^q = g_k(x)(x^q + 1) = g_k(x) x^n \mod f(x).$$

The theorems stated here can be found in the authorative guide into the topic of Lidl and Niederreiter (1983) for example, see also MacWilliams and Sloane (1981), Collings and Hembree (1986).

A3. The distribution function of the χ_n^2 distribution

Let χ_n^2 be a χ^2 distributed random variable with n degree of freedom and F_n its distribution function:

$$F_n(x) = P\{\chi_n^2 < x\} = \int_0^x f_n(y) \, \mathrm{d}y,$$

where

$$f_n(y) = \frac{1}{2^{n/2}\Gamma(n/2)} y^{n/2-1} \, \mathrm{e}^{-y/2}, \ y \geqq 0, \ n \geqq 1.$$

The function value $F_n(x)$ can be determined by the aid of the function $G_n(x) = 1 - F_n(x)$, for which we have the following equation

$$G_n(x) = P\{\chi_n^2 > x\} = P\{\chi_{n-2}^2 > x\} + \frac{(x/2)^{n/2-1} \, \mathrm{e}^{-x/2}}{\Gamma(n/2)}.$$

This recursive equation when applied $[n/2]$ times gives us the function $G_n(x)$ as a sum of finite number of terms, where the terms can directly be evaluated (the second term on the right-hand side of the last equation) or a remainder term. This remainder is either

(i) a distribution function of the χ^2 distribution with 2 degrees of freedom, if the original value n was even, or

(ii) a distribution function value with 1 degree of freedom if the initial value of n was odd.

In case of (i) the remainder takes on the form

$$G_2(x) = \frac{1}{2} \int_x^\infty e^{-y/2}\, dy = e^{-x^2/2},$$

while in case of (ii) we have

$$G_1(x) = P\{\chi_1^2 > x\} = \frac{1}{\sqrt{2\pi}} \cdot \int_x^\infty y^{-1/2}\, e^{-y/2}\, dy = \frac{2}{\sqrt{2\pi}} \cdot \int_{\sqrt{x}}^\infty e^{-y^2/2}\, dy,$$

that can be rewritten by the aid of distribution function ϕ of the standard normal distribution as

$$G_1(x) = 2\phi(-\sqrt{x}).$$

Using these expressions the value of $F_n(x)$ can be evaluated, but this procedure is replaced by an approximate expression, the Wilson–Hilferty approximation formula, to be described later, if the degree of freedom n is great (if $n > 60$).

Summarizing all the above mentioned details after some streamlining the following formal algorithm can be obtained to evaluate the distribution function F_n of the χ^2 distribution.

S10 Subalgorithm. The $F_n(x)$ distribution function (χ^2)

1. If $n > 60$, then set $y \leftarrow (x/n)^{1/3}$, $s \leftarrow 2/(9n), y \leftarrow (y - (1 - s))/\sqrt{s}$ and deliver the probability $p \leftarrow \phi(y)$ as the value of $F_n(x)$.
2. Set $a \leftarrow x/n$. If $a < \varepsilon$ (we can initialize $\varepsilon \sim 10^{-8}$ for instance), then set $p \leftarrow 0$ and go to Step 7. If $a > 10$ then set $p \leftarrow 1.0$ and go to Step 7.
3. Let $c \leftarrow \sqrt{x}$, if n is odd then $m \leftarrow 1$, otherwise $m \leftarrow 0$. If $m = 1$, then set $t \leftarrow c$, $d \leftarrow 1$, otherwise $t \leftarrow x/2$, $d \leftarrow 2$.
4. [Start summation.] Let $\ell \leftarrow (n-1)/2$, $s \leftarrow 0$. Repeat for indices $i = 1, 2, \ldots, \ell$ the following operations $s \leftarrow s + t$, $d \leftarrow d + 2$, $t \leftarrow tx/d$.
5. If $m = 1$, then set $q \leftarrow \phi(c)$ and $p \leftarrow 2(1 - q + \exp\{-x/2\} * s/2.5066283)$ and deliver the value $p \leftarrow 1 - p$.
6. If $m = 0$, then evaluate $p \leftarrow \exp(-x/2) \cdot (s + 1.0)$, $p \leftarrow 1 - p$.
7. Deliver the value p as the distribution function value $F_n(x)$.

In less demanding computations approximate procedures are used to compute $F_n(x)$, that have somewhat greater error than the above Algorithm S10. The best known approximation is that of Fischer:

$$P\{\chi_n^2 < x\} \sim \phi(\sqrt{2x} - \sqrt{2n-1}).$$

A little bit more accurate is the Wilson–Hilferty approximation formula:

$$P\{\chi_n^2 < x\} \sim \phi\left(\left(\left(\frac{x}{n}^{1/3}\right) - 1 + \frac{2}{9n}\right)\sqrt{\frac{9n}{2}}\right);$$

the Fischer formula generally gives two accurate digits while this last expression has an error of approximately 10^{-3}.

In our algorithms for evaluating the distribution function of the multidimensional normal distribution we have to evaluate the distribution function H_n of the χ distribution with n degree of freedom ($H_n(x) = F_n(x^2)$ holds). To achieve satisfactory speed the following linear interpolation technique was employed. First the two constants c_1 and c_2 were determined for which the inequalities

$$H_n(x) < 10^{-7}, \text{ if } x < c_1,$$

and

$$H_n(x) > 1 - 10^{-7}, \text{ if } x > c_2$$

hold, and then the interval $[c_1, c_2]$ is divided into 1000 equal parts. The function value $H_n(x)$ at these splitting points is stored in an array A with 1000 components. Any value $H_n(x)$, $x \in [c_1, c_2]$ is determined henceforth by using a linear interpolation between the values stored in the vector A.

S11 Subalgorithm, Linear interpolation for $H_n(x)$

1. If $x < c_1$ then let $p \leftarrow 0$ and go to Step 4. If $x > c_2$, then let $p \leftarrow 1$ and go to Step 4.
2. Set $y \leftarrow x - c_1$ and $i \leftarrow [10^4 y/(c_2 - c_1)]$, where $[\cdot]$ is the integer part function.
3. Evaluate $p \leftarrow A(i) + [A(i+1) - A(i)]\dfrac{y - i(c_2 - c_1)10^{-4}}{(c_2 - c_1)10^{-4}}$.
4. Deliver the value p.

Finally we give an approximate expression for evaluating the one-dimensional standard normal distribution function with an error of 10^{-6}–10^{-7}, this expression may be used in Algorithm S10.

The formula is

$$\phi(x) \sim \begin{cases} p & \text{if } x \geqq 0, \\ 1-p & \text{if } x \leqq 0, \end{cases}$$

where the probability p is computed from

$$p = 1 - \phi(x) \cdot t \cdot (a_1 + a_2 t + a_3 t^2 + a_4 t^3 + a_5 \cdot t^4),$$

where the constants $a_1, \ldots, a_5$ and other notations are defined by the following equations

$$t = 1/(1 - 0.2316419|x|),$$

$$\phi(c) = 0.3989423 \exp\{-x^2/2\},$$

$$a_1 = 0.3193815, \qquad a_2 = 0.3565638, \qquad a_3 = 1.781478,$$

$$a_4 = -1.821256, \qquad a_5 = 1.330274.$$

A4. Examples with known results and exercises with unknown probabilities in evaluating the distribution function of the multidimensional normal distribution

The details of examples and exercises used in the computer testing of algorithms in Chapter 7 are listed here. In constructing the examples we used the following three types of problems (or their combinations): two-dimensional correlated distributions, distributions with equal correlation coefficients and with independent components, for each of these types there is an exact method to evaluate the distribution function value or a fast computational method is known, in the independent case the probabilities are to be multiplied.

The examples are given by the aid of the following quantities: probability p, number of dimensions n, integration limit $\mathbf{h}$ and the correlation coefficients r_{ij}, $i = 1, \ldots, n-1$, $j = i+1, \ldots, n$. Of course we have $r_{ii} = 1$, $i = 1, \ldots, n$ and because of the symmetry of the correlation matrix $\mathbf{R}$ $r_{ij} = r_{ji}$. Coefficients r_{ij}, $i \neq j$ not appearing in the examples are assumed to be 0; thus the data given for an example completely determine an integral of form (7.1).

Example 1. $p = 0.64064279$, $n = 4$; $\mathbf{h} = (2.1;\ 1.2;\ 0.6;\ 3.8)$, $r_{12} = 0.8$, $r_{34} = -0.9$.

Example 2. $p = 0.23241$, $n = 5$, $h_1 = \ldots = h_5 = 0.3$, $r_{ij} = 0.4$, $i = 1, \ldots, 4$, $j = i+1, \ldots, 5$.

Example 3. $p = 0.58300606$, $n = 10$, $\mathbf{h} = (1.7;\ 0.8;\ 5.1;\ 3.2;\ 2.4;\ 1.8;\ 2.7;\ 1.5;\ 1.2;\ 2.6)$, $r_{12} = -0.6$, $r_{34} = 0.9$, $r_{56} = 0.4$, $r_{78} = 0.2$, $r_{9,10} = -0.8$.

Example 4. $p = 0.85276$, $n = 10$, $h_1 = \ldots = h_{10} = 1.5$, $r_{ij} = 0.9$, $i = 0.9$, $i = 1,\ldots,9$, $j = i+1,\ldots,10$.

Example 5. $p = 0.11054$, $n = 15$, $h_1 = \ldots = h_7 = 0.5$,
$\mathbf{h} = (0.5\ldots0.5;\ 1.2;\ 2.7;\ 1.8;\ 0.4;\ 2.4;\ 0.6;\ 1.5;\ 0.9)$,
$r_{ij} = 0.625$, $i = 1,\ldots,6$, $j = i+1,\ldots,7$, $r_{89} = 0.8$, $r_{10,11} = 0.7$, $r_{12,13} = -0.8$, $r_{14,15} = -0.9$.

Example 6. $p = 0.92295$; $n = 20$; $h_1 = \ldots = h_8 = 3.1$;
$\mathbf{h} = (3.1\ldots3.1;\ 2.8;\ 3.2;\ 4.3;\ 3.8;\ 2.3;\ 1.9;\ 2.6;\ 2.2;\ 3.1;\ 2.3;\ 2.7;\ 3.4)$,
$r_{ij} = 0.875$, $i = 1,\ldots,7$, $j = i+1,\ldots,8$, $r_{9,10} = 0.1$, $r_{11,12} = 0.8$, $r_{13,14} = -0.2$, $r_{15,16} = 0.5$, $r_{17,18} = 0.4$, $r_{19,20} = -0.9$.

In the following Examples 7–10 the correlation matrix was the unit matrix (independence of components), the integration limit $\mathbf{h}$ was given as the first n components of the vector $\mathbf{h}^{(1)}$, where

$\mathbf{h}^{(1)} =(2.84;\ 2.06;\ 2.64;\ 3.44;\ 3.73;\ 2.55;\ 2.99;\ 2.24;\ 2.16;\ 2.97;\ 1.73;\ 1.43;\ 2.39;\ 2.02;\ 1.74;\ 2.51;\ 2.54;\ 1.95;\ 1.66;\ 2.53;\ 1.53;\ 2.34;\ 2.46;\ 2.69;\ 1.88;\ 1.99;\ 2.21;\ 2.70;\ 2.88;\ 1.57;\ 2.15;\ 2.31;\ 2.02;\ 1.93;\ 1.55;\ 3.20;\ 2.33;\ 1.60;\ 1.77;\ 3.20;\ 3.10;\ 2.37;\ 1.50;\ 1.43;\ 2.84;\ 3.34;\ 2.17;\ 2.21;\ 3.30;\ 2.98)$.

Example 7. $p = 0.70443223$, $n = 20$.
Example 8. $p = 0.56617742$, $n = 30$.
Example 9. $p = 0.44324325$, $n = 40$.
Example 10. $p = 0.36586192$, $n = 50$.

The six examples with unknown probabilities are described next. The correlation matrix was the left upper $n \times n$ part of the matrix $\mathbf{R}^{(1)}$ ($\mathbf{R}^{(2)}$, resp.), and the integration limit was given by the first n components of the vector $\mathbf{h}^{(2)}$ ($\mathbf{h}^{(3)}$ resp.).

Example 11. $n = 5$, $\mathbf{h}^{(2)}$, $\mathbf{R}^{(1)}$.
Example 12. $n = 10$, $\mathbf{h}^{(2)}$, $\mathbf{R}^{(1)}$.
Example 13. $n = 20$, $\mathbf{h}^{(2)}$, $\mathbf{R}^{(1)}$.
Example 14. $n = 5$, $\mathbf{h}^{(3)}$, $\mathbf{R}^{(2)}$.
Example 15. $n = 10$, $\mathbf{h}^{(3)}$, $\mathbf{R}^{(2)}$.
Example 16. $n = 20$, $\mathbf{h}^{(3)}$, $\mathbf{R}^{(2)}$.

$\mathbf{h}^{(2)}$ =(1.03; 2.59; 2.20; 2.71; 1.09; 2.67; 1.54; 2.85; 2.22; 3.42; 1.19; 3.77; 2.62; 2.55; 2.45; 3.86; 2.05; 1.30; 1.65; 3.40),

$\mathbf{h}^{(3)}$ =(2.95; 2.97; 3.78; 3.00; 4.02; 2.85; 4.34; 2.17; 2.81; 4.03; 4.37; 2.59; 3.10; 4.18; 4.34; 4.03; 2.72; 3.89; 3.11; 2.01).

The matrices $\mathbf{R}^{(1)}$ and $\mathbf{R}^{(2)}$ are given with the elements above the diagonal

$\mathbf{R}^{(1)}$ *rowwise*

Row 1. 0.039; 0.097; −0.041; 0.040; 0.144; 0.095; 0.553; −0.223; −0.035; −0.215; 0.113; −0.409; −0.435; 0.519; −0.024; −0.416; 0.209; 0.129; −0.013;

Row 2. 0.474; 0.398; 0.221; 0.344; 0.134; 0.385; −0.009; 0.030; −0.257; −0.268; 0.069; 0.037; 0.072; 0.155; −0.070; 0.002; −0.192; 0.257;

Row 3. 0.054; −0.048; −0.080; −0.145; 0.446; −0.043; −0.228; −0.230; 0.177; 0.210; −0.054; −0.145; −0.092; −0.120; −0.050; 0.161; 0.307;

Row 4. −0.099; −0.272; −0.003; 0.232; −0.013; 0.138; −0.099; −0.036; 0.103; −0.002; 0.115; 0.484; −0.277; 0.274; −0.239; 0.062;

Row 5. 0.087; −0.145; −0.241; −0.141; −0.173; 0.119; 0.024; −0.016; 0.051; 0.490; 0.335; −0.102; 0.258; 0.022; 0.225;

Row 6. 0.397; 0.002; −0.282; 0.022; −0.202; 0.144; −0.293; −0.091; 0.096; −0.087; 0.264; −0.228; 0.018; −0.020;

Row 7. −0.132; 0.179; 0.226; −0.120; −0.024; −0.173; −0.142; −0.140; −0.121; 0.180; −0.404; 0.080; 0.052;

Row 8. −0.186; −0.085; −0.219; −0.257; −0.138; −0.134; −0.067; 0.113; −0.130; 0.147; −0.219; 0.079;

Row 9. 0.072; 0.195; −0.468; −0.035; 0.362; −0.354; −0.224; 0.114; −0.113; 0.065; −0.009;

Row 10. −0.218; 0.148; −0.446; −0.102; 0.056; −0.012; 0.281; −0.243; 0.176; −0.191;

Row 11. −0.016; 0.113; 0.185; −0.213; 0.259; 0.146; −0.130; −0.208; −0.188;

Row 12. 0.133; −0.427; 0.410; −0.189; 0.106; −0.299; 0.488; −0.046;

Row 13. 0.0; −0.003; −0.203; −0.123; −0.159; 0.194; 0.129;

Row 14. −0.337; 0.127; 0.139; 0.200; 0.110; −0.107;

Row 15. 0.193; −0.259; 0.070; 0.224; 0.024;

Row 16. −0.167; 0.257; −0.412; −0.180;

Row 17. −0.475; 0.094; −0.0117;
Row 18. −0.092; 0.242;
Row 19. 0.137;

$\mathbf{R}^{(2)}_{rowwise}$

Row 1. −0.006; −0.198; −0.412; −0.352; −0.169; 0.251; 0.003; −0.213; 0.012; 0.094; 0.115; 0.017; 0.129; −0.195; 0.123; −0.010; 0.247; 0.149; 0.554;
Row 2. −0.140; 0.115; 0.246; 0.290; 0.209; −0.077; 0.197; −0.106; 0.039; −0.195; 0.401; −0.220; 0.158; −0.296; 0.108; 0.259; −0.212; −0.098;
Row 3. 0.295; 0.409; −0.163; −0.210; −0.141; −0.471; 0.203; 0.288; 0.045; 0.212; −0.020; −0.026; 0.436; −0.164; −0.178; 0.272; −0.058;
Row 4. 0.387; −0.115; −0.233; 0.127; −0.289; −0.068; 0.235; −0.055; −0.039; −0.006; 0.194; −0.203; 0.379; 0.146; 0.017; 0.183;
Row 5. −0.314; −0.106; −0.075; 0.088; 0.210; 0.218; −0.438; 0.090; 0.078; 0.222; −0.128; −0.168; 0.107; 0.097; 0.130;
Row 6. 0.367; 0.043; 0.065; 0.237; −0.204; 0.252; 0.056; −0.279; 0.086; −0.343; 0.298; 0.238; 0.140; −0.402;
Row 7. 0.133; 0.287; 0.225; −0.076; 0.168; −0.185; 0.134; −0.056; −0.168; −0.048; 0.064; 0.191; −0.055;
Row 8. 0.076; 0.006; −0.025; 0.266; −0.572; 0.235; 0.240; 0.102; 0.404; 0.017; 0.180; 0.103;
Row 9. 0.071; −0.146; −0.197; −0.213; −0.287; 0.286; −0.146; −0.070; 0.262; −0.392; −0.367;
Row 10. 0.169; 0.031; −0.200; 0.001; 0.091; 0.230; 0.016; 0.212; 0.125; −0.159;
Row 11. −0.143; −0.195; 0.285; −0.024; 0.261; −0.092; −0.235; −0.041; 0.236;
Row 12. 0.001; 0.016; −0.065; 0.116; 0.222; 0.053; −0.121; −0.118;
Row 13. −0.245; 0.031; 0.022; −0.391; 0.297; −0.043; −0.196;
Row 14. −0.005; 0.143; −0.177; −0.112; 0.366; 0.440;
Row 15. −0.239; 0.002; 0.374; −0.214; −0.140;
Row 16. −0.236; −0.321; 0.045; −0.132;
Row 17. 0.164; −0.123; 0.085;
Row 18. −0.033; 0.158;
Row 19. 0.371;

Finally numerical data concerning the examples that were used to illustrate the speed of the algorithms for computing probabilities of rectangles are described here. The n-dimensional rectangle is given by $a \leqq x \leqq b$, $r_{ij}, i, j = 1, \ldots, n$ are the elements of the correlation matrix R.

Example 17. $n = 2$, $a = (-2.1\ 0.6)$, $b = (-1.2\ 3.8)$, $P_{ex} = 0.02665$, $r_{ij} = 0$, $i \neq ji$, $j = 1, 2$, $r_{ii} = 1$, $i = 1, 2$.

Example 18. $n = 4$, $a = (-2.1\ -1.2\ -0.6\ -3.8)$, $b = (1.5\ 1.5\ 1.5\ 1.5)$, $P_{ex} = 0.46044$, $r_{ij} = 0$, $i \neq j$, i, $j = 1, \ldots, 4$, $r_{ii} = 1$, $i = 1, \ldots, 4$.

Example 19. $n = 6$, $a = (-1.5\ -2.3\ -2.17\ -1.28\ -2.1\ -0.6)$, $b = (3.8\ 4.1\ -0.5\ 3.2\ 1.2\ 3.8)$, $P_{ex} = 0.15327$, $r_{ij} = 0$, $i \neq j$, i, $j = 1, \ldots, 6$, $r_{ii} = 1$, $i = 1, \ldots, 6$.

Example 20. $n = 10$, $a = (-3.2 \ldots -3.2)$, $b = (2.5 \ldots 2.5)$, $P_{ex} = -$, $r_{ij} = 0.9$, $i \neq j$, $r_{ii} = 1$, $i = 1, \ldots, 10$.

Example 21. $n = 10$, $a = (-1.9\ -2.7\ -2.0\ -2.5\ -1.9\ -2.0\ -1.5\ -2.0\ -1.4\ -2.2)$, $b = (0.3\ 1.1\ 0.4\ 0.9\ 0.3\ 0.4\ -0.1\ 0.4\ -0.2\ 0.6)$, $P_{ex} = -$, $r_{ij} = 0.9$, $i =\neq j, i$, $j = 1, \ldots 10$, $r_{ii} = 1$, $i = 1, \ldots, 10$.

Example 22. $n = 20$, $a = (-2.4\ -0.4\ -1.8\ -1.0\ -2.6\ -1.6\ -2.1\ +0.1\ -2.7\ -1.0\ -2.5\ -1.6\ -2.5\ -1.4\ -2.2\ -3.9\ -2.5\ -1.5\ -1.8\ 1.1)$, $b = (1.5\ 2.4\ 2.1\ 2.9\ 1.3\ 2.3\ 1.8\ 3.8\ 1.2\ 2.9\ 1.4\ 2.3\ 1.4\ 2.5\ 1.7\ 0.5\ 1.4\ 2.4\ 2.1\ 2.8)$, $P_{ex} = -$, $r_{ij} = 0.8$, $i \neq j = 1, \ldots, 20$, $r_{ii} = 1$, $i = 1, \ldots, 20$.

A5. Simulation related subroutines in a subroutine library

Here we give a list of subroutines in the IMSL (International Mathematical Subroutine Library) that are related or useful in simulation. Of course in other subroutine libraries (like SSP, PL-MATH) we can find similar subroutines.

Generally we recommend at least an elementary test of the random number generators found in subroutine libraries, but in ideal cases we should test them against problems, whose solutions are known.

GFIT	χ^2 fit test
GGAMR	Gamma generator (one parameter)
GGAMS	Gamma generator (two parameters, rejection)
GGAMT	Gamma generator (two parameters)

GGBIR	Binomial generator
GGBNR	Beta generator (rejection)
GGCAY	Cauchy generator
GGCHS	χ^2 generator
GGEOT	Geometrical generator
GGEXN	Exponential generator
GGHPR	Hypergeometric generator
GGMLT	Multinomial generator
GGNLG	Lognormal generator
GGNML	Normal generator
GGNPM	Normal generator (polar)
GGNSM	Normal vector generator (given correlation matrix)
GGPON	Poisson generator (changing parameter)
GGPOS	Poisson generator (not frequently changing parameter)
GGSPH	Uniform vector generator on a sphere ($n = 3$ and $n = 4$ dimensions)
GGTRA	Triangular generator
GGUBS	Basic uniform [0.1) generator
GGUW	Uniform [0.1) generator (shuffling)
GGVCR	General continuous generator
GGWIB	Weibull generator
GTNOR	Normality test
GTPOK	Poker test
GTPST	Pairs' test
MDBETA	Beta distribution function
MDBETI	Beta distribution function's inverse
MDCH	χ^2 distribution function
MDCHI	χ^2 distribution function's inverse
MDNOR	Normal distribution function
MDNRIS	Normal distribution function's inverse

LIST OF ALGORITHMS

Chapter 2

Chapter 3

C12	Continuous variable by acceptance-complement
C13	Continuous variable by inversion-rejection
C14	Continuous variable by series
C15	Continuous variable, by series+alternating sequence
C16	Continuous variable, by series-alternating-squeeze
D2	Discrete variable, by lineary search
D3	Discrete variable, by table method
D4	Discrete variable, by decomposition
D5	Discrete variable, by modified decomposition
D6	Preparation to D5
D7	Discrete variable, by economical (preliminary)
D8	Preparation to D9
D9	Discrete variable, by economical

Chapter 4

N1	Normal, approximate
N2	Normal, polar transformation
N3	Normal, modified polar method
S1	Subalgorithm, almost linear density
S2	Subalgorithm, normaltail
N4	Normal, economical triangle
S3	Normal, by comparison, preliminary
N5	Normal, comparison
S4	Subalgorithm, exponential, truncated density
E4	Exponential, modified decomposition
G1	Gamma, $n > 0$ integer parameter
G2	Gamma, fractional part parameter $0 < b < 1$
G3	Gamma, arbitrary parameter $b > 0$
G4	Gamma, by Cauchy, parameter $b > 1$
G5	Gamma, normal-exponential upper bound $(b > 2.53)$
G6	Generalized gamma, log logistic upper bound $(b > 1)$
G7	χ^2 distribution integer parameter $n > 0$
ST	Student parameter
B1	Beta, b, c integer
B2	Beta, arbitrary parameters (b, c)
B3	Beta, symmetrical density, $b = c > 1.5$
BI1	Binomial, count off
BI2	Binomial, large parameter n

PO1	Poisson, count-down
PO2	Poisson, from gamma
WE	Weibull
BU	Burr
DE	Double exponential
PE	Power exponential via double exponential, $a \geqq 1$
FR	Frechet
RA	Rayleigh
NB1	Negative binomial
NB2	Negative binomial, from Poisson

Chapter 5

V1	Vector generation, conditional distribution function
V2	Vector generation, acceptance-rejection
VRE	Uniform vector in rectangle
VSI	Uniform vector in simplex
S5	Subalgorithm, converting an interior algorithm to surface algorithm
S6	Subalgorithm, converting a surface algorithm to an interior algorithm
VSA	Uniform vector in a sphere, acceptance-rejection
S7	Subalgorithm, polar coordinates, preliminary
VSP	Uniform vector in sphere, polar coordinates
VSN	Uniform vector on sphere, normal approach
VEN	Uniform vector in ellipsoid, normal approach
VSC	Uniform vector in sphere, chord method
VEC	Uniform vector in ellipsoid, chord method (changes in VSC)
VNT	Normal vector, matrix transformation
VNP	Normal vector, polar transformation
VSS	Uniform vector in sphere, stationary
VNE	Normal vector, ellipsoid decomposition

Chapter 6

MC1	Crude method
MC2	Acceptance-rejection method
MC3	Importance sampling
MC4	Stratified sampling
MC5	Control variates
MC6	Antithetic variates

Chapter 7

NP1	Normal probability, by direct integration
NP2	Normal probability, by indirect integration
NP3	Normal probability, by importance sampling ("acceptance-rejection")
NP4	Normal probabiltiy, "acceptance-rejection" for small probability
NP5	Normal probability, "acceptance-rejection" for great probability
NP6	Normal probability, integration along a line
NP7	Normal probability, orthonormalized-2 estimator for distribution function (preliminary version)
NP8	Normal probability, importance sampling ("acceptance-rejection") for a rectangle
NP9	Normal probability, orthonormalized-2 estimator for the probability of a rectangle (preliminary version)
S8	Subalgorithm, computation of the function $e_{2+}(\mathbf{y})$
SE2	Subalgorithm, computation of $c_1, c_2(r\mathbf{y} \in D$ if $c_1 \leqq r \leqq c_2)$
S9	Subalgorithm, probability content of a line in a rectangle (function $e_3(\mathbf{y})$)
NP10	Normal probability, orthonormalized-2 estimator ($h \geqq 0$ only)

Chapter 8

A1	Solving a system of linear equations
A2	Solving an integral equation
A3	Solving a differantial equation
A4	Interpolation of a function of very many variables
A5	Maximizing a function (crude method)
A6	Maximizing a function (random search)
A7	Stochastic quasi-gradient algorithm
A8	Thermodynamical system stabilization
A9	Single server queue

Appendix

S10	Subalgorithm, the $F_n(x)$ distribution function (χ^2)
S11	Subalgorithm, linear interpolation for $H_n(x)$

REFERENCES

Ahrens, J.H. and Dieter, U. (1972). Computer methods for sampling from the exponential and normal distributions. *Comm. ACM* **15**, 873–882.

Ahrens, J.H. and Dieter, U. (1973). Extensions of Forsythe's method for random sampling from the normal distributions. *Math. Comp.* **27**, 927–937.

Ahrens, J.H and Dieter, U. (1974). Computer methods for sampling from gamma, beta, Poisson and binomial distributions. *Computing* **12**, 223–246.

Ahrens, J.H. and Dieter, U. (1974). *Non-uniform random numbers.* Technische Hochschule, Graz, 192 pp.

Ahrens, J.H. and Dieter, U. (1981a). Generating gamma variates by comparison of probability densities, Manuscript, Technical University, Graz.

Ahrens, J.H., Kohrt, K.D. (1981b). Computer methods for efficient sampling from largely arbitrary statistical distributions. *Computing* **26**, 19–31.

Ahrens, J.H. and Dieter, U. (1982). Computer generation of Poisson deviates from modified normal distributions. *ACM Trans. on Math. Softw.* **8**, 163–179.

Ahrens, J.H., Kohrt, K.D., Dieter, U. (1983). Algorithm 599. Sampling from gamma and Poisson distributions. *ACM Trans. on Math. Softw.* **9**, 255–257.

Ahrens, J.H., Dieter, U. (1985). Sequential random sampling. *ACM Trans. on Math. Softw.* **11**, 157–169.

Anderson, T.W. (1974). *Introduction to multidimensional statistical analysis.* New York: J. Wiley.

Apostolopoulos, N., Schuff, S. (1979). Initializing algorithms: note to the article Ahrens, J.H., Dieter, U.: Computer methods for sampling from gamma, beta, Poisson and binomial distributions, *Computing* **22**, No. *2*, 185–189.

Arvillias, A.C. and Maritsas, D.G. (1978). Partitioning the period of a class of m-sequences and application to pseudorandom number generation. *J. ACM* **25**, 675–686.

Arvillias, A.C. and Maritsas, D.G. (1979). Toggle registers generating in parallel k^{th} decimations of m sequences $x^p + x^k + 1$ design tables. *IEEE Trans. on Computers* **C-28**, 89–101.

Atkinson, A.C. and Pearce, M.C. (1976). The computer generation of beta, gamma and normal variables (with discussion). *J. Roy. Stat. Soc.* **A 139**, 431–461.

Atkinson, A.C. (1979). The computer generation of Poisson random variables. *Applied Statistics*, **28**, 29–35.

Baba, N., Uchida, H. and Samaragi, Y. (1984). A gaming approach to the acid rain problem. *Simulation and Games* **15**, 305–314.

Bahvalov, N.S. (1977) *Numerical Methods of Computer Mathematics.* (In Hungarian.) Műszaki Könyvkiadó, Budapest.

Bank, R., Rose, D.J. and Fichtner, W. (1983). Numerical methods for semiconductor device simulation. *SIAM J. Sci. Stat. Comp.* **4**, 416–435.

Barbu, Gh. (1982). On computer generation of a random variable by transformations of uniform variables. *Bull. Math. Soc. Sci. Math. Romanie* **26**, *2*, 129–139.

Best, D.J. (1978). A simple algorithm for the computer generation of random samples from a Student's *t* or symmetric beta distributions, in: COMPSTAT 1978, Proceedings in Computational Statistics. Physics Verlag, Wien, 341–347.

Best, D.J. (1983). A note on gamma variate generators with shape parameters less than unity. *Computing* **30**, 185–188.

Beyer, J.E. and Madsen, O.B.G. (1979). Computer simulations of combined ecological and hydrodynamic processes in sill fjord. Preprint.

Borosh, I., Niederreiter, H. (1983). Optimal multipliers for pseudorandom number generation by the linear congruential method. *BIT* **23**, No. *1*, 65–74.

Braaten, E. (1980). An improved measure of the nonuniformity of the distribution of a sequence. *SIAM J. Numer. Anal.* **17**, 31–32.

Bratley, P., Fox, B.L., Schrage, L.E. (1983). *A guide to simulation.* Springer Verlag, New York–Berlin–Heidelberg–Tokyo.

Bright, H.S., Enison, R.L. (1979). Quasi-random number sequences from a long period TLP generator with remarks on application to cryptography. *Computing Surveys* **11**, 357–370.

Burkard, R.E. and Rendl, F. (1983). A thermodynamically motivated simulation procedure for combinatorial optimization problems. TU Graz, Bericht 83–12.

Burr, I.W. (1967). A useful approximation to the normal distribution function, with application to simulation. *Technometrics* **9**, 647–651.

Buslenko, N.P., Golenko, D.I., Sreiger, J.A., Sobol, I.M. and Sragovich, V.G. (1965). *Monte Carlo Methods.* (In Hungarian.) Műszaki Könyvkiadó, Budapest.

Buslenko, N.P. (1972). *Simulation of Sophisticated Systems.* (In Hungarian.) Műszaki Könyvkiadó, Budapest.

Butler, J.W. (1956). Machine sampling from given probability distributions. In: *Symp. on Monte Carlo methods.* H.A. Mayer, ed. J. Wiley, New York, 249–264.

Chaitin, G. (1966). On the length of programs for computing finite binary sequences. *J.ACM* **13**, 547–569.

Chambers, J.M., Mallows, C.L., Stuck, B.W. (1976). A method for simulating stable random variables. *J. Amer. Stat. Ass.* **71**, 340–344.

Cheng, R.C.M. (1977). The generation of gamma random variables with non-integral shape parameters. *Appl. Stat.* **26**, 71–75.

Cheng, R.C.M.(1978). Generating beta variates with non-integral shape parameters. *Comm. ACM* **21**, 317–322.

Cheng, R.C.H., Feast, G.M. (1980). Gamma variate generators with increased shape parameter range, *Comm. ACM* **23**, 389–398.

Choi, M.Y. and Huberman, B.A. (1983). Digital dynamics and the simulation of magnetic systems. *Physical Review* **828**, 2547–2554.

Collings, B.J., Hembree, G.B. (1986). Initializing generalized feedback shift register pseudorandom number generators. *J.ACM* **33**, 706–711.

Davis, P.J. and P. Rabinowitz (1975). *Methods of numerical integration.* Academic Press, New York.

Deák, I. (1976). Cold rolling mills in the Danube Iron Works, Research Report.

Deák, I. (1978). Monte Carlo methods for determining the probabilities of sets in the n-dimensional space in case of normal distribution. *Alkalmaz. Mat. Lapok* **4**, 35–94. (In Hungarian.)

Deák, I. (1979). Computation of multiple normal probabilities. Symposium on Stochastic Programming, Oberwolfach. P. Kall, A. Prékopa, eds. In *Springer Lecture Notes in Economics and Mathematical Systems.* Springer Verlag, Berlin, New York, 107–120.

Deák, I. (1979). Comparison of methods for generating uniformly distributed random points in and on a hypersphere. *Probl. of Control and Inform. Theory* **8**, 105–113.

Deák, I. (1980). The ellipsoid method for generating normally distributed random vectors. *Zastosowania Mat.* **17**, 95–117.

Deák, I. (1980). Three digit accurate multiple normal probabilities. *Num. Math.* **35**, 369–380.

Deák, I. (1980). Fast procedures for generating stationary normal vectors. *J. Stat. Comp. and Simul.* **10**, 225–242.

Deák, I. and Bene, B. (1980). Random number generation: a bibliography. MTA SzTAKI Working Paper MO/5, 53 pp.

Deák, I. (1981). An economical method for random number generation and a normal generator. *Computing* **27**, 113–121.

Deák, I. (1984). On random number generation procedures. Proceedings of the VII Brasov Symp. on Probability. 541–545.

Deák, I. (1986). Computing probabilities of rectangles in case of multinormal distribution. *J. Stat. Comp. and Simul.* **26**, 101–114.

Deák, I. (1986). *Random number generators and their applications.* in: Mathematical Methods of Operations Research, Series ed.: A. Prékopa. Publishing House of the Hungarian Academy of Sciences, 235 pp. (In Hungarian.)

Deák, I. (1986). The economical method for generating random samples from discrete distributions. *ACM Trans. on Math. Software* **12**, 34–36.

Denning, D.E.R. (1983). Cryptography and data security. Addison-Wesley, Reading, Mass., pp. 400.

DeRaedt, H. (1985). Monte Carlo simulation of quantum statistical lattice models. Physics Reports (Review section of Physics Letters) **127**, 233–307.

Devroye, L. (1980a). The computer generation of binomial random variables. Technical Report, McGill University.

Devroye, L. (1980b). Generating the maximum of independent, identically distributed random variables. *Computers and Mathematics with Applications.* **6**, 305–315.

Devroye, L., Naderisamani, A. (1980). A binomial random number generator. Manuscript, McGill University.

Devroye, L. (1981). The computer generation of Poisson random variables. *Computing* **26**, 197–207.

Devroye, L. (1981). On the computer generation of random variables with a given characteristic function. *Comp. and Maths. with Applications* **7**, 547–552.

Devroye, L. (1981). The series method for random variate generation and its application to the Kolmogorov-Smirnov distribution. *Amer. J. Math. Management Sci.* **1**, 359–379.

Devroye, L. (1984a). Random variate generation for unimodal and monotone densities. *Computing* **32**, No. *1*, 43-68.

Devroye, L. (1984b). On the use of probability inequalities in random variate generation. *J. Statist. Comput. and Simul.* **20**, 91–100.

Devroye, L. (1986). *Non-uniform random variate generation.* Springer Verlag, New York-Berlin-Heidelberg-Tokyo. 843 pp.

Dieter, U., Ahrens, J.H. (1983). A combinatorial method for the generation of normally distributed random numbers, *Computing* **11**, 137–146.

Dieter, U. and Ahrens, J.H. (1974). *Pseudo-random Numbers.* Graz: Technische Hochschule, 216.

Dieter, U. (1982). An alternate proof for the representation of discrete distributions by equiprobable mixtures. *J. Applied Probability* **19**, 869–872.

Donelly, T.G. (1973). Bivariate normal distribution algorithm 462. *Comm. ACM* **16**, 638.

Dudewicz, E., Ralley, T.G. (1981). *The handbook of random number generation and testing with TESTRAND computer code,* American Science Press Series in Mathematical and Management Sciences 4, American Sciences Press, Columbus, Ohio, 634 pp.

Ermoliev, Yu.M. (1969). On the stochastic quasi-gradient method and stochastic quasi-Feyer sequences. *Kibernetika* **2**.

Ermoliev, Yu.M. (1983). Stochastic quasi-gradient methods and their applications to system optimization. *Stochastics* **9**, 1–36.

Fichtner, W., Rose, D.J. and Bank, R. (1983). Semiconductor device simulation. *SIAM J. Sci. Stat. Comp.* **4**, 391–415.

Fishman, G.S. (1975). Sampling from the gamma distribution on a computer. *Comm. ACM* **19**, 407–409.

Fishman, G.S., Moore, L.R. (1978). A statistical analysis of multiplicative congruential generators with modulus 2 3*k*. Technical Report, 78-11. Curriculum in Operations Research and Systems Analysis, University of North Carolina, Chapel Hill.

Fishman, G.S. (1979). Sampling from the binomial distribution on a computer. *J. Amer. Stat. Ass.* **74**, 418–423.

Fishman, G.S. (1979). *Principles of Discrete Event Simulation.* Wiley Interscience, New York.

Fishman, G.S., Huang, B.D. (1980). Antithetic variate revisited. Technical Report 80-4, University of North Carolina, Chapel Hill.

Forsythe, G.E. (1972). Von Neumann's comparison method for random sampling from the normal and other distributions. *Math. of Comp.* **26**, 817–826.

Fox, B.L. (1963). Generation of random samples from the beta and *F* distributions. *Technometrics*, **5**, 269–270.

Fox, B.L. (1980). Monotonicity, external correlations and sinchronization; implications for nonuniform random numbers, Technical Report, Universite de Montreal.

Fox, B.L. (1983). Counterparts of variance reduction techniques for quasi Monte Carlo integration, manuscript, Universite de Montreal.

Fushimi, M. (1983). Increasing the orders of equidistribution of the leading bits of the Tausworthe sequence, *Inform. Processing Letters* **16**, No. *4*, 189–192.

Golder, E.R. (1976). The spectral test for the evaluation of congruential pseudorandom generators, Algorithm AS 98. *Appl. Stat.* **25**, 173–180. See also ibid. **25** (1976), 324.

Hamidi-Noori, A. (1984). Scheduling a high contact service organization. *Simulation and Games* **15**, 315–327.

Hammersley, J.M. and Handscomb, D.C. (1964). *Monte Carlo Methods.* Methuen, London.

Hicks, J.S. and Wheeling, R.F. (1959). An efficient method for generating uniformly distributed points on the surface of an *n*-dimensional sphere. *Comm. ACM* **2**, 17–19.

Jávor, A. and Benkő, T. (1979). *Simulation of Discrete Systems.* (In Hungarian.)

Johnson, N.L. and Kotz, S. (1972). *Distributions in statistics.* Vols. I-IV. New York: J. Wiley.

Kendall, M.G., Moran, P.A.P. (1963). *Geometrical probability*, London.

Kinderman, A.J. and Ramage, J.G. (1976). Computer generation of normal random variables. *J. Amer. Stat. Assn.* **17**, 893–896.

Kinderman, A.J. and Monahan, J.F. (1977). Computer generation of random variables using the ratio of uniform deviates. *ACM Trans. on Math. Software* **3**, 257–260.

Kirkpatrick, S. Stoll, E.P. (1981). A very fast shift register sequence random number generator, *J. Comput. Phys.* **40**, No. *2*, 517–526.

Knuth, D.E. (1969). The art of computer programming. Vol. II. In: *Seminumerical algorithms.* Addison-Wesley, Reading, Mass.

Kolmogorov, A.M. (1963). On tables of random numbers. *Sankhya Ser. A* **25**, 369–376.

Kramosil, I. (1982). Pseudorandom Monte Carlo methods. *Fund. Inform.* **5**, No. *3–4*, 301–312.

Kronmal, R.A. and Peterson, A.V. (1979). On the alias method for generating random variables from a discrete distribution. *Amer. Statist.* **33**, 214–223.

Kronmal, R.A. and Peterson, A.V. (1981). A variant of the acceptance-rejection method for computer generation of random variables. *J. Amer. Stat. Assn.* **76**, 446–451.

Lehmer, D.H. (1951). Mathematical methods in large-scale computing units. *Ann. Comp. Lab. Harvard Univ.* **26**, 141–146.

Lewis, T.G. and Payne, W.H. (1973). Generalized feedback shift register pseudo-random number algorithm. *J.ACM* **20**, 456–468.

Lidl, R., Niederreiter, H. (1983). *Finite fields,* in: Encyclopedia of Mathematics and its Applications, 735, Addison-Wesley.

Lux, I. (1978). A special method to sample some probability density functions. *Computing* **20**, 183–188.

MacLaren, M.D. and Marsaglia, G. (1965). Uniform random number generators. *J.ACM* **12**, 83–89.

MacWilliams, F.J., Sloane, N.J.A. (1981). *The theory of error-correcting codes,* North-Holland, p. 762.

Maisel, A. and Gnugnoli, G. (1972). *Simulation of Discrete Stochastic Systems.* Chicago: Science Research Associates.

Maritsas, D.G., Arvillias, A.C. and Bounas, A.C. (1978). Phase shift analysis of linear feedback shift register structures generating pseudorandom sequences. *IEEE Trans. on Computers* **C-27**, 660–669.

Marsaglia, G. (1961). Expressing a random variable in terms of uniform random variables. *Ann. Math. Stat.* **32**, 894–898.

Marsaglia, G. (1963). Generating discrete random variables in a computer. *Comm. ACM* **6**, 37–38.

Marsaglia, G. (1964). Random variables and computers. Trans. Third Prague Conf. on Inf. Theory, Statistical Decision Functions. Publishing House of the Czechoslovak Academy of Sciences, 499–512.

Marsaglia, G., MacLaren, M.D., Bray, T.A. (1964). A fast procedure for generating normal random variables. *Comm. ACM* **7**, 4–10.

Marsaglia, G. (1964). Generating a variable from the tail of the normal distribution. *Technometrics* **6**, 101–102.

Marsaglia, G. (1968). Random numbers fall mainly in the planes. *Proc. Nat. Acad. Sci. USA* 61, 25–28.

Marsaglia, G. (1970). Regularities in congruential random number generators. *Numer. Math.* 16, 8–10.

Marsaglia, G. (1977). The squeeze method for generating gamma variates. *Comp. and Maths. with Applications* **3**, 321–325.

Marsaglia, G., Olkin, I. (1984). Generating correlation matrices. *SIAM J. Sci. Statist. Comput.* **5**, No. *2*, 470–475.

Martin-Lof, P. (1966). The definition of random sequences. *Information and Control* **9**, 602–619.

Metropolis, N., Rosenbluth, A., Rosenbluth, M., Teller, A. and Teller, E. (1953). Equation of state calculations by fast computing machines. *J. Chem. Phys.* **21**, 1087–1092.

Miyatake, O., Inoue, M. and Yoshizawa, J. (1975). Generation of physical random numbers. *Math. Japonica* **20**, 207–217.

Muller, M.E. (1959). A comparison of methods for generating normal deviates on digital computers. *J. ACM* **6**, 376–383.

Muller, M.E. (1959). A note on a method for generating points uniformly on n-dimensional spheres. *Comm. ACM* **2**, 19–20.

Nance, R.E., Overstreet, C. Jr. (1972). A bibliography on random number generation. *Computing Reviews*, 495–508.

Niederreiter, H. (1978). The serial test for linear congruential pseudorandom numbers. *Bull. Amer. Math. Soc.* **84**, 273–274.

Niederreiter, H. (1978). Quasi Monte Carlo methods and pseudorandom numbers. *Bull. Amer. Math. Soc.* **84**, 957–1041.

Neumann, J. von (1951). Various techniques used in connection with random digits. *J. Res. N.B.S. Appl. Math. Series* **3**, 36–38.

Partridge, S.E., Sculli, D. and Wong, S.H. (1984). Designing management games for production management training. *Simulation and Games* **15**, 328–345.

Pearson, R.B. (1983). An algorithm for pseudorandom number generation suitable for large scale integration. *J. Comput. Phys.* **48**, No. *3*, 478–489.

Peart, P. (1982). The dispersion of the Hammersley sequence in the unit square. *Monatsh. Math.* **94**, No. *3*, 249–261.

Peskun, P.H. (1980). Theoretical tests for choosing the parameters of the general mixed linear congruential pseudorandom number generator. *J. Stat. Comput. Simulation* **11**, No. *3-4*, 281–305.

Prékopa, A. and Szántai, T. (1978). On optimal regulation of a storage level with application to the water level regulation of a lake. Computer and Automation Institute, Hungarian Academy of Sciences. *Tanulmányok* **80**, 119–154.

Prékopa, A., Szántai, T. (1978). A new multivariate gamma distribution and its fitting to empirical streamflow data. *Water Resources Research* **14**, 19–24.

Prékopa, A. (1974). *Probability Theory.* Budapest: Műszaki Könyvkiadó. (In Hungarian.)

Prékopa, A., Ganczer, S., Deák, I. and Patyi, K. (1980). The STABIL stochastic programming model and its experimental application to the Hungarian electricity sector. In: *Stochastic Programming.* M.A.H. Dempster, ed. Oxford: Academic Press.

Relles, D. (1972). A simple algorithm for generating binomial random variables when *N* is large. *J. Amer. Stat. Ass.* **67**, 612–613.

Rényi, A. (1966). *Probability Theory.* Budapest: Tankönyvkiadó. (In Hungarian.)

Ripley, B.D. (1983a). Computer generation of random variables: a tutorial, *Internat. Statist. Rev.* **51**, No. *3*, 301–319.

Ripley, B.D. (1983b). The lattice structure of pseudorandom generators, *Proc. Royal Stat. Soc. London Ser.* A 389, No. 1796, 197–204.

Rizzi, A. (1982). Generation of pseudorandom binary digits by primitive polynomials, Statistics, Bologna **42**, No. *2*, 193–207.

Rosenberg, R.S. (1970). Simulation of genetic populations with biochemical properties: I. The model. *Mathematical Biosciences* **7**, 223–257. II. Selection of crossover probabilities. Ibid. **8**, 1–37.

Rubin, P.A. (1984). Generating random points in a polytope, *Comm. Statist. B-Simulation and Computation* **13**, No. *3*, 375–376.

Rubinstein, Y.R. (1981). *Simulation and the Monte Carlo Method.* New York: J. Wiley, 278.

Ruszczynski, A. Feasible direction methods for stochastic programming problems, *Math. Prog.* **19** (1980), 220–229.

Sahai, H. (1979). A supplement to Sowey's bibliography on random number generation and related topics. *J. Stat. Comp. Simul.* **10**, 31–52.

Samoridnitsky, G., Weisman, I. and Rubinstein, R.Y. (1982). An efficient mixture method. *Operations Research Letters* 1:5, 198–200.

Schmeiser, B.W. (1978). The generation of order statistics in digital computer simulation. In: Proc. of the 1978 Winter Simulation Conference, 137–140.

Schmeiser, B.W. (1978). Generation of the maximum (minimum) value in digital computer simulation, *J. Statist. Comp. and Simul.* B 103–115.

Schmeiser, B.W. and Ram Lal (1979). Squeeze Methods for Generating Gamma Variates. Technical Report OREM 78009, Southern Methodist University.

Schmeiser, B.W. (1980). Random variate generation: A survey. In: *Simulation with discrete models: A state-of-the-art view.* T.I. Oren, C.M. Shub and P.F. Roth, eds., 79–104.

Schmeiser, B.W., Babu, A.J.B. (1980). Data variate generation via exponential majorizing functions, *Oper. Res.* **28**, 917–926. *Errata: Oper. Res.* **31** (1983) No. *4*, 802.

Schmeiser, B.W. (1980). Generation of variates from distribution tails, *Oper. Res.* **28**, 1012–1017.

Schmeiser, B.W., Lal, R. (1980). Squeeze methods for generating gamma variates, *J. Amer. Stat. Ass.* **75**, 679–682.

Schmeiser, B.W. (1980). Random variate generation: a survey, in: Proceedings of the 1980 Winter Simulation Conference, Orlando, Florida, 79–104.

Schmeiser, B.W. (1980). Multivariate modelling in simulation: a survey, in: ASQC technical Conference Transactions, Atlanta, 252–261.

Schmeiser, B.W., Babu, A.J.B. (1980). Beta variate generation via exponential majorizing functions, *Oper. Res.*, **28**, 1012–1017.

Schmeiser, B.W., Katchitvichyanukul, V. (1981). Poisson random variate generation. Research Memorandum 81–4, School of Industrial Engineering, Purdue University, West Lafayette, Indiana, 38.

Schmeiser, B.W., Lal, R. (1982). Bivariate gamma random vectors, *Oper. Res.* **30**, 355–374.

Sós, V.T. (1976). On the discrepancy of the sequence $\{n\alpha\}$. *Coll. Math. Soc. J. Bolyai* **13**, 359–367.

Sowey, E.R. (1972). A chronological and classified bibliography on random number generation and testing, *Int. Stat. Rev.* **40**, 255–371.

Sowey, E.R. (1978). A second classified bibliography on random number generation and testing, *Int. Stat. Rev.* **46**, 89–102.

Sowey, E.R. (1987). A third classified bibliography on random number generation and testing, *J. Roy. Stat. Soc. Ser. A.*

Szántai, T. (1976). A procedure for determining the distribution function and its gradient of the multinormal distribution. *Alk. Mat. Lapok* **2**, 27–39. (In Hungarian.)

Szántai, T. (1986). Evaluation of a special multivariate gamma distribution function, *Math. Prog. Study.*

Szobol, I.M. (1981). *Basics of the Monte Carlo Methods.* Budapest: Műszaki Könyvkiadó (In Hungarian.)

Tadikamalla, P.R. (1978a). Computer generation of gamma random variables, *Comm. ACM* **21**, 419–422.

Tadikamalla, P.R. (1978b). Computer generation of gamma random variables II, *Comm. ACM* **21**, 925–929.

Tadikamalla, P.R. (1979). Random sampling from the generalized gamma distribution, *Computing* **23**, 199–263.

Tadikamalla, P.R. (1980). Random sampling from the exponential power distribution, *J. Amer. Stat. Ass.* **75**, 683–686.

Tadikamalla, P.R., Johnson, M.E. (1981). A complete guide to gamma variate generation, *Amer. J. Math. Management Sci.* **1**, No. *3*, 213–236 and ibid. **2** (1982) No. *1*, 93.

Tausworthe, R.C. (1965). Random numbers generated by linear recurrence modulo two. *Math. Comp.* **19**, 201–209.

Tootill, J.P.R., Robinson, W.D., Adams, A.G. (1971). The runs up-and-down performance of Tausworthe pseudo-random number generators, *J. ACM* **18**, 381–399.

Vaduva, I. (1983) Computer generation of random vectors based on transformation of uniform distributed vectors, Proceedings of the VII Brasov Conference on Probability Theory.

Walker, A.J. (1977). An efficient method for generating discrete random variables with general distributions. *ACM Trans. on Math. Software* 3, 253–256.

Wallace, N.D. (1974). Computer generation of gamma random variates with nonintegral shape parameters. *Comm. ACM* **17**, 691–695.

Wilson, J.R. (1983). Antithetic sampling with multivariate inputs, *Amer. J. Math. Management Sci.* **3**, No. *2*, 121–144.

Whittaker, J. (1974). Generating gamma and beta variables with nonintegral shape parameters, *Applied Statistics*, **23**, 210–214.

Wyman, F.P. (1970). *Simulation Modelling: A Guide to Using SIMSCRIPT*. New York: John Wiley.

Zaremba, S.K. (1968). The mathematical basis of Monte Carlo and quasi-Monte Carlo methods. *SIAM Rewiew* **10**, 303–314.

Zielinski, R. (1972). *Generatory liczb losowych*. Warsawa: Wyd. Naukowo-Techniczne.

Zierler, N. (1959). Linear recurring sequences, *J. SIAM* **7** (1959), 31–48.

Zierler, N. and Brillhart, J. On primitive trinomials (mod 2). *Inform. Control* **13** (1968), 541–554, **14** (1969), 566–569.

ADDITIONAL REFERENCES
(Added in proof)

Dantzig, G.B., Glynn, P.W. (1990). Parallel processors for planning under uncertainty, Annals of Operations Research (forthcoming).

Deák, I. (1988). Multidimensional integration and stochastic programming, in: *"Numerical techniques for stochastic optimization 1984"*, Springer, ed. Yu. Ermoliev, R. Wets, 187–200.

Deák, I. (1989). Uniform random number generators on parallel computers, Research Report 89–1, Industrial Engineering, University of Wisconsin-Madison, pp. 17.

Deák, I. (1990). Procedures to solve STABIL on a parallel computer, Annals of Operations Research (Proceedings of the Vth Int. Conference on Stochastic Programming) (forthcoming).

Gassman, H. (1988). Conditional probability and conditional expectation of a random vector, in: *"Numerical techniques for stochastic optimization 1984"*, Springer, ed. Yu. Ermoliev, R. Wets, 237–254.

L'Ecuyer, P., Blouin, F. (1988). Generalized linear congruential generators, Reserach Report DIUL–RR–8814, Department d'Informatique, Univ. Leval, pp. 25.

Ripley, B. D. (1987). *Stochastic simulation*, Wiley, New York, 237 pp.

Standlober, E., Dieter, U. (1989). A bibliography on random number generation, University of Graz, Austria (in preparation).

Suri, R., Zazanis, M. A. (1988). Perturbation analysis gives strongly consistent sensitivity estimates for the M/G/1 queue, *Management Science*, Vol. **34**, 39–64.

Suri, R., Leung, Y. T. (1989). Single run optimization of discrete event simulation: an empirical study using the M/M/1 queue, *IIE Transactions*, March 1989.

Thesen, A., Travis, L. (1988). Introduction to simulation, in: Proceedings of the Winter Simulation Conference, 1988, ed. M. Abrams, P. Haigh, J. Comfort, 7–14.

SUBJECT INDEX